HYDROSTATICS

HYDROSTATICS

[Written according to the syllabi of BA & BSc (General & Honours) of all Indian universities and all Competitive Examinations of equivalent standard]

Dr M M Rahman

Double Gold Medallist
Reader, PG Department of Mathematics
A N College, Patna 13

New Central Book Agency (P) Ltd

LONDON
HYDERABAD ERNAKULAM BHUBANESWAR
NEW DELHI KOLKATA PUNE GUWAHATI

NCBA

REGD OFFICE
8/1 Chintamoni Das Lane, Kolkata 700 009, India
email: ncbapvtltd@eth.net

OVERSEAS
NCBA (UK) Ltd, 149 Park Avenue North, Northampton, NN3 2HY, UK
email: ncbauk@yahoo.com

EXPORT
NCBA Exports Pvt. Ltd, 212 Shahpur Jat, New Delhi 110 049
email: ncbaexports@ncbapvtltd.com

BRANCHES

2/27 Ansari Road, 1st Floor
Daryaganj, **New Delhi** 110 002
email: ncbadel@ncbapvtltd.com

House No. 3-1-315, 1st Floor
Nimboliadda, Kachiguda, **Hyderabad** 500 027
email: ncbahydb@ncbapvtltd.com

Shop Nos. 3 & 4, Vinayak Towers
681/B Budhwar Peth, Appa Balwant Chowk
Pune 411 002
email: ncbapune@ncbapvtltd.com

GSS Shopping Complex, 1st Floor
Opposite College Post Office, Convent Road
Ernakulam 682 035
0484-4021054 / 53

Shop No. 15, Bharati Towers, Ground Floor
Forest Park
Bhubaneswar 751 001
email: ncbabub@ncbapvtltd.com

Radhamayee Complex [Second Floor]
Opp. Axis Bank, Panbazar, Jaswanta Road
Guwahati 781 001
email: ncbaguwahati@ncbapvtltd.com

HYDROSTATICS • Dr M M Rahman

First Published: 1995
Thoroughly Revised and Updated Second Edition: 2005
Reprinted: January 2009, July 2012, September 2013, July 2017, 2019, **2021**

PUBLISHER AND TYPESETTER
New Central Book Agency (P) Ltd
8/1 Chintamoni Das Lane, Kolkata 700 009

PRINTER
New Central Book Agency (P) Ltd
Web-Offset Division, Dhulagarh, Sankrail, Howrah

PROOF CHECKING
P K Chowdhury

PROJECT TEAM
Abhijit Sen and Subir Pal

ISBN: 978 81 7381 036 7

Price ₹ 415.00

Preface to the Second Edition

The book has been thoroughly revised. Several new examples from various examination papers have been added and few of them are solved. The book has been revised with an aim so that it may be more useful to students. At some places alternative easier solutions have replaced the old ones. The university papers have been added uptodate.

I'm indebted to the teachers and students on account of whose patronage the book is appearing in its second edition. I'm also thankful to may wife Mrs. N. Jahan and sons Tauseef and Tauqeer for their encouragement and inspiration in the completion of this book.

Every effort has been made to avoid the errors and misprints and to improve it. Even then there is always a scope for improvement and every suggestion for the further improvement of the book as well as pointing out of errors will be thankfully received.

Lastly but not the least I'm thankful to Mr. Amitabha Sen, Director of New Central Book Agency (P) Ltd., for their cordial cooperation during the preparation of second edition of this book.

Patna, 20 Sept., 2005 MD. MOTIUR RAHMAN

Preface to the First Edition

This book HYDROSTATICS is intended to cover the syllabi of B.A. and B.Sc.(H) of all Indian Universities as well as competitive examinations like I.A.S. and P.S.C. After teaching more than twelve years to the Honours Students, I have experienced that average students found some existing books too difficult while the others too inadequate to meet their requirements. This book is complete as far as the subject matter is concerned and I hope that nothing of importance has been left out.

Articles have been arranged in a systematic order and fundamental principles involved have been amply illustrated by diagrams and worked out examples. A large number of worked out examples are given corresponding to each article, so that the students may understand the methods clearly. Most of the questions have been taken from the question papers of various Universities and Public Service Examinations.

I'm grateful to my students whose constant inspiration and help insisted me to complete the present book. I also want to express my sincere thanks to my wife Mrs. N. Jahan and sons Tauseef and Tauqeer whose encouragement and inspiration forced me to complete the book in time.

It is hoped that the book is relatively free from errors and misprints but concrete suggestions for further improvement of the book as well as pointing out of errors will be thankfully received.

MD. MOTIUR RAHMAN

A few lines about the book

This book covers the syllabi of B.A. and B.Sc(H) of all Indian Universities and all competitive examinations of equivalent standard. The stress has been given on fundamental ideas, so that students may grasp it easily. Each topic has been introduced and developed in as simple and straight-forward manner as possible and well within the comprehension of students of all categories. Articles are arranged in a systematic order and the fundamental principles involved have been amply illustrated by diagrams and worked out examples. The three-dimensional diagrams have been drawn as would be viewed from a distance in a glass container showing the vertical section through the body. A special feature in the chapter on CENTRE OF PRESSURE is the treatment of liquids with variable density.

Syllabus

I. New Syllabus of C.U.

1. Definition of fluid, perfect fluid, pressure.

2. To prove that the pressure at a point in a fluid in equilibrium is the same in every direction. Transmissibility of liquid pressure. Pressure of heavy fluids. To prove—

   (a) In a fluid at rest under gravity the pressure is the same at all points in the same horizontal plane.

   (b) In a homogeneous fluid at rest under gravity the difference between the pressures at two points is proportional to the difference of their depths.

   (c) In a fluid at rest under gravity horizontal planes are surfaces of equal density.

   (d) When two fluids of different densities at rest under gravity do not mix their surface of separation is a horizontal plane. Pressure in heavy homogeneous liquid. Thrust of heavy homogeneous liquid on plane surfaces.

3. Definition of centre of pressure. Formula for the depth of the centre of the pressure of a plane area. Position of the centre of pressure. Centre of pressure of a triangular area whose angular points are at different depths. Centre of pressure of a circular area. Position of the centre of pressure referred to coordinate axe through the centroid of the area. Centre of pressure of an elliptical area when its major axis is vertical or along the line of greatest slope. Effect of addition depth on centre of pressure.

4. Thrust on curved surfaces. Floating bodies : the principle of Archimedes. Force of buoyancy, centre of buoyancy. To find the thrust on any surface exposed to the pressure of heavy fluid at rest under gravity. Horizontal and vertical thrust on a surface exposed to the pressure of heavy fluid. To find the resultant pressure on any surface of a fluid at rest under the action of any given forces. Conditions of equilibrium of a .floating body.

5. Equilibrium of fluids in given fields of force: Definition of field of force, line of force, pressure derivative in terms of force. Surface of equi-pressure. To find the necessary and sufficient conditions of equilibrium of a fluid under the action of a force whose components are X, Y, Z along the coordinate axes. To prove:

   (a) that surfaces of equal pressure are the surfaces intersecting orthogonally the lines of force,

   (b) when the force system is conservative, the surfaces of equal pressure are equipotential surfaces and are also surfaces of equal density. To find the differential equations of the surfaces of equal pressure and density.

6. Rotating fluids: to determine the pressure at any point and the surfaces of equal pressure when a mass of homogeneous liquid contained in a vessel, revolves uniformly about a vertical axis.

7. The stability of the equilibrium of floating bodies: definition, stability of equilibrium of a floating body, metacentre, plane of flotation, surface of buoyancy. General propositions about small rotational displacements. To derive the condition for stability.

8. Pressure of gases: the atmosphere. Relations between pressure, density and temperature. Pressure in an isothermal atmosphere. Atmosphere in convective equilibrium.

II. Syllabus for I.A.S. (Main Examination)

Pressure of heavy fluids, Equilibrium of fluids under given system of forces, Centre of pressure, Thrust on curved surfaces, Equilibrium of floating bodies, Stability of equilibrium, Metacentre, Pressure of gases.

Contents

Abbreviations

B.B.A.B.U.	—	Babasaheb Bhim Rao Ambedkar Bihar University
B.U.	—	Bihar University
C.G.	—	Centre of Gravity
C.P.	—	Centre of Pressure
C.U.	—	Calcutta University
L.N.M.U. (Mith)	—	Lalit Narayan Mithila University
M.U.	—	Magadh University
P.U.	—	Patna University
R.U.	—	Ranchi University
Sp.Gr.	—	Specific Gravity
Sp.Wt.	—	Specific Weight
T.M.B.U.	—	Tilka Manjhi Bhagalpur University

SOME USEFUL GEOMETRICAL RESULTS

Areas

Triangle.

1. If h be the height and a the length of the base, then its area $= \frac{1}{2}ah$.
2. If a and b be the two sides having angle θ between them, then its area $= \frac{1}{2}ab\sin\theta$.

Hexagon. If a be the side of a regular hexagon, then its area $= 3\frac{\sqrt{3}}{2}a^2$.

Octagon. If a be the side of a regular octagon, then its area $= 2(1+\sqrt{2})a^2$.

Circle. If r be the radius of the circle, then its area $= \pi r^2$ and circumference $= 2\pi r$.

Sector of a circle. The area of a sector of a circle of radius r subtending an angle 2θ of the centre $= \frac{1}{2}r^2\theta$.

Ellipse. If $2a$ and $2b$ be the lengths of the major and minor axes of an ellipse, then its area $= \pi ab$.

Volumes and Surface

Rectangular solid. If a, b and c be the length, breadth and height of a rectangular solid, then

its volume $= abc$

its whole surface $= 2(ab+bc+ca)$

and its diagonal $= \sqrt{a^2+b^2+c^2}$.

Cube. If a be the length of each edge, of the cube, then

its volume $= a^3$

its whole surface $= 6a^2$

its diagonal $= \sqrt{3}a$.

Prism. If the base of the prism be a regular polygon, then

its volume = (area of the base) × height

its lateral face = (perimeter of the base) × height.

Cylinder. If r be the radius of the base of the cylinder of height h, then

its curved surface $= 2\pi rh$

its whole surface $= 2\pi r(h + r)$

its volume $= \pi r^2 h$.

Pyramid. Its volume $= \frac{1}{3}$ (area of base) × height.

Its slant surface $= \frac{1}{2} \times$ (perimeter of its base) × slant height.

Cone. If h be the height, r the radius, l the slant height, α the semivertical angle of the cone, then

$r = h \tan a, l = h \sec \alpha$

its volume $= \frac{1}{3}\pi r^2 h = \frac{1}{3}\pi h^3 \tan^2 \alpha$

its curved surface $= \pi r l = \pi h^2 \tan \alpha \sec \alpha$.

Frustum of a cone. If r_1, r_2 be the radii of its circular ends and d the perpendicular distance between them, then

its volume $= \frac{1}{3}\pi d(r_1^2 + r_1 r_2 + r_2^2)$.

Sphere. If r be the radius of the sphere, then

its volume $= \frac{4}{3}\pi r^3$

its curved surface $= 4\pi r^2$.

Zone or portion of a sphere. If r be the radius of the sphere of which the zone is given between two parallel planes of distances h_1 and h_2 from the centre of the sphere, then

its volume $= \frac{\pi}{3}(h_1 - h_2)\{3r^2 - (h_1^2 + h_1 h_2 + h_2^2)\}$

its curved surface $= 2\pi rd$,

where d is the perpendicular distance between the two parallel planes.

Paraboloid of revolution. Solid formed by the revolution of a parabola about its axis is known as paraboloid of revolution.

Its volume cut off by a plane perpendicular to its axis $= \frac{1}{2}$ (area of its plane base) × its height.

Centres of Gravity

C.G. of a **cylinder** is the mid-point of its axis.

C.G. **of a solid cone** of height h is the point on the axis at a distance $\frac{3h}{4}$ from the vertex.

C.G. **of the curved surface of a circular cone** of height h is the point on axis at a distance $\frac{3h}{4}$ from the vertex.

C.G. of a solid half cone: base of radius r, is a point whose distance from the axis of the whole cone $= \frac{r}{\pi}$.

C.G. of solid hemisphere of radius r, is a point on its axis at a distance $\frac{3r}{8}$ from the centre of the base.

C.G. of the curved surface of a hemisphere of radius r, is a point on its axis at a distance $\frac{r}{2}$ from the centre of this base.

C.G. of a semicircle of radius r, is a point on the radius perpendicular to the bounding diameter at a distance $\frac{4r}{3\pi}$ from the centre of the circle.

C.G. of a sector of a circle of radius r subtending an angle 2θ of the centre of the circle is a point on the bisector of the angle 2θ, whose distance from the centres $= \frac{2r}{3}\frac{\sin\theta}{\theta}$.

C.G. of a circular arc of radius r subtending an angle 2θ of the centre is a point on the bisector of the angle 2θ, whose distance from the centre $= \frac{r\sin\theta}{\theta}$.

C.G. of a semicircular arc of radius r is a point on the radius perpendicular to the bounding diameter at a distance $\frac{2r}{\pi}$ from the centre of the circle.

Radius of Gyration

If Ak^2 is the moment of inertia, where k is the radius of gyration, value of k^2 is given below for some cases:

1. For a thin rod of length l, about an axis through the middle point perpendicular to the rod, $k^2 = \frac{l^2}{12}$.
2. For a rectangular lamina of sides a and b, about a line through its centre parallel to the side a, $k^2 = \frac{b^2}{12}$, and parallel to the side b, $k^2 = \frac{a^2}{12}$.
3. For a circular area of radius r; about a diameter, $k^2 = \frac{r^2}{4}$.
4. For an elliptic area of major and minor axes $2a$ and $2b$,

   $k^2 = \frac{a^2}{4}$ about the axis $2b$

   $k^2 = \frac{b^2}{4}$ about the axis $2a$.

5. For triangular area ABC, about a straight line through A in the plane of the area

$$k^2 = \frac{1}{6}(\beta^2 + \beta\gamma + \gamma^2),$$

where β and γ being the distances of B and C from the axis of rotation through A.

CHAPTER 1

Introduction

1.1 Matter

Anything which occurs in nature and occupies space is called *Matter* or *Substance.* Matter is divided into two classes: (*i*) *Solids* and (*ii*) *Fluids.*

- *Solid:* A solid is a substance which preserves its shape and size when left to itself without the application of a certain amount of external force.

- *Fluid:* A fluid is a substance which does not preserve its shape when left to itself. In other words, the fluids are substances which flow or are capable of flowing.

Examples of solids are *lumps of iron, wood, stone,* etc., and the examples of fluid are *air, water, oil, honey,* etc.

The property by virtue of which a substance offers resistance to forces tending to change its shape is called *rigidity*. Hence, solid may also be defined as a substance which has rigidity and fluid as a substance having no rigidity.

Fluids are further divided into two classes, namely, (*i*) *Liquids* and (*ii*) *Gases.*

- *Liquids:* A liquid is a substance which offers a great resistance to forces tending to change its volume. Its shape is the same whatever be the shape of the vessel containing it.

- *Gases:* A gas is a substance which offers very small resistance to forces tending to change its volume, i.e., gases can easily be made to change its volume.

Examples of liquids are *alcohol, water, oil,* etc., and examples of gases are *nitrogen, oxygen and vapour,* etc.

The property by virtue of which the application of even a very large force does not change the volume of a substance, is called *incompressibility.* Hence, liquids are practically incompressible substances and gases are compressible substances.

For this reason, liquids are sometimes called *inelastic fluids* and the gases *elastic fluids.*

1.2 Distinction between Solid, Liquid and Gas

It may be noted that there is no substance which is perfectly in compressible. Thus, we have three states of matter solid, liquid and gas. Their main distinguish features are as follows:

- A solid is incompressible and has rigidity, i.e., it has a definite size and definite shape.
- A liquid is incompressible but has no rigidity, i.e., it has a definite size but no definite shape.
- A gas is easily compressible and has no rigidity, i.e., it has neither definite size and nor definite shape.

1.3 Conversion of Different States of Matter

It has experimentally been seen that a substance can be changed from one of the three states (*solid, liquid* and *gas*) to another, i.e., by suitable devices of changing the temperature and pressure, the substance (*matter*) can be transformed from one state to another. Thus, if a lump of ice is heated it changes into water. By supplying further heat to water, it can be changed into water vapours or steam. Similarly, a gas like air or oxygen can be reduced to a liquid state, by lowering the temperature and increasing the pressure. There are substances like camphor, which change directly from solid state to gaseous state and *vice versa* without entering into the intermediate liquid state.

After a certain level of temperature, the increase in temperature, however great it is, with no increase in pressure, causes the gas to condense and also when the temperature is below this level of temperature a sufficient increase of pressure causes condensation takes place is called the *critical temperature.* Critical temperature varies for different gases. A gas below the critical temperature is called the *vapour* and above of the gas is called *permanent gas.*

1.4 Hydrostatics

Hydrostatics, a Greek word, means that branch of Mathematics which is concerned with the conditions of equilibrium of fluids at rest and their reactions upon solids in contact with them.

1.5 Types of Fluids

There are two types of fluids: (*i*) *Perfect* and (*ii*) *Viscous.*

- *Perfect fluid:* We learn from our common experience that fluids like water and air are characterised by the property that the portions of their mass can be very easily separated, that is, they are extremely divisible. Thus, if a thin plate be immersed in water or air, the resistance offered to the division is very slight and almost inappreciable; or, in other words, the only force exerted on the plate is normal to the plate, the tangential force being almost negligible. This leads to conceive of a fluid for which there is no force tangential to its different portions. Though such fluids do not exist in nature such a fluid is called a *perfect fluid*.

  Thus, a perfect fluid is an aggregate of particles (*a fluid*) the portions of which can be separated from each other by slightest effort, there being no force tangential to the different portions. In other words, a perfect fluid is a substance which does not exert any tangential force but the only direction along which it can exert any force is along the normal to the surface with which it is in contact.

  A perfect fluid is an incompressible fluid. Its volume cannot be altered. But its shape can be changed with the slightest force that is applied to it.

  We have assumed a perfect fluid as having no property of the nature of friction between the different particles, but there are no fluids which are known to satisfy this definition completely. For example, we set water revolving in a tumbler, the water after some time comes to rest. The forces which are responsible for stoppage of motion are the tangential forces, or the functional forces between different particles and between the water and the tumbler at the common surface. Had these forces been absent the water would have continued to revolve. Thus, a perfect fluid is only a hypothetical fluid. We postulate such an ideal fluid, because it greatly simplifies the mathematical discussion connected with the fluids.

- *Viscous fluid:* Contrary to perfect fluids there are fluids which offer resistance to forces tending to change their shape. In these fluids the internal forces of the nature of friction come into play and retard the relative motion of particles. Such fluids are called *viscous fluids*.

  Thus, a viscous fluid is an aggregate of particles (*a fluid*) which exert friction on another particles, when in motion, i.e., it offers resistance to the motion of a body in it.

  Examples of viscous fluids are *honey, coaltar* and *treacle,* etc.

  The property by virtue of which the particles exert friction on one another is called *viscosity.* It may be noted that viscosity comes into play only when the fluid is in motion. Hence, for fluids at rest, there is no tangential action or shearing stress, whatever degree of viscosity it may possess.

  As hydrostatics deals with fluid at rest, therefore, all the propositions in hydrostatics are true for fluids, irrespective of their viscosity.

  Just as perfect fluids are not found in nature, in the same way perfectly viscous fluids are also impossible to be found. Even a very viscous fluid will change its shape under a force, however, small it may be if the force be continued for a long time.

1.6 Fluids are also classified according to the Distribution of Matter in its Different Parts

If the masses of equal volumes of a substance, taken from anywhere on it, be equal, it is said to be homogeneous, otherwise, it is heterogeneous.

1.7 Shearing Stress

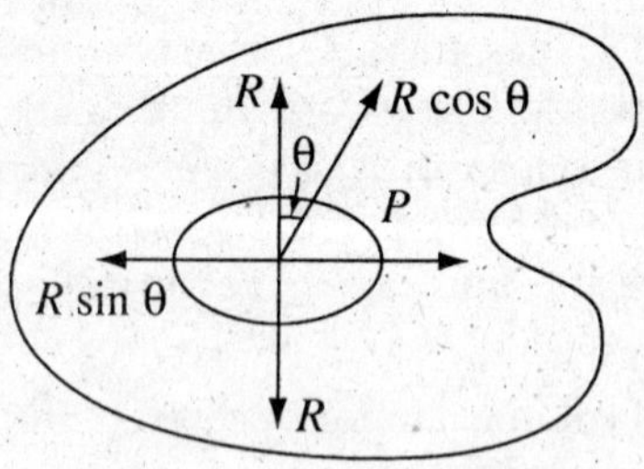

Fig. 1.1

Let us imagine a small plane surface in a material substance surrounding the point P. By Newton's third law of motion, two portions of the substance on opposite sides of the plane surface exert on one another equal and opposite to forces. Any one of the two forces may be called *stress.* Let this stress be denoted by R, which can be resolved into two components normally and tangentially to this plane, viz., $R\cos\theta$ and $R\sin\theta$. $R\cos\theta$ is called the *thrust* or *tension* and $R\sin\theta$ is known as *shearing stress.*

CHAPTER 2

Density and Specific Gravity

2.1 Mass

The mass of a body is defined as the quantity of matter that is contained in the body.

2.2 Density

The density at a point of a substance is defined as the limit to which the ratio of the mass of a small region of substance containing the point and the volume of that region tends, as the linear dimensions of the region are indefinitely diminished.

Analytically, density of a point of a substance is expressed as

$$\lim_{\delta V \to 0} \frac{\delta m}{\delta V} = \frac{dm}{dV},$$

where δm and δV are the mass and the volume of the small region containing the point.

In case, the substance is homogeneous, for every point of the substance, the density ρ is given by

$$\rho = \frac{\delta m}{\delta V} \quad \text{or,} \quad \rho = \frac{m}{V},$$

where m is the mass and V the volume of the substance.

Hence, the density of a homogeneous substance is its mass per unit volume.

The mean density of a portion of a substance is the ratio of its mass to its volume.

From the definition of density and mean density it follows that the mean density of all parts of a homogeneous substance is the same and is equal to its density. The dimension of density

$$= \frac{M}{V} = \frac{M}{L^3} = ML^{-3}.$$

Also from the definition of density and mean density it follows that a homogeneous substance is said to be of uniform density and a heterogeneous substance to be of variable density.

The mass of one cubic centimeter of water at 4°C is 1 gm, hence density of water in C.G.S. system is 1; and mass of one cubic feet of water at 4°C is 62.5 lbs, hence density of water in F.P.S. system is 62.5

2.3 Weight in Terms of Density

If the weight of a homogeneous substance of mass M be W gms, g be the acceleration due to gravity measured in C.G.S. system; ρ, its density in gms per cubic cm and V, its volume in cubic centimetre, then

$$W = V\rho g = Mg \qquad (\because\ M = V\rho).$$

Similar results will be true if the units are taken in F.P.S. system.

2.4 Specific Gravity or Relative Density or Relative Weight

The specific gravity of any substance is the ratio of the weight of any volume of that substance to the weight of an equal volume of some standard substance.

For solids and liquids the standard substance is generally taken to be pure distilled water at 4°C and for gases air or hydrogen is taken as the standard substance.

If ρ and ρ' be the densities of the given and the standard substance, then

$$\begin{aligned}\text{Specific gravity} &= \frac{\text{Wt. of volume } V \text{ of the body}}{\text{Wt. of a volume } V \text{ of the standard substance}}\\ &= \frac{V\rho g}{V\rho g} \qquad (\because\ W = V\rho g)\\ &= \frac{\rho}{\rho'}.\end{aligned}$$

So, specific gravity of a substance can also be defined as the ratio of the density of that substance to the density of a standard substance. The specific gravity is sometimes called *relative weight* or *relative density*. It is written in short form as sp. gr. or R.D. The specific gravity has no dimensions, it is simply a number.

2.5 Weight in Terms of Specific Gravity

If W be the weight of a volume V of a substance whose specific gravity is S and w be the weight of a unit volume of the standard substance, then

$$S = \frac{W}{V_w} \quad \text{or,} \quad W = S \times V_w.$$

Hence,

$$W = VS_w$$

w being a constant, we may say $W \propto SV$.

Thus, weight of a body varies as its volume and its specific gravity conjointly.

Also, from the definition of specific gravity, density of homogeneous substance

$$= \text{sp. gr. of homogeneous substance} \times \text{density of a standard substance},$$

i.e., $V\times$ density of homogeneous substance

$$= \text{sp. gr. of homogeneous substance} \times V \times \text{density of a standard substance},$$

i.e., $W = V\rho S$.

In case the standard substance is taken to the water at 4°C, then in C.G.S. system the numbers expressing the density and the specific gravity of the substance are the same.

From this it does not, however, follow that the density and specific gravity are the same.

2.6 Specific Gravity of Mixtures

2.6.1 Specific gravity of a mixture of number of substances whose volumes and specific gravities are given

Let $V_1, V_2, V_3, \ldots, V_n$ be the volumes of indifferent substances whose specific gravities are $S_1, S_2, S_3, \ldots, S_n$ and let W be the weight of a unit volume of the standard substance, so that weights of these different substances are V_1S_1W, $V_2S_2W, V_3S_3W, \ldots, V_nS_nW$. Hence, sum of the weights of different substances is

$$V_1S_1W + V_2S_2W + V_3S_3W + \cdots + V_nS_nW.$$

Assuming no expansion or shrinkage in volume on account of mixture, the volume of the mixtures is $V_1 + V_2 + V_3 + \cdots + V_n$.

Let S be the sp. gr. of the mixture, then the weight of the mixture is

$$(V_1 + V_2 + V_3 + \cdots + V_n)SW.$$

Now, the weight of the mixture must be equal to the sum of the weights of different substances be

$$\begin{aligned}(V_1 + V_2 + V_3 + \cdots + V_n)SW &= V_1S_1W + V_2S_2W + V_3S_3W + \cdots + V_nS_nW \\ \text{or,}\quad S &= \frac{V_1S_1 + V_2S_2 + V_3S_3 + \cdots + V_nS_n}{V_1 + V_2 + V_3 + \cdots + V_n}.\end{aligned}$$

If on mixing the different substances, the volume of the mixture is not equal to the sum of the volumes of the different substances but is different, say V, then the sp. gr. of the mixture is given by

$$S = \frac{V_1S_1 + V_2S_2 + \cdots + V_nS_n}{V}.$$

Corollary 2.6.1 *If the volume of the mixture of n different substances is reduced to K times the sum of their individual volumes as a result of mixing, then the specific gravity of the mixture is*

$$\frac{V_1S_1 + V_2S_2 + \cdots + V_nS_n}{K(V_1 + V_2 + \cdots + V_n)}.$$

Corollary 2.6.2 *If n different substances of volumes $V_1, V_2, V_3, \ldots$ and densities $\rho_1, \rho_2, \rho_3, \ldots$ be mixed together, then the density of the mixture is*

$$\frac{V_1\rho_1 + V_2\rho_2 + \cdots + V_n\rho_n}{V_1 + V_2 + \cdots + V_n}$$

$$\text{or,} \quad \frac{V_1\rho_1 + V_2\rho_2 + \cdots + V_n\rho_n}{K(V_1 + V_2 + \cdots + V_n)}$$

according as there is no change of volume or a change of volume $K(V_1 + V_2 + \cdots)$ as a result of mixing.

Corollary 2.6.3 *If $V_1 = V_2 = V_3 = \cdots = V_n = V$ (say), then the specific gravity S of the mixture is given by*

$$\begin{aligned} S &= \frac{V(S_1 + S_2 + \cdots + S_n)}{V + V_2 + \cdots \text{ to } n \text{ terms}} \\ &= \frac{V(S_1 + S_2 + \cdots + S_n)}{nV} \\ &= \frac{S_1 + S_2 + \cdots + S_n}{n}. \end{aligned}$$

Thus, the specific gravity of a mixture of equal volumes of different liquids when there is no change of volume on account of mixing is the arithmetic mean of the specific gravities of the given liquids.

2.6.2 Specific gravity of a mixture of fluids when their weights and specific gravities are given

Let $W_1, W_2, W_3, \ldots, W_n$ be the weights of n different substances whose specific gravities are $S_1, S_2, \ldots, S_n$. Let S be the specific gravity of the mixture and let W be the weight of a unit volume of the standard substance; therefore, respective volumes of different substances are

$$\frac{W_1}{S_1W}, \frac{W_2}{S_2W}, \ldots, \frac{W_n}{S_nW}.$$

Supposing that there is no change of volume when the mixture is formed, the volume of the mixture is

$$\left(\frac{W_1}{S_1W} + \frac{W_2}{S_2W} + \cdots + \frac{W_n}{S_nW}\right).$$

∴ Weight of the mixture is

$$\left(\frac{W_1}{S_1W} + \frac{W_2}{S_2W} + \cdots + \frac{W_n}{S_nW}\right) SW.$$

But the weight of the mixture is

$$W_1 + W_2 + \cdots + W_n.$$

Now, the weight of the mixture must be equal to the sum of the weights of the different substances of which the mixture is made. Hence,

$$\left(\frac{W_1}{S_1 W}+\frac{W_2}{S_2 W}+\cdots+\frac{W_n}{S_n W}\right) SW = W_1+W_2+\cdots+W_n$$

$$\text{or,}\quad S = \frac{W_1+W_2+\cdots+W_n}{\frac{W_1}{S_1}+\frac{W_2}{S_2}+\cdots+\frac{W_n}{S_n}}.$$

When the volume of the mixture is not the sum of the volumes of its constituents but a changed volume V, say, we have

$$S = \frac{W_1+W_2+\cdots+W_n}{WV}.$$

Corollary 2.6.4 *If the volume of the mixture of n different substances is reduced, as a result of mixing, to K times the sum of the volumes of the substances, then the specific gravity of the mixture is*

$$= \frac{W_1+W_2+\cdots+W_n}{K\left(\frac{W_1}{S_1}+\frac{W_2}{S_2}+\cdots+\frac{W_n}{S_n}\right)}.$$

Corollary 2.6.5 *If n different substances of weights $W_1, W_2, \ldots, W_n$ and densities $\rho_1, \rho_2, \ldots, \rho_n$ be mixed together, then the density ρ of the mixture is given by*

$$\rho = \frac{W_1+W_2+\cdots+W_n}{\frac{W_1}{\rho_1}+\frac{W_2}{\rho_2}+\cdots+\frac{W_n}{\rho_n}}$$

$$\text{or,}\quad \rho = \frac{W_1+W_2+\cdots+W_n}{K\left(\frac{W_1}{\rho_1}+\frac{W_2}{\rho_2}+\cdots+\frac{W_n}{\rho_n}\right)}$$

according as there is no change of volume or a change of volume to K times the sum of the original volumes as a result of mixing.

Corollary 2.6.6 *If $W_1 = W_2 = \cdots = W_n = W$ (say), then the specific gravity S of the mixture is given by*

$$S = \frac{W+W+\cdots \text{to } n \text{ terms}}{\frac{W}{S_1}+\frac{W}{S_2}+\cdots+\frac{W}{S_n}}$$

$$= \frac{nW}{W\left(\frac{1}{S_1}+\frac{1}{S_2}+\cdots+\frac{1}{S_n}\right)}$$

$$= \frac{n}{\frac{1}{S_1}+\frac{1}{S_2}+\cdots+\frac{1}{S_n}}.$$

Thus, the specific gravity of a mixture of equal weights of different liquids when there is no change of volume on account of mixing is the harmonic mean of the specific gravities of the component substances.

2.7 An Important Result

The specific gravities of a mixture of n liquids is greater when equal volumes are taken than when equal weights are taken, assuming no change in volume as a result of mixing.

Let the specific gravities of the mixture when equal volumes and equal weights of n liquids are mixed, be S and S' and let $S_1, S_2, \ldots, S_n$ be the specific gravities of different substances. Then we have

$$\begin{aligned} S &= \frac{VS_1 + VS_2 + \cdots + VS_n}{V + V + \cdots \text{to } n \text{ terms}} \\ &= \frac{V(S_1 + S_2 + \cdots + S_n)}{nV} \quad \text{(V is the volume of each liquid)} \\ &= \frac{S_1 + S_2 + \cdots + S_n}{n} \end{aligned}$$

$$\text{i.e.,} \quad S = \text{Arithmetic Mean of } S_1, S_2, \ldots, S_n \tag{2.1}$$

$$\begin{aligned} \text{and} \quad S' &= \frac{W + W + \cdots \text{to } n \text{ terms}}{\frac{W}{S_1} + \frac{W}{S_2} + \cdots + \frac{W}{S_n}} \\ &= \frac{nW}{W\left(\frac{1}{S_1} + \frac{1}{S_2} + \cdots + \frac{1}{S_n}\right)} \quad \text{(W is the weight of each liquid)} \\ &= \frac{n}{\frac{1}{S_1} + \frac{1}{S_2} + \cdots + \frac{1}{S_n}} \end{aligned}$$

$$\text{i.e.,} \quad S' = \text{Harmonic Mean of } S_1, S_2, \ldots, S_n \tag{2.2}$$

Now we know that the arithmetic mean of n unequal positive numbers is greater than their harmonic mean. Hence, from (2.1) and (2.2), we conclude that

$$S > S'.$$

Note 1. *Cases of three and four liquids can be treated in the similar manner.*

Note 2. *The theorem is also true when instead of specific gravities, densities are taken into consideration.*

Illustrative Examples

► **Ex. 1.** A mixture is made of 7 cu. cms. of sulphuric acid (sp. gr. 1.843) and 3 cu. cms. of distilled water and its specific gravity when cold, is found to be 1.615. What contraction has taken place?

Solution: Let V be the volume of the mixture. Now the specific gravity of the mixture

$$= \frac{V_1S_1 + V_2S_2}{V_1 + V_2}. \tag{1}$$

Here, $S = 1.615$, $V_1 = 7$, $S_1 = 1.843$, $V_2 = 3$, $S_2 = 1$.

Hence from (1), we get

$$1.615 = \frac{7 \times 1.843 + 3 \times 1}{V}$$

$$V = \frac{12.901 + 3}{1.615} \text{c.c.} = \frac{15.901\text{c.c.}}{1.165} \text{c.c.}$$

Original volume $= (7 + 3)$c.c. $= 10$ c.c.

Hence, amount of contraction

$$= 10 - \frac{15901}{1615} = \frac{16150 - 15901}{1615} = \frac{249}{1615} \text{ c.c.} \quad \textbf{Ans.}$$

▶ **Ex. 2.** The specific gravities of pure gold and copper are 19.3 and 8.62. Find the specific gravity of standard gold which is an alloy of gold and copper in the proportion of 1 : 1.

Solution: Let the volumes of pure gold and copper be 11V and V, respectively. Let S be the specific gravity of standard gold, then

$$S = \frac{V_1 S_1 + V_2 S_2}{V_1 + V_2}. \tag{1}$$

Here, $V_1 = 11V$, $V_2 = V$, $S_1 = 19.3$, $S_2 = 8.62$.

Hence, from (1)

$$S = \frac{11V \times 19.3 + V \times 8.62}{11V + V} = \frac{212.3 + 3.62}{12}$$

$$= \frac{220.92}{12} = 18.41 \quad \textbf{Ans.}$$

▶ **Ex. 3.** When equal volumes of two substances are mixed, the specific gravity of the mixture is S_1; when equal weights of the same substances are mixed, the specific gravity is S_2. Find the specific gravity of the substance.

[Cal 45, Bih 56S, Ran 67A, 71A, Mag 68A, L.N.M.U. 75A]

Solution: Let the specific gravities of two substances be S and S' respectively, where $S > S'$. Then

$$S_1 = \frac{VS + VS'}{V + V} = \frac{S + S'}{2} \tag{1}$$

$$\text{and} \quad S_2 = \frac{W + W}{\frac{W}{S} + \frac{W}{S'}} = \frac{2SS'}{S + S'}. \tag{2}$$

From (1), we have

$$S + S' = 2S_1. \tag{3}$$

Hence from (2), we get

$$S_2 = \frac{2SS'}{2S_1}. \quad \text{or,} \quad S_1 S_2 = SS'.$$

Now, $(S - S')^2 = (S + S')^2 - 4SS' = 4S_1^2 - 4S_1S_2 = 4(S_1^2 - S_1S_2)$

$$\therefore\ S - S' = 2\sqrt{S_1^2 - S_1S_2}. \tag{4}$$

Adding (3) and (4), we get

$$2S = 2\left[S_1 + \sqrt{S_1^2 - S_1S_2}\right] \quad \text{or,} \quad S = S_1 + \sqrt{S_1^2 - S_1S_2}.$$

Subtracting (4) from (3), we get

$$2S' = 2\left[S_1 - \sqrt{S_1^2 - S_1S_2}\right]$$

$$S' = S_1 - \sqrt{S_1^2 - S_1S_2} \quad \textbf{Ans.}$$

▶ **Ex. 4.** Three equal vessels, A, B and C are half full of liquids of densities ρ_1, ρ_2 and ρ_3, respectively. If now B be filled from A and then C from B, prove that the density of the liquid now contained in C is $\frac{\rho_1+\rho_2+2\rho_3}{4}$, the liquids being supposed to mix completely.

Solution: Let V be the volume of either vessel. Now, when B is filled with A; if contains a mixture of liquids of densities ρ_1 and ρ_2 (*mixed in equal volumes V each*). Let its density be x, then

$$2Vx = V\rho_1 + V\rho_2 \quad \text{or,} \quad x = \frac{\rho_1 + \rho_2}{2}.$$

Now when the mixture is poured into C, the vessel C contains liquid of densities ρ_3 and x mixed in equal volume V each. Let y be the density of the mixture, then

$$2Vy = V\rho_3 + Vx$$

$$\text{or,} \quad 2y = \rho_3 + x = \rho_3 + \frac{\rho_1 + \rho_2}{2} = \frac{\rho_1 + \rho_2 + 2\rho_3}{2}$$

Hence,

$$y = \frac{\rho_1 + \rho_2 + 2\rho_3}{4} \quad \textbf{Proved.}$$

▶ **Ex. 5.** How much water must be added to 2703 of a salt solution whose specific gravity is 1.08 so that specific gravity of the mixture be 1.05?

Solution: Let Wozs of water be added to the salt solution. Also the specific gravity of water is:

Using the formula

$$S = \frac{W_1 + W_2}{\frac{W_1}{S_1} + \frac{W_2}{S_2}}. \tag{1}$$

Here, $S = 1.05$, $W_1 = 27$, $W_2 = W$, $S_1 = 1.08$, $S_2 = 1$.

Hence, from (1), we get

$$1.05 = \frac{27+W}{\frac{27}{1.08}+\frac{W}{1}} \quad \text{or,} \quad \frac{105}{100} = \frac{27+W}{\frac{27\times 100}{108}+W}$$

$$\text{or,} \quad \frac{105}{100} = \frac{(27+W)108}{2700+108W} \quad \text{or,} \quad \frac{21}{20} = \frac{(27+W)108}{(25+W)108}$$

$$\text{or,} \quad \frac{21}{20} = \frac{27+W}{25+W} \quad \text{or,} \quad 525+21W = 540+20W$$

or, $W = 15$ ozs **Ans.**

▶ **Ex. 6.** Equal volumes of three liquids are mixed and the mixture separated into three parts; to each of these parts is added its own volume of the original liquids, and the densities so formed are in the ratio 3 : 4 : 5. Prove that the densities of the liquids are as 1 : 2 : 3.

Solution: Let the volume of each liquid be V, let the densities of the three liquids be ρ_1, ρ_2, ρ_3 and the density of the mixture be ρ. Then

$$\rho = \frac{\rho_1 V + \rho_2 V + \rho_3 V}{V+V+V} = \frac{\rho_1+\rho_2+\rho_3}{3}.$$

Let the liquid of density ρ_1 be added to one part of the mixture. Then density of this mixture

$$= \frac{\frac{\rho_1+\rho_2+\rho_3}{3}+\rho_1 V}{V+V} = \frac{4\rho_1+\rho_2+\rho_3}{6}.$$

Similarly, the densities of other two mixtures are

$$\frac{\rho_1+4\rho_2+\rho_3}{6} \quad \text{and} \quad \frac{\rho_1+\rho_2+4\rho_3}{6}.$$

According to the question

$$\frac{4\rho_1+\rho_2+\rho_3}{\rho_1+4\rho_2+\rho_3} = \frac{3}{4} \qquad (1)$$

$$\text{and} \quad \frac{\rho_1+4\rho_2+\rho_3}{\rho_1+\rho_2+4\rho_3} = \frac{4}{5}. \qquad (2)$$

From (1), we get $16\rho_1+4\rho_2+4\rho_3 = 3\rho_1+12\rho_2+3\rho_3$, or, $13\rho_1-8\rho_2+\rho_3=0$.

From (2), we get $5\rho_1+20\rho_2+5\rho_3 = 4\rho_1+4\rho_2+16\rho_3$, or, $\rho_1+16\rho_2-11\rho_3=0$.

Hence, we get

$$13\rho_1-8\rho_2+\rho_3=0$$

$$\rho_1+16\rho_2-11\rho_3=0$$

$$\therefore \quad \frac{\rho_1}{88-16} = \frac{\rho_2}{1+143} = \frac{\rho_3}{208+8}$$

$$\text{or,} \quad \frac{\rho_1}{72} = \frac{\rho_2}{144} = \frac{\rho_3}{216}.$$

Hence,

$$\frac{\rho_1}{1} = \frac{\rho_2}{2} = \frac{\rho_3}{3}.$$

▶ **Ex. 7.** Pure water is added, drop by drop, to a vessel of volume V filled with a salt solution of specific gravity S which is allowed to overflow. Find the specific gravity of the solution when a volume V of water has been poured in.

[L.N.M.U. 75H, Bih. 57H, R.U. 62H, M.U. 64H, Bhag. 72H]

Solution: Let σ denote the specific gravity when a volume ν of water has been added and $\sigma + d\sigma$ the specific gravity when a volume $\nu + d\nu$ has been added, i.e., $d\nu$ may denote the volume of a drop of water.

The addition of a drop at this stage means that a volume $d\nu$ of specific gravity 1 is added to a volume V of specific gravity σ and forms a total volume $V + d\nu$ of specific gravity $\sigma + d\sigma$, where of a drop overflows. Hence, equating the weight of the mixture to the sum of the weights of the constituents, we have

$$(V + d\nu)(\sigma + d\sigma) = d\nu + V\sigma$$
$$\text{or,} \quad V\sigma + Vd\sigma + d\nu\sigma + d\nu d\sigma = d\nu + V\sigma.$$

As $d\nu d\sigma$ are very small, we can neglect their product. Then

$$Vd\sigma + d\nu\sigma = d\nu$$
$$\text{or,} \quad Vd\sigma + d\nu(\sigma - 1) = 0$$
$$\text{or,} \quad \frac{d\nu}{V} + \frac{d\sigma}{\sigma - 1} = 0.$$

Integrating, we get

$$\frac{\nu}{V} + \log(\sigma - 1) = C. \qquad (1)$$

But, when $\nu = 0$, then $\sigma = S$ so that $C = \log(S - 1)$.

Hence, from (1), we get

$$\frac{\nu}{V} + \log(\sigma - 1) = \log(S - 1)$$
$$\log(\sigma - 1) = \log(S - 1) - \frac{\nu}{V}$$
$$\text{or,} \quad \log(\sigma - 1) = \log(S - 1) + \log_e^{-(\nu/V)}$$
$$\text{or,} \quad \log(\sigma - 1) = \log\left[(S - 1)e^{-(\nu/V)}\right]$$
$$\text{or,} \quad \sigma - 1 = (S - 1)e^{-(\nu/V)}$$
$$\text{or,} \quad \sigma = 1 + (S - 1)e^{-(\nu/V)} \quad \textbf{Ans.}$$

Exercises 2(I)

1. When equal volumes of two liquids are mixed the specific gravity is 4, when equal weights of the same substances are mixed the specific gravity is 3. Find the specific gravities of the substances.

   [Bih. 80, R.U. 67A, M.U. 68A]

   [*Ans.* 2, 6]

2. Two liquids of specific gravities S_1, S_2 and of volumes V_1, V_2 having been mixed, the specific gravity of the mixture is found to be σ. Find the volume of the mixture. [M.U. 84]

   [*Ans.* $\frac{V_1S_1+V_2S_2}{\sigma}$]

3. When a vessel is filled by means of equal volumes of two liquids, the specific gravity of the compound is $\frac{4}{3}$ of what it would have been. If the vessel would have been filled by means of equal weights of fluids, compare the specific gravities of the two liquids. [*Ans.* 3 : 1 or 1 : 3]

4. A mixture is formed of two fluids. The density ρ of the mixture, the ratio $m : 1$ of the volumes and the ratio $n : 1$ of the densities are given. Prove that the densities of the fluids are

$$\frac{mn+n}{mn+1}\rho \quad \text{and} \quad \frac{m+1}{mn+1}\rho.$$

5. The density of mixture of equal volumes of two liquids is a and that of a mixture of equal weights of the same liquids is b. Show that the specific gravies of the liquids are given by $x^2 - 2ax + ab = 0$.

   [R.U. 77]

6. Equal weights of gold, silver and an alloy of gold and silver are dipped successively in a cylinder of water and they cause water to rise a, b, c centimetres, respectively. Prove that the alloy contains gold and silver in the ratio $b - c : c - a$ by weight.

7. The mixture of a quantity of a liquid A with x lbs of B has a specific gravity S, with $2x$ lbs of B a specific gravity S', with $3x$ lbs of B a specific gravity S'', show that

$$2(S - S_1)(S_2 - S') = (S_2 - S)(S' - S_1)$$
$$3(S - S_1)(S_2 - S'') = (S_2 - S)(S'' - S_1),$$

   where S_1 and S_2 are specific gravities of the liquids.

8. The density of milk in the vessel of a milkman was found to be ρ. If the density of the pure milk is σ, show that the ratio of weights of water and milk on the adulterated liquid is $\sigma - \rho : \sigma(\rho - 1)$. [M.U. 82]

   Also prove that the volumes of water and milk in the mixture bear a ratio $\sigma - \rho : \rho - 1$. [R.U. 82]

9. Having given the specific gravity ρ of a mixture formed of equal volumes of two fluids and also the specific gravity σ of a mixture formed by taking a volume of one fluid double that of the other, prove that the specific gravities of the fluids are

$$(3\sigma - \rho) \text{ and } (4\rho - 3\sigma).$$

10. From a vessel full of liquid of specific gravity S_1 is removed $\frac{1}{m}$th of the contents and it is filled up with a liquid of specific gravity S_2. If

this operation be repeated n times, find the resulting specific gravity of the liquid in the vessel. If n is very large, prove that the limiting specific gravity is $\frac{S_1 - S_2}{\rho} + S_2$.

[*Ans.* $\left(1 - \frac{1}{m}\right)^n (S_1 - S_2) + S_2$]

11. A mixture is formed of equal volumes of n liquids, the densities of which are in the ratio of the numbers 1, 2, 3, ..., n; find the density of the mixture. Also find the density of the mixture when the volumes are in the ratio:

    (a) of the numbers 1, 2, 3, ..., n and

    (b) of the numbers n, $n - 1$, ..., 3, 2, 1.

    [*Ans.* $(n + 1) : 2$, (*a*) $(2n + 1) : 3$, (*b*) $(n + 2) : 3$]

12. A vessel contains a certain volume of liquid of specific gravity ρ and another contains an equal volume of liquid of specific gravity σ, $\frac{1}{n}$th of each is taken out and put into the other and well mixed; this process is repeated m times. Show that the final specific gravities are

$$\rho + \frac{\sigma - \rho}{2}\left\{1 - \left(1 - \frac{2}{n}\right)^m\right\}$$

and

$$\sigma + \frac{\rho - \sigma}{2}\left\{1 - \left(1 - \frac{2}{n}\right)^m\right\}.$$

13. A mixture is formed of equal volumes of n liquids, the specific gravities of which are S, 2S, 3S, ..., nS. Find the specific gravity of the mixture, assuming that at no change in volume takes place. [*Ans.* $\frac{n+1}{2}S$]

14. A mixture of oxygen and nitrogen is found to consist of 21 parts of oxygen to 79 nitrogen by volume, or by weight 23 parts to 77. Compare the densities of the gases.

    [*Ans.* 1817 : 1617]

15. A given body weights four times as much in vacuum as in water and one-third as much again in water as in another liquid. Find the specific gravity of the latter liquid. [*Ans.* $\frac{13}{12}$]

16. The specific gravity of cork is 24. Find the volume of water that weights as much has a cubic yard of cork. [*Ans.* 24 cu. yd]

CHAPTER 3

Fluid Pressure

3.1 Fluid Pressure

Let a vessel with a horizontal base and vertical sides be filled with any fluid and a hole be made in the side of the vessel and further let this hole be covered by a plate, which exactly fits in it.

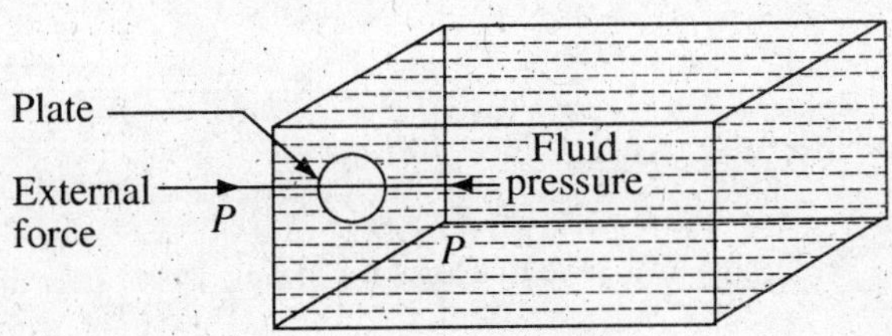

Fig. 3.1

Now it will be observed that the plate can be kept at rest only if some external force is applied to it, which shows that the fluid in side the vessel is also exerting some force on the plate. If P be the force necessary to keep the plate in equilibrium, then P is called the *fluid pressure* or the *fluid thrust* on the whole area of the plate.

- *Uniform Pressure:* If the fluid pressure is the same for each equal element of the area which can be on the horizontal base of the vessel or in other words the fluid pressure or any portion of the plane area is proportional to the area of that portion, the pressure on that area is said to be uniform.

  On the other hand, if the pressure is not the same for different equal areas as it would be when the areas are taken on the vertical side of the vessel, the fluid pressure is said to be varying or non-uniform.

- *Mean Pressure:* The mean pressure on a given plane area is the uniform pressure on it which will give the same resultant thrust on that area as the actual total pressure. Thus, if A be the area upon which a fluid thrust P is acting, then the mean pressure is $\frac{P}{A}$.

3.2 Pressure at a Point

The pressure at a point of a surface subjected to a fluid thrust is defined as the limit to which the ratio of the thrust on a very small area of the surface enclosing the point and the elementary area tends, as the elementary area is indefinitely diminished. In other words, the pressure at any point of an area is the limit of the mean pressure on a very small area enclosing that point.

Thus, if δP be the fluid pressure on a small area δA enclosing the given point in the fluid, the mean pressure per unit area

$$= \frac{\delta P}{\delta A}.$$

Now, let δA be diminished indefinitely, i.e., when $\delta A \to 0$, then

$$\lim_{\delta A \to 0} \frac{\delta P}{\delta A} = p$$

$$\text{or,} \quad p = \frac{dp}{dA}.$$

This p is called the *pressure* at the given point. Sometimes pressure at any point is also called the *intensity of pressure* at that point.

3.3 Equality of Pressure in All Directions

The pressure at a point in a fluid in equilibrium is the same in all directions.

[M.U. 84A, 97H, L.N.M.U. 86H, C.U. 84H, 93H, 95H, 04H, R.U. 74H, 91H; 93H]

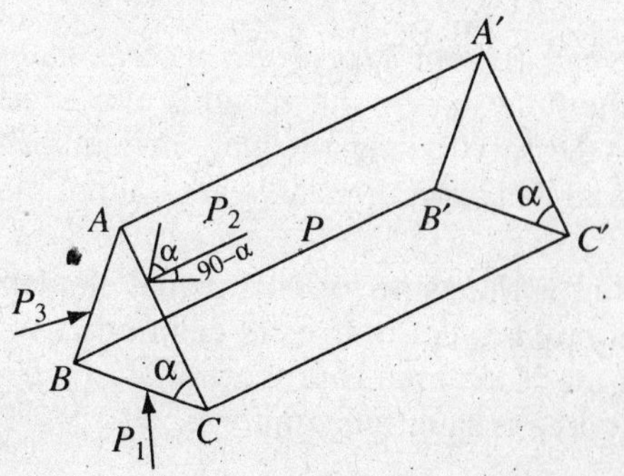

Fig. 3.2

Let the point P be enclosed by a small triangular prism $(ABCA'B'C')$ such that $\angle B = 90^\circ$ and hence $\angle B' = 90^\circ$, BC is horizontal and hence $B'C'$ is horizontal, AB is vertical and hence $A'B'$ is vertical.

Let

$$\angle C = \angle C' = \alpha.$$

Let P_1, P_2, P_3 be the average pressures on the rectangular faces $BCC'B'$, $CAA'C'$, $ABB'A'$, respectively and p_1, p_2, p_3 be the pressures at point P in directions normal to these faces.

Let W be the weight per unit volume of the fluid and l, the length of the prism.

We consider the equilibrium of the fluid prism and resolving the forces horizontally, we get

$$\begin{aligned} P_3(AB\cdot l) &= P_2(CA\cdot l)\cos(90-\alpha) \\ \text{or,}\quad P_3AB &\doteq P_2CA\sin\alpha. \end{aligned} \tag{3.1}$$

In $\triangle ABC$

$$\sin\alpha = \frac{AB}{AC}$$

$$\therefore\ AB = AC\sin\alpha.$$

Hence, from (3.1), we get

$$P_3AB = P_2AB \quad \text{or,} \quad P_3 = P_2.$$

When the prism shrinks to point P,

$$P_2 \to p_2 \quad \text{and} \quad P_3 \to p_3$$

$$\therefore \quad p_3 = p_2 \tag{3.2}$$

Again, resolving the forces vertically we have

$$P_1\cdot(BC\cdot l) - P_2\cdot(AC\cdot l)\cos\alpha - W\left(\frac{1}{2}AB\cdot BC\right)l = 0$$

$$\text{or,}\quad P_1\cdot BC - P_2(AC\cos\alpha) - W\left(\frac{1}{2}AB\cdot BC\right) = 0 \tag{3.3}$$

In $\triangle ABC$

$$\cos\alpha = \frac{BC}{AC},\ \text{or, } BC = AC\cos\alpha.$$

Hence, from (3.3), we get

$$P_1BC - P_2BC - W\left(\frac{1}{2}AB\cdot BC\right) = 0$$

$$\text{or,}\quad P_1 - P_2 = W\left(\frac{1}{2}AB\right).$$

When the prism reduces to point P

$$P_1 \to p_1, P_2 \to p \text{ and } AB \to 0$$

$$\therefore \quad p_1 - p_2 = 0 \tag{3.4}$$

From (3.2) and (3.4), we have

$$p_1 = p_2 = p_3$$

which proves the proposition.

3.4 Transmissibility of Fluid Pressure

Any pressure or increase of pressure applied to any point of a fluid at rest is transmitted equally to every part of the liquid. This principle of Transmissibility of Fluid Pressure is known as *Pascal's Law.* [R.U. 92H, 94H]

This property of fluid is a direct result of experiment. We shall deduce it from the theoretical considerations as follows:

Let A and B be two points in a fluid at rest. Join A and B. Then two cases arise line AB may be wholly inside the fluid or it may not lie entirely in the fluid. We will consider these cases separately.

Case I. *The line AB is entirely in the fluid.*

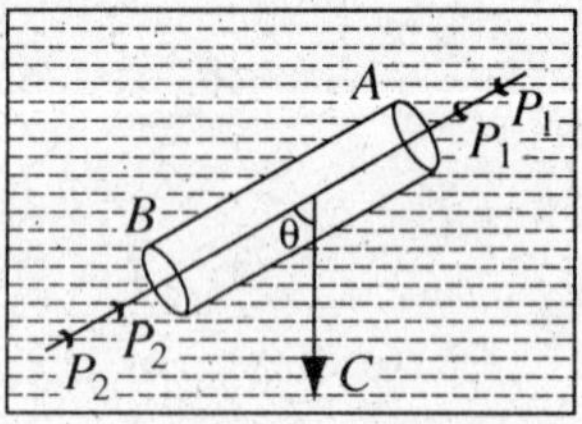

Fig. 3.3

We construct a cylinder of small cross-sectional area α with AB as the axis and the end planes at A and B, at right angles to AB. Let p_1 and p_2 be the pressures at points A and B, respectively. Since the ends are very small, the pressures, there may be assumed to be uniform.

The fluid contained in the cylinder is in equilibrium under the action of the following forces:

- Fluid pressure $p_1\alpha$ on the end plane at A in the direction of AB.
- Fluid pressure $p_2\alpha$ on the end plane at B in the direction of BA.
- Weight W of the fluid, acting vertically downwards.
- Pressure on the curved surface, of the cylinder which is at every point perpendicular to AB.
- The resultant external force on the fluid inside the cylinder. Let X be its component along AB.

Considering the equilibrium of the fluid-cylinder and resolving along AB, we have

$$p_1\alpha - p_2\alpha + W\cos\theta + X = 0, \tag{3.5}$$

where θ is the angle at which the axis of cylinder is inclined to the vertical.

Let an additional pressure P_1 be given at A and in consequence let P_2 be the correspondinğ increase of pressure at B. Then similarly, we have

$$(p_1 + P_1)\alpha + W\cos\theta - (p_1 + P_2)\alpha + X - 0. \tag{3.6}$$

Subtracting (3.5) from (3.6), we have

$$P_1\alpha - P_2\alpha = 0 \quad \text{or,} \quad P_1 = P_2,$$

i.e., Increase of pressure at A = Increase of pressure at B.

Case II. *The line AB dc°s not lie entirely in the fluid.*

Join A and B by a series of straight lines AC, CD, DB such that each of these lies entirely inside the fluid.

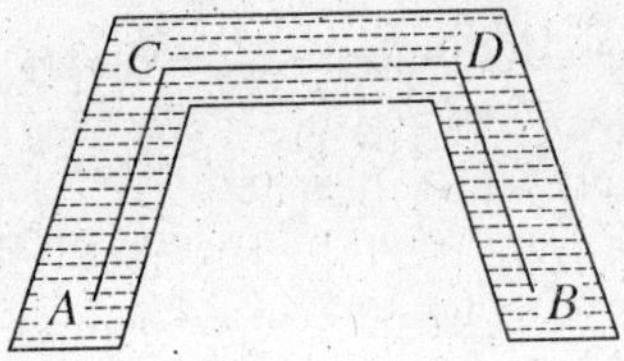

Fig. 3.4

Then by case I, we have

$$\text{Increase of pressure at A} = \text{Increase of pressure at C}$$
$$\text{Increase of pressure at C} = \text{Increase of pressure at D}$$
$$\text{Increase of pressure at D} = \text{Increase of pressure at B}$$

From these we conclude that

$$\text{Increase of pressure at A} = \text{Increase of pressure at B}$$

which proves the proposition.

3.5 Fluids at Rest under the Action of Gravity

From this point onwards we consider more specially the case of the portion of a fluid which is in equilibrium due to the weight of the portion of the fluid considered and the thrusts due, either the action of the surrounding fluid or, to solids with which the fluid is in contact. The case, where fluid is light, i.e., gravity does not act on it, will be expressly indicated.

3.6 Pressure at Points in a Horizontal Plane

Theorem 3.6.1 *In a fluid, at rest under gravity, the pressure is the same at all points in the same horizontal plane.* [R.U. 80H, 94H, L.N.M.U. 93H]

Let A and B be any two points in fluid in the same horizontal plane. Join A and B. Then two cases arise: line AB may be wholly inside the fluid or it may not lie entirely in the fluid. We will consider the two cases separately.

Case I. *When the straight line AB lies wholly inside the fluid.*

With AB as axis, we construct a right circular cylinder of small cross-sectional area α. Let p_1 and p_2 be the pressures at A and B, respectively.

Since the ends of the cylinder are small, we can assume the pressure on them to be uniform and equal to the pressures at A and B, i.e., p_1 and p_2, respectively.

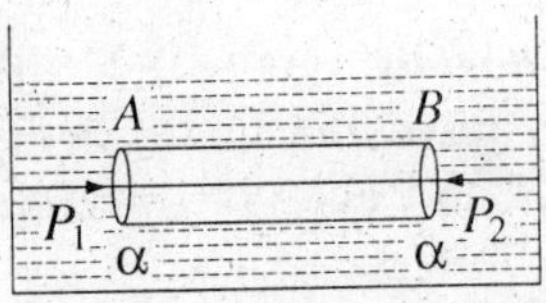

Fig. 3.5

Now the fluid in the cylinder is in equilibrium under the action of the following forces:

- The pressure $p_1\alpha$ on the end at A along AB.
- The pressure $p_2\alpha$ on the end at B along BA.
- The weight of the fluid in the cylinder acting vertically downwards.
- The pressure at each point on the curved surface of the cylinder, which are perpendicular to AB.

Hence, for equilibrium resolving all the forces along AB, we get

$$p_1\alpha - p_2\alpha = 0 \quad \text{or,} \quad p_1 = p_2,$$

i.e., Pressure at A = Pressure at B.

Case II. *When the straight line AB does not lie wholly inside the fluid.*

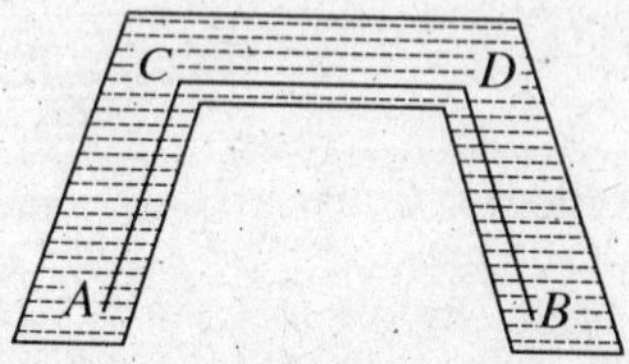

Fig. 3.6

Join A and B by a series of straight lines AC, CD, DB such that each of them lie wholly in the fluid. Then by case I

Pressure at A = Pressure at C
= Pressure at D
= Pressure at B.

Hence, the pressures at A and B are same.

3.7 Pressure at a Point in a Liquid

[M.U. 96H, 99H]

Case I. *When the liquid is heterogeneous.*

Let P and Q be two points on the same vertical in the liquid at the depths z and z + dz beneath the free surface. Then PQ = dz. With PQ as axis, describe a right circular cylinder of small cross-sectional area α.

Let p and p + dp be the pressures at points P and Q, respectively. Since the cylinder is small, the variation of the density of the liquid in it may be neglected and so let the density be ρ.

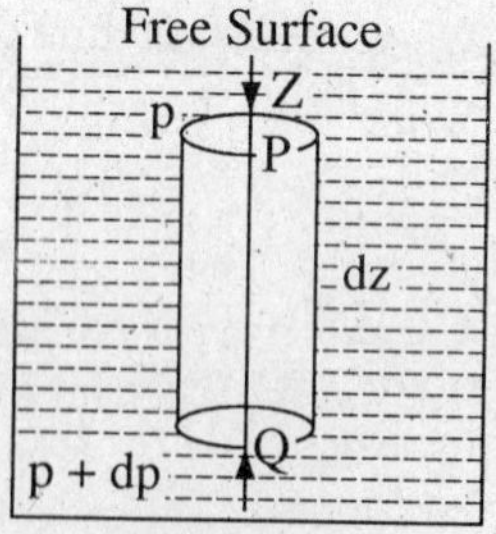

Fig. 3.7

The forces acting on the cylinder of liquid are:

- The thrust $p\alpha$ on the plane face at P, acting vertically downwards.

- The thrust $(p+\delta p)\alpha$ on the plane face at Q, acting vertically downwards.
- The thrust on the curved surface acting horizontally.
- The weight $\rho g\alpha dz$ of the liquid inside the cylinder acting vertically downwards.

Hence, for equilibrium resolving all the forces; acting on the cylinder, vertically. We get

$$\alpha(p+dp) - p\alpha = \rho g dz\alpha$$

$$\text{or,} \quad dp = \rho g dz \quad \text{or,} \quad \frac{dp}{dz} = \rho g.$$

This is known as *pressure equation.*
Integrating, we get

$$p = g\int \rho dz + A,$$

where A is the constant of integration.

Case II. *When the liquid is homogeneous.*

From the pressure equation we have

$$\frac{dp}{dz} = \rho g, \quad \text{or,} \quad dp = \rho g dz.$$

Integrating, we get

$$p = g\int \rho dz.$$

In a homogeneous liquid ρ is constant

$$p = \rho g\int dz$$

$$p = \rho g z + C \qquad (3.7)$$

Let π be the atmospheric pressure. On the free surface $z = 0$, $p = \pi$, so from (1) we get $C = \pi$.

Substituting this value of C in (3.7), we get

$$p = \pi + \rho g z. \qquad \text{[T.M.B.U. 2001H]}$$

If the atmospheric pressure be neglected, then $p = \rho g z$, i.e., $p \propto z$.

That is, *the pressure at a point varies as the depth of the point below the surface when the atmospheric pressure is neglected.*

If we take into account the atmospheric pressure, let us imagine the atmospheric pressure to be equivalent to the pressure due to a stratum of the liquid of thickness $\frac{\pi}{\rho g} = h$ (*say*). Then the pressure at a depth h below the original surface of the liquid $= \rho g(h+z)$.

The upper surface of the superimposed liquid is termed the *effective surface.* Hence, *the pressure at any point in a homogeneous liquid is proportional to the depth below the effective surface.*

3.8 In a Homogeneous Fluid at rest under Gravity the Difference between the Pressures at Two Points is Proportional to the Difference of Their Depths

The following three cases arise: [P.U. 95H, L.N.M.U. 94H]

Case I. *When the line joining the two points is vertical.*

Let A and B be the two points, in the same vertical line in a homogeneous fluid of density ρ. Produce BA to meet the free surface at O.

Let $AB = h$ and p_A, p_B be the pressures at A and B, respectively.

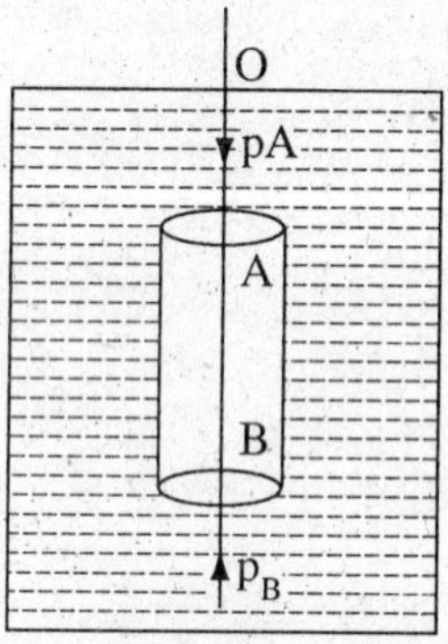

Fig. 3.8

Then by the pressure equation

$$\frac{dp}{dz} = \rho g \quad \text{or,} \quad dp = \rho g dz$$

$$\text{or,} \quad \int_{p_A}^{p_B} dp = \rho g \int_{OA}^{OB} dz$$

$$\text{or,} \quad p_B - p_A = \rho g(OB - OA)$$

$$= \rho g AB = \rho g h$$

i.e., the difference of pressures at B and A $\propto$ difference of the depths of B and A.

Case II. *When the line joining the two points is not vertical.*

Let C be the point in the horizontal plane through B and vertically below A.

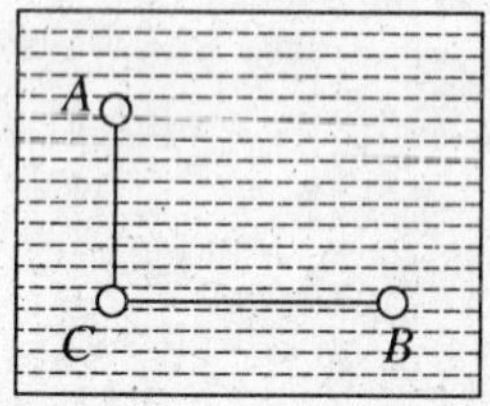

Fig. 3.9

Since A and C are in the same vertical line, so by case I

$$p_C - p_A = \rho g AC. \qquad (3.8)$$

Also since C and B are in the same horizontal plane, we have

Pressure at C = Pressure at B, i.e., $p_C = p_B$.

Hence from (3.8), we get $p_B - p_A = \rho g h$,

where $AC = h = \rho g \times$ difference of depths of B and A.

Case III. *When the line joining the two points is neither vertical nor lies entirely in the fluid.*

Let C and D be two points in the same horizontal plane and vertically below A and B, respectively so that CD completely lies in the fluid. Then by case I, we have

$$p_C - p_A = \rho g AC \qquad (3.9)$$

$$p_D - p_B = \rho g BD \qquad (3.10)$$

Subtracting (3.10) from (3.9), we get

$$(p_C - p_D) + (p_B - p_A) = \rho g(AC - BD). \quad (3.11)$$

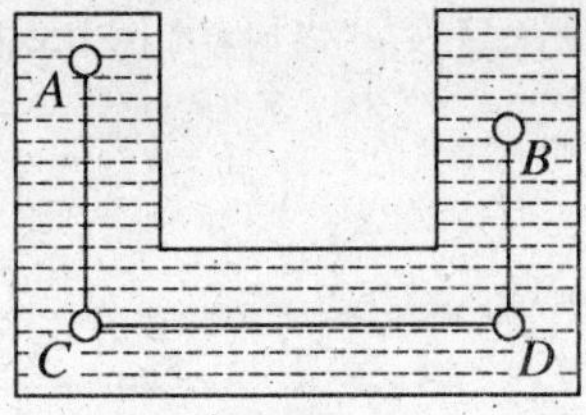

Fig. 3.10

Also since C and D are in the same horizontal plane; we have $p_C = p_D$.

Hence from (3.11), we have

$$p_B - p_A = \rho g(AC - BP) = \rho g \text{ (difference of depths of B and A).}$$

If the shape of the vessel containing the liquid is as shown in the figure, by case I, we have

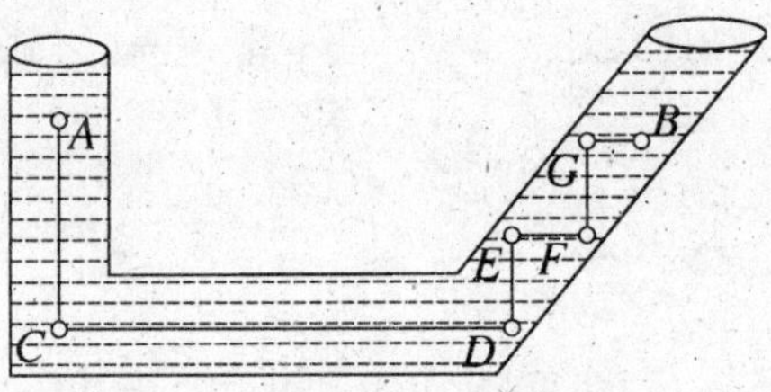

Fig. 3.11

$$p_C - p_A = \rho g AC \quad (3.12)$$

Also $$p_C = p_D \quad (3.13)$$

Again, $$p_D - p_E = \rho g ED \quad (3.14)$$

Also $$p_E = p_F \quad (3.15)$$

$$p_F - p_G = \rho g GF \quad (3.16)$$

Also $$p_G = p_B \quad (3.17)$$

Adding (3.14) and (3.16) and subtracting the sum from (3.12), we get

$$p_C - p_A - [p_D - p_E + p_F - p_G] = \rho g(AC - ED - GF)$$

or, $P_G - p_A = \rho g[AC - (ED + GF)$ [From (3.13) and (3.15)]

or, $p_B - p_A = \rho g$ (difference of depths of B and A) [From (3.17)].

Corollary 3.8.1 *If A and B are in the same horizontal line; $p_B - p_A = 0$ in all cases, i.e., $p_B = p_A$.*

Thus, *in a fluid at rest under gravity, horizontal planes are surfaces of equal pressures.* **[B.B.A.B.U. 92H]**

Corollary 3.8.2 *The difference of pressures between two points in a homogeneous fluid at rest under gravity is equal to the weights of a column of fluid whose cross-sectional area is unity and whose height is equal to the difference of depths of the two points.*

3.9 Pressure at a Point in the Lowest Layer of Different Liquids which do not Mix

[R.U. 79H, L.N.M.U. 93H]

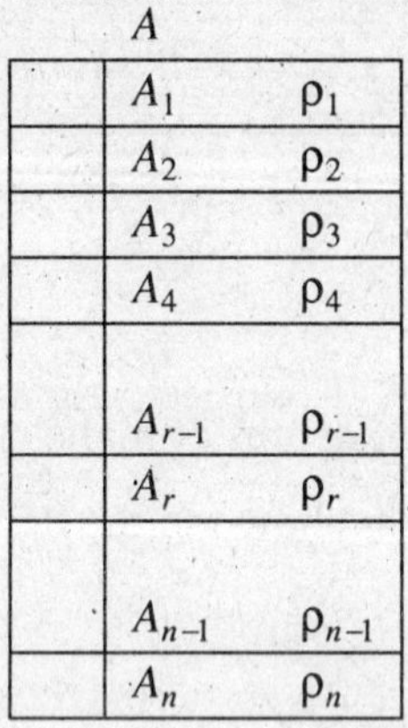

Fig. 3.12

When different liquids are placed, one over the other, it has been seen by common experience that the liquid having lowest density occupies the uppermost position and the liquid having the highest density, the lowermost. The positions of the strata of various liquids from the top to bottom are in the order of their increasing density.

Suppose that there are n liquids in-equilibrium of densities $\rho_1, \rho_2, \ldots, \rho_n$, such that $\rho_1 < \rho_2 < \rho_3 < \cdots < \rho_n$.

Let a vertical line AA_n be drawn, such that A is in the surface of the uppermost liquid and the point A_n in the lower surface of the lowermost liquid, at which the pressure is to be determined. Suppose AA_n intersects the planes of separation of liquids at A_1, A_2, A_r, A_{n-1} and $AA_1 = h_1, A_1A_2 = h_2, \ldots, A_{r-1}A_r = h_r, \ldots$

Also let the atmospheric pressure be π and the pressure at the point A_r is ρ_r, then we have

$$p_1 - \pi = g\rho_1 h_1$$
$$p_2 - p_1 = g\rho_2 h_2$$
$$\ldots\ldots\ldots\ldots\ldots\ldots$$
$$\ldots\ldots\ldots\ldots\ldots\ldots$$
$$p_r - p_{r-1} = g\rho_r h_r$$
$$\ldots\ldots\ldots\ldots\ldots\ldots$$
$$\ldots\ldots\ldots\ldots\ldots\ldots$$
$$p_n - p_{n-1} = g\rho_n h_n.$$

Adding all these, we get

$$p_n - \pi = g(\rho_1 h_1 + \rho_2 h_2 + \cdots + \rho_n h_n)$$
$$p_n = \pi + g\sum_{r=1}^{n} \rho_r h_r,$$

3.10 Surfaces of Equal Pressure and Density

In a fluid at rest under gravity horizontal planes are surfaces of equal density.

[C.U. 90H, M.U. 2001H]

Let A and B be any two points in the same horizontal plane. Let A′ and B′ be two points close to A and B, and vertically above A and B, respectively so that $AA' = BB'$.

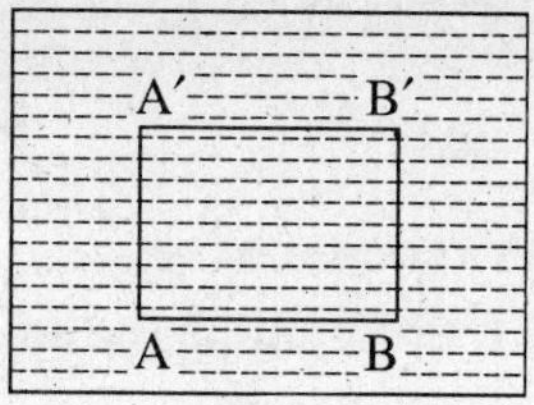

Fig. 3.13

Let ρ and ρ′ be the mean density of the fluid between, A and A′, B and B′.

If p_A, p_B, p'_A, p'_B denote the pressures at A, B, A′B′, respectively, then

$$p_A - p'_A = g\rho AA' \tag{3.18}$$

$$p_B - p'_B = g\rho' BB' \tag{3.19}$$

Since AB and A′B′ are horizontal. So

$$p_A = p_B,\ p'_A = p'_B. \tag{3.20}$$

Subtracting (3.19) from (3.18), we get

$$(p_A - p_B) - (p'_A - p'_B) = \rho g AA' - p' g BB'$$

$$0 = \rho g AA' - \rho' g BB' \quad \text{[From (3.20)]}$$

$$\text{or,} \quad \rho g AA' = \rho' g BB'$$

$$\text{or,} \quad \rho = \rho' \quad \text{[Since } AA' = BB'\text{]}.$$

It follows that density at A = density at B.

Hence, the density at points A and B are equal. But A and B are any two points in this horizontal plane. Hence, the densities at all points in the same horizontal plane are same.

Thus, *when a fluid is at rest under gravity, horizontal planes are surface of equal density. But in fluid at rest under gravity horizontal planes are also surfaces of equal pressure.*

Hence, it follows that in case of a fluid at rest under gravity horizontal planes are surfaces of equal density and equal pressure.

3.11 Free Surface of a Heavy Homogeneous Liquid at Rest under Gravity is Horizontal

[C.U. 90H, 94H, R.U. 95H, 98H]

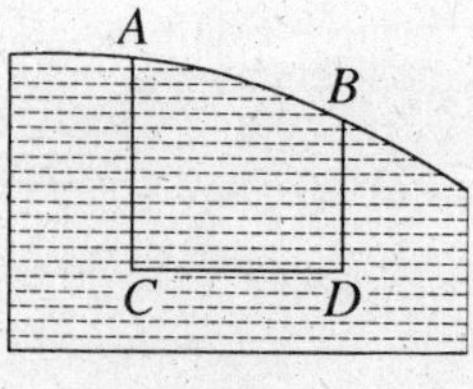

Fig. 3.14

Let C and D be any two points in the same horizontal plane in a liquid of density ρ and let the verticals through C and D meet the free surface of the liquid at A and B, respectively.

Since C and D are in the same horizontal plane.

$$\text{Pressure at C} = \text{Pressure at D}$$
$$\text{or,}\quad \pi + \rho g AC - \pi + \rho g BD,$$

where π is the atmospheric pressure. Hence,

$$AC = BD$$

which shows that A and B are in the same horizontal plane.

Again, since C and D are any two points in the same horizontal line, any line AB drawn in the free surface is horizontal.

Thus, *the free surface of a heavy homogeneous liquid at rest under gravity is horizontal.*

3.12 Common Surface of Two Heavy Homogeneous Fluids

When two fluids of different densities at rest under gravity do not mix their surface of separation is a horizontal plane. [R.U. 81H, 94H, 2000H, M.U. 2000H]

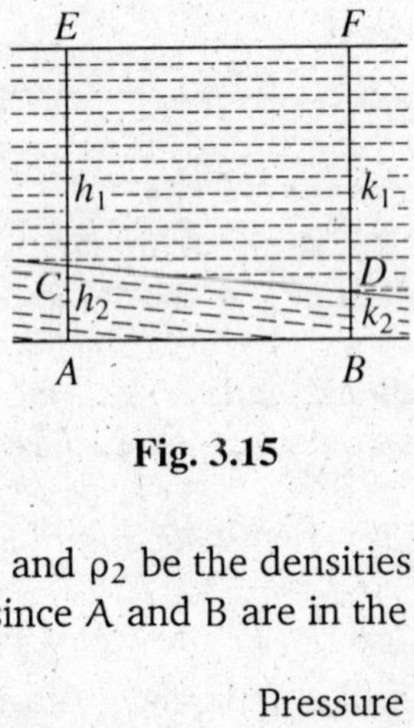

Fig. 3.15

Let A and B be any two points in the lower fluid and in the same horizontal plane. Through A and B we draw verticals to meet the common surface at C and D, and the free surface at E and F, respectively so that $AC = h_2$, $BD = k_2$ and $CE = h_1$, $DF = k_1$.

Let ρ_1 and ρ_2 be the densities of the upper and lower fluids, respectively. Now since A and B are in the same horizontal plane, we have

$$\text{Pressure at A} = \text{Pressure at B}$$
$$\text{or,}\quad \pi + \rho_1 g h_1 + \rho_2 g h_2 = \pi + \rho_2 g k_2 + \rho_1 g k_1$$
$$\text{or,}\quad g\rho_1 h_1 + g\rho_2 h_2 = g\rho_1 k_1 + g\rho_2 k_2$$
$$\text{or,}\quad \rho_1 h_1 + \rho_2 h_2 = \rho_1 k_1 + \rho_2 k_2 \qquad (3.21)$$

Since AB and EF are both horizontal, we have

$$AF = BF$$
$$\text{or,}\quad AC + CE = BD + DF$$
$$\text{or,}\quad h_1 + h_2 = k_1 + k_2$$
$$\text{or,}\quad \rho_1 h_1 + \rho_1 h_2 = \rho_1 k_1 + \rho_1 k_2 \qquad (3.22)$$

Subtracting (3.22) from (3.21), we have

$$h_2(\rho_2 - \rho_1) = k_2(\rho_2 - \rho_1)$$
$$\text{or,} \quad h_2 = k_2 \quad [\because \rho_2 - \rho_1 \neq 0].$$

Hence, CD, the common surface is horizontal.

3.13 Dimension of Pressure

Since pressure is thrust per unit area, its dimensions are

$$\frac{MLT^{-2}}{L^2} = ML^{-1}T^{-2}.$$

3.14 Head of Liquid

The pressure of a point due to head of h feet of liquid means that the pressure at that point is the same as if the effective surface were h feet above it.

Illustrative Examples

▶ **Ex. 1.** A vessel full of water is fitted with a light cork. How is it that a slight blow on the cork may be sufficient to break the vessel?

Solution: Let A be the area of cross-section of the cork and S the area of the surface of the vessel. Let p be the intensity of the slight blow on the cork, then as the water is incompressible. So by Pascal's law, the pressure $\frac{p}{A}$ per unit area will be transmitted to each point on the surface of the vessel, and the total pressure thus transmitted will be $\frac{p}{A}S$, which is very large as S is large. This pressure on the surface of the vessel, being very large may be sufficient to break the vessel.

Hence, a slight blow on the cork may be sufficient to break the vessel. The chances of breaking increases if the surface of the vessel S is large and the area of the cork, viz., A is small.

▶ **Ex. 2.** A glass tube of uniform section is bent into the form of a U-tube, whose arms are vertical; it contains mercury of specific gravity 13.6; water is poured into one arm to occupy a height of 40.8 cms of it. Find the rise of the level of mercury in the other arm. **[M.U. 69S, 72A]**

Solution: Let the initial levels of mercury in the two arms of the U-tube be at A and B in the same horizontal. When water has been poured in, let the mercury levels, then being at A′ and B′, respectively in the two arms. As the tube being of uniform cross-section AA′ = BB′.

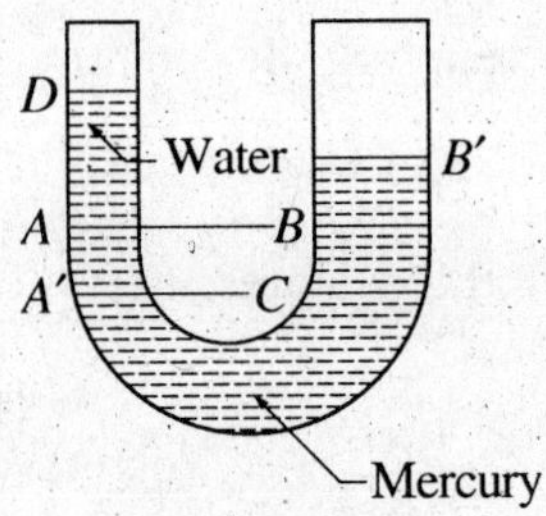

Fig. 3.16

Suppose the length occupied by water is DA' and C being a point in the same horizontal level as A'. Then, we have

$$\text{Pressure at } A' = \text{Pressure at } C$$

$$\text{i.e.,} \quad \pi + DA' \times \rho g = \pi + B'C \times 13.6\rho g,$$

where ρ is the density of water and π is the atmospheric pressure

$$\therefore \; DA' = B'C \times 13.6 \tag{1}$$

Now $DA' = 40.8$, $BB' = AA' = BC$, $B'C = B'B + BC = 2BB'$

Hence, from (1), $40.8 = 2BB' \times 13.6$

$$BB' = \frac{40.8}{2 \times 13.6} = \frac{40.8}{2 \times 13.6} = \frac{204}{136} = \frac{12}{8} = \frac{3}{2} = 1.5 \text{ cm } \textbf{(Ans.)}.$$

► **Ex. 3.** In the lower half of a uniform circular tube one quadrant is occupied by a liquid of density 2ρ and the other by two liquids which do not mix, of densities 3ρ and ρ. Prove that the volume of the lower of the two latter liquids is twice that of the other. **[M.U. 76A, R.U. 68A]**

Solution: The quadrant BD is occupied by the liquids of densities ρ and 3ρ and the quadrant AB by the liquid of density 2ρ.

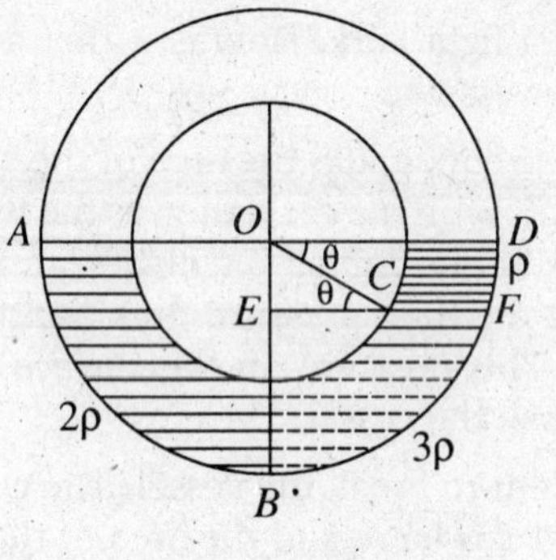

Fig. 3.17

Let CF be the surface of separation of the liquids of densities ρ and 3ρ. Let the radius OC make an angle θ with the horizontal diameter AD. Let a be the radius of the circular tube.

Let $OC = a$.

In ΔOCE

$$\sin\theta = \frac{OE}{OC} = \frac{OE}{a}$$

$$OE = a\sin\theta$$

Now

$$EB = OB - OE = a - a\sin\theta.$$

Now pressure B due to the liquid in $AB = g2\rho OB = 2g\rho a$.

Also pressure of B due to the liquid in

$$DB = \rho g OE + 3\rho g EB = \rho g a\sin\theta + 3\rho g a(1 - \sin\theta).$$

Equating the two pressures at B, we have

$$2\rho g a = \rho g a[\sin\theta + 3 - 3\sin\theta]$$

$$\text{or,} \quad 2 = (3 - 2\sin\theta) \quad \text{or,} \quad 2\sin\theta = 1$$

$$\text{or,} \quad \sin\theta = \frac{1}{2} \quad \therefore \; \theta = 30°.$$

Hence, $\angle DOC = 30°$

$$\angle COB = 90° - \theta = 90° - 30° = 60° = 2\angle COD$$

$$\therefore \quad \text{arc } BC = 2 \text{ arc } CD.$$

Thus, the volume of the liquid in arc BC = 2 (volume of the liquid in arc CD).

► **Ex. 4.** Three fluids, whose densities are in AP, fill a semi-circular tube whose bounding diameter is horizontal. Prove that the depth of one of the common surfaces is double that of the other.

[Pat. 74A, Cal. 53H, R.U. 65A, 70A, 74S, M.U. 84, Mith. 82, Bhag. 75, B.B.A.B.U. 93H, 95H]

Solution: Let (O, ABD) be the semi-circle in which columns AB, BC and CD contain fluids of densities $\rho - K$, $\rho + K$ and ρ, respectively.

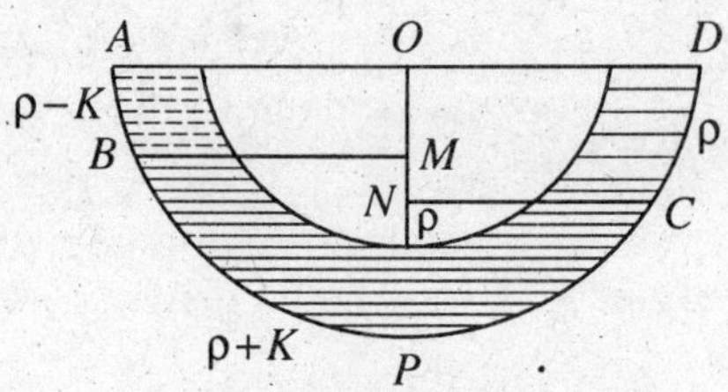

Fig. 3.18

From B and C we draw BM and CN perpendicular to OP.

Now pressure at P due to fluids in

$$ABP = (\rho - K)gOM + (\rho + K)gMP.$$

Also pressure at P due to fluids in

$$DCP = (\rho + K)gNP + \rho gON.$$

Equating the two pressures at P, we have

$$(\rho - K)gOM + (\rho + K)gMP = (\rho + K)gNP + \rho gON$$

or, $(\rho - K)OM + (\rho + K)(OP - OM) = (\rho + K)(OP - ON) + \rho ON$

or, $OM(\rho - K - \rho - K) + (\rho + K)OP = (\rho + K)OP + ON(\rho - \rho - K)$

or, $-2KOM = -KON$

or, $2OM = ON.$

Hence, the proposition.

► **Ex. 5.** A find tube bent in the form of an ellipse is held with its plane vertical and is filled with n liquids whose densities are $\rho_1, \rho_2, \ldots, \rho_n$ taken in order round the elliptic tube. If $r_1, r_2, \ldots, r_n$ be the densities of the points of separation from either focus, prove that $r_1(\rho_1 - \rho_2) + r_2(\rho_2 - \rho_3) + \cdots + r_n(\rho_n - \rho_1) = 0.$

[R.U. 69H, 66H, M.U. 63H, Mith. 88H]

Solution: Let S be the focus and ZZ_1 be the directrix of the ellipse. Let the portions $A_1A_2, A_2A_3, A_3A_4, \ldots$ be occupied by liquids of densities $\rho_2, \rho_3, \ldots$, etc.

Then

$$SA_1 = r_1,\ SA_2 = r_2,\ \ldots,\ SA_n = r_n.$$

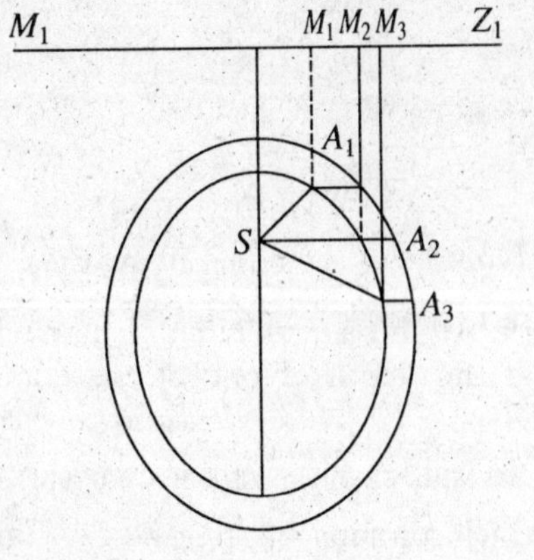

Fig. 3.19

From $A_1, A_2, A_3, \ldots$ we draw $A_1M_1, A_2M_2, \ldots$ perpendiculars to the directrix ZZ_1.

Let

$$A_1M_1 = z_1,$$
$$A_2M_2 = z_2, \ \ldots,$$
$$A_nM_n = z.$$

Then by the definition of ellipse, the focal distance of any point of the ellipse $= e \times$ distance of the point from the directrix, where e is the eccentricity of the ellipse.

Hence,

$$SA_1 = eA_1M_1, \quad \text{or,} \quad r_1 = eA_1M_1,$$
$$\text{or,} \quad A_1M_1 = \frac{r_1}{e}.$$

$$\therefore \ z_1 = \frac{r_1}{e}.$$

Similarly

$$z_2 = \frac{r_2}{e}, \ \ldots, \ z_n = \frac{r_n}{e}.$$

Let the pressures of $A_1, A_2, A_3, \ldots$ be $p_1, p_2, p_3, \ldots$, respectively. Then

$$p_2 - p_1 = \rho_2 g(z_2 - z_1)$$
$$p_3 - p_2 = \rho_3 g(z_3 - z_2)$$
$$p_4 - p_3 = \rho_4 g(z_4 - z_3)$$
$$\cdots\cdots\cdots\cdots\cdots\cdots$$
$$p_n - p_{n-1} = \rho_n g(z_n - z_{n-1})$$
$$p_1 - p_n = \rho_1 g(z_1 - z_n)$$
$$\cdots\cdots\cdots\cdots\cdots\cdots$$

Adding all these, we get

$$0 = g\,[\rho_2(z_2 - z_1) + \rho_3(z_3 - z_2) + \cdots + \rho_n(z_n - z_{n-1}) + \rho_1(z_1 - z_n)]$$

$$\text{or, } 0 = g\left[\rho_2\left(\frac{r_2}{e} - \frac{r_1}{e}\right) + \rho_3\left(\frac{r_3}{e} - \frac{r_2}{e}\right) + \cdots + \rho_n\left(\frac{r_n}{e} - \frac{r_{n-1}}{e}\right) + \rho_1\left(\frac{r_1}{e} - \frac{r_n}{e}\right)\right]$$

$$\text{or, } 0 = \rho_2(r_2 - r_1) + \rho_3(r_3 - r_2) + \cdots + \rho_n(r_n - r_{n-1}) + \rho_1(r_1 - r_n).$$

Hence, $r_1(\rho_1 - \rho_2) + r_2(\rho_2 - \rho_3) + \cdots + r_n(\rho_n - \rho_1) = 0.$

▶ **Ex. 6.** A tube in the form of a parabola held with its vertex downwards and axis vertical, is filled with two different liquids of densities δ and δ'. If the distances of the free surfaces of the liquids from the focus be r and r', respectively, show that the distance of their common surface from the focus is

$$\frac{r\delta - r'\delta'}{\delta - \delta'}.$$

[Bhag. 68H, R.U. 74H, 79H, 81H, 82H]

Solution: Let OY be the axis and OX the directrix of the parabola.

Let P, Q be the free surfaces of the liquids of densities δ and δ' respectively and R, their common surface.

From P, Q, R we draw PL, QM and PN perpendicular to OX respectively.

Let F be the focus

$$PL = FP = r,$$
$$QM = FQ = r',$$
$$RN = FR = z \text{ (let)}$$

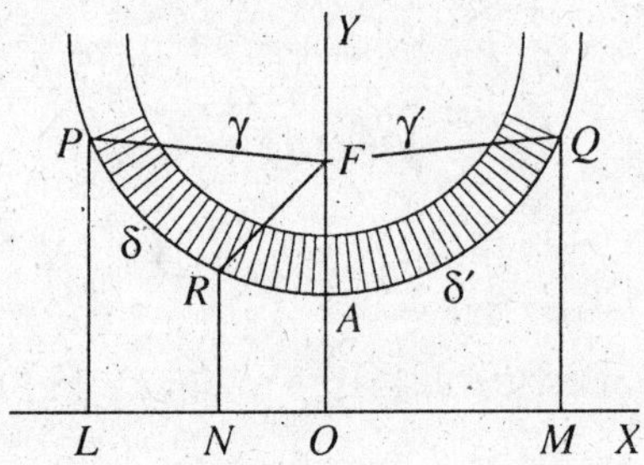

Fig. 3.20

Since in a parabola focal distance of a point = distance of the point from the directrix.

Equating pressures from both sides of A, we have

$$(PL - RN)\delta g + (RN - OA)\delta' g = (QM - OA)\delta' g$$

or, $(r - z)\delta + (z - OA)\delta' = (r' - OA)\delta'$

or, $(r - z)\delta + z\delta' = r'\delta'$

or, $z(\delta' - \delta) = r'\delta' - r\delta$

or, $z = \dfrac{r\delta - r'\delta'}{\delta - \delta'}$ Proved.

▶ **Ex. 7.** A fine circular tube in the vertical plane contains a column of liquid of density ρ, which subtends a right angle at the centre and a column of density σ subtending an angle α. Prove that the radius through the common surface makes with the vertical an angle $\tan^{-1}\frac{\rho - \sigma + \sigma\cos\alpha}{\rho + \sigma\sin\alpha}$

[P.U. 67A, 89A, M.U. 64H, 69A, 72S, R.U. 75S, 95H, M.U. 93H]

Solution: Let the tube ADC be occupied by the liquid of density ρ, such that $\angle AOC = 90°$. The liquid of density σ is contained in BC and $\angle BOC = \alpha$.

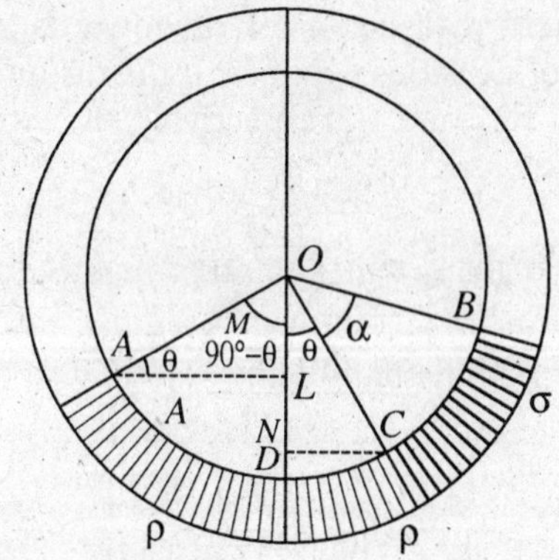

Fig. 3.21

Let θ be the angle that common radius OC makes with the vertical, then $\angle AON = 90° - \theta$, $\angle BOM = \alpha + \theta$.

From A, B, C we draw AL, BM and CN perpendiculars on OD. Let a be the radius of the circle.

$$\therefore \quad OL = a\sin\theta,$$
$$OM = a\cos(\alpha+\theta),$$
$$ON = a\cos\theta.$$

Now pressure at D due to liquid in

$$AD = \rho g LD = \rho g(OD - OL)$$
$$= \rho g(a - a\sin\theta).$$

Also pressure at D due to liquids in BCD

$$= \rho g ND + \sigma g MN$$
$$= \rho g(OD - ON) + \sigma g(ON - OM)$$
$$= \rho g(a - a\cos\theta) + \sigma g[a\cos\theta - a\cos(\alpha+\theta)].$$

Since D is the lowest point, therefore, the pressures at D due to the liquids on both the sides of the tube must be equal.

$$\therefore \quad \rho g(a - a\sin\theta) = \rho g(a - a\cos\theta) + \sigma g[a\cos\theta - a\cos(\alpha+\theta)]$$
$$\Rightarrow \quad \rho(1-\sin\theta) = \rho(1-\cos\theta) + \sigma[\cos\theta - \cos(\alpha+\theta)]$$
$$\Rightarrow \quad \rho - \rho\sin\theta = \rho - \rho\cos\theta + \sigma\cos\theta - \sigma\cos(\alpha+\theta)$$
$$\Rightarrow \quad \rho\cos\theta - \rho\sin\theta = \sigma\cos\theta - \sigma(\cos\alpha\cos\theta - \sin\alpha\sin\theta)$$
$$\Rightarrow \quad \rho\cos\theta - \rho\sin\theta = \sigma\cos\theta - \sigma\cos\alpha\cos\theta + \sigma\sin\alpha\sin\theta$$
$$\Rightarrow \quad \rho\cos\theta - \sigma\cos\theta + \sigma\cos\alpha\cos\theta = \sigma\sin\alpha\sin\theta + \rho\sin\theta$$
$$\Rightarrow \quad \cos\theta(\rho - \sigma + \sigma\cos\alpha) = \sin\theta(\sigma\sin\alpha + \rho)$$
$$\Rightarrow \quad \tan\theta = \frac{\rho - \sigma + \sigma\cos\alpha}{\rho + \sigma\sin\alpha} \quad \text{Proved.}$$

► **Ex. 8.** A closed tube in the form of an equilateral triangle contains equal volumes of three liquids which do not mix and is placed with the lower side horizontal. Prove that if the density of the liquids are in AP, their surfaces of separation will be at points of trisection of the sides of the triangle.

[P.U. 76A, 80H, Bih. 65A, M.U. 63H, 94H, L.N.M.U. 93H, B.B.A.B.U. 94H]

Solution: Let ABC be an equilateral triangular tube of side a each. Let D, E, F be the surfaces of separations of liquid of densities $\rho - \delta$, ρ, $\rho + \delta$.

Let $AD = BF = CE = x$ as the liquids are of equal volumes. Then

$$AF = CD = BE = a - x.$$

Pressure at B due to the liquids in AFB

$$= (\rho-\delta)gAK + (\rho+\delta)gKL$$
$$= (\rho-\delta)gAF\sin 60^\circ + (\rho+\delta)gBF\sin 60^\circ$$
$$= [(\rho-\delta)(a-x) + (\rho+\delta)x]g\sin 60^\circ.$$

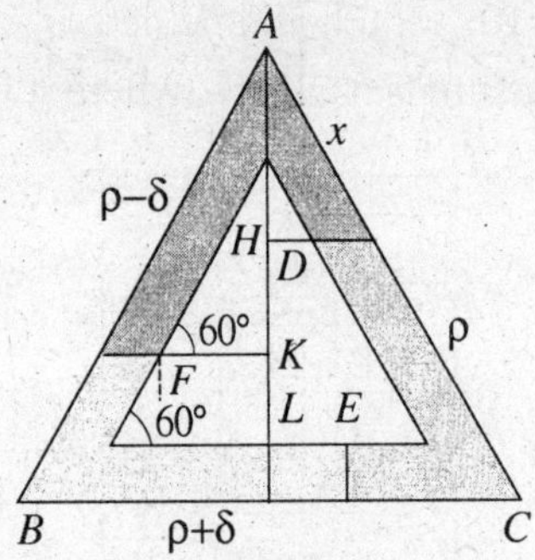

Fig. 3.22

Pressure at C due to the liquids in ADC

$$= (\rho-\delta)gAH + \rho gDL$$
$$= (\rho-\delta)gAD\sin 60^\circ + \rho gDC\sin 60^\circ$$
$$= [(\rho-\delta)x + \rho(a-x)]g\sin 60^\circ.$$

Since B and C are points in the same level.

$\therefore$ pressure at B = pressure at C.

$\Rightarrow \quad [(\rho-\delta)(a-x) + (\rho+\delta)x]g\sin 60^\circ = [(\rho-\delta)x + \rho(a-x)]g\sin 60^\circ$

$\Rightarrow \quad \rho(a-x) - \delta(a-x) + \rho x + \delta x = \rho x - \delta x + \rho(a-x)$

$\Rightarrow \quad -\delta a + \delta x + 2\delta x = 0$

$\Rightarrow \quad 3\delta x = \delta a$

$\Rightarrow \quad x = \frac{a}{3}$ Proved.

▶ **Ex. 9.** A vertical cylinder of height $2h$ and radius r closed at the top is just filled by equal volumes of two liquids of densities ρ and σ, show that if the axis be gradually inclined to the vertical, the pressure at the lowest point of the base will never exceed $g(\rho+\sigma)\cdot(r^2+h^2)^{1/2}$.

[B.U. 57H, 62H, P.U. 62H, 65H, Bh.U. 62H, 66H, 75H, R.U. 68H, 71H, C.U. 92H]

Solution: Let one of the plane sections of the cylinder through the axis be ABCD. When the axis is gradually inclined to the vertical, let C always touch the ground. Evidently the pressure at the lowest point C will be maximum, when the diagonally opposite point A is vertically above it. In this position let axis PQ make the vertical AC an angle θ.

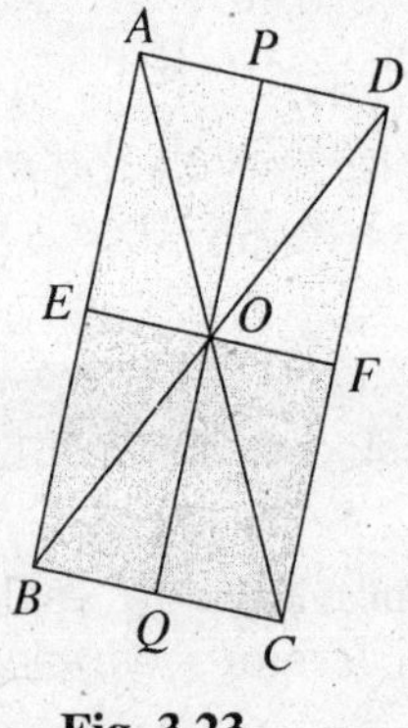

Fig. 3.23

As the liquids of densities ρ and σ occupy enual volumes ($\rho > \sigma$), the pressure p at C will be due to a height OC of liquid of density ρ and a height OA of liquid of density σ, where O is the midpoint of AC.

$$\therefore p = \rho gOC + \sigma gOA$$
$$= g(\rho+\sigma)OA \quad (\because OC = OA)$$
$$= g(\rho+\sigma)\sqrt{OP^2 + AP^2}$$
$$= g(\rho+\sigma)\sqrt{h^2 + r^2} \text{ Proved.}$$

▶ **Ex. 10.** If a liquid be heterogeneous and of density $\frac{\rho z}{a}$ at a depth z, show that the pressure is $\pi + \frac{\rho g z^2}{2a}$, where π is the atmospheric pressure. **[L.N.M.U. 94H]**

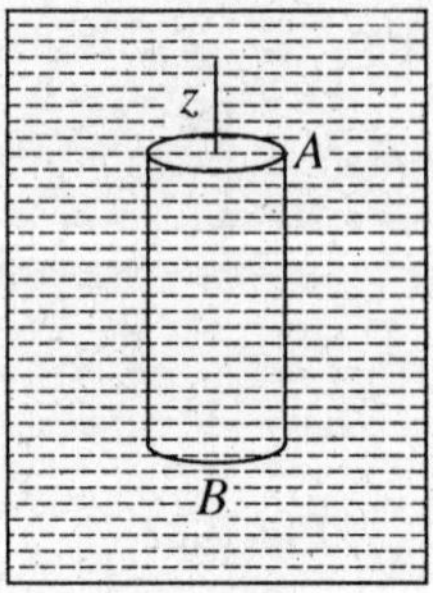

Fig. 3.24

Solution: Let p be the pressure at a given point A at a depth z below the surface. We take another point B vertically below A and very close to it such that $AB = \delta z$. Let $p + \delta p$ be the pressure at B. We consider the equilibrium of the elementary cylinder of liquid of very small cross-section α.

Since the cylinder is very small, the density of the liquid within it may be taken to be constant and let it be σ, where σ is not the same at every point but varies from point to point.

Now the forces acting on the elementary cylinder are

- the thrust $p\alpha$ at the end A acting along AB;
- the thrust $(p + \delta p)\alpha$ at the end B acting along AB;
- the weight of the liquid $\sigma g \alpha z \delta z$ acting vertically downward; and
- the thrust on the curved surface acting in the direction normal to the curved surface, i.e., horizontal.

Hence, for equilibrium, resolving vertically we have

$$p\alpha + \sigma g \alpha \delta z = (p + \delta p)\alpha \quad \Rightarrow \quad \frac{\delta p}{\delta z} = \sigma g.$$

Now, if $\delta z \to 0$, $\frac{\delta p}{\delta z}$ tends to the limit $\frac{dp}{dz}$ and σ becomes the density at the point A.

$$\therefore \quad \frac{dp}{dz} = \sigma g.$$

Here $\sigma = \frac{\rho z}{a}$

$$\therefore \quad \frac{dp}{dz} = \frac{\rho z}{a} g \quad \Rightarrow \quad dp = \frac{\rho z}{a} g dz.$$

Integrating we get $p = \frac{\rho g}{a}\frac{z^2}{2} + C$. At the free surface $z = 0$, $p = \pi$.

$$\therefore \quad C = \pi$$

$$\therefore \quad p = \pi + \frac{\rho g z^2}{2a} \quad \text{Proved.}$$

▶ **Ex. 11.** Liquids of densities, ρ and δ occupy areas subtending angles α and β at the centre of a fine vertical circular tube. If the surface of separation be at the lowest point, prove that $\sin \frac{\beta}{2} = \sqrt{\frac{\rho}{\delta}} \sin \frac{\alpha}{2}$. **[R.U. 93H]**

Solution: Let the tube AB be occupied by the liquid of density ρ, such that $\angle AOB = \alpha$ and the tube AC by the liquid of density δ, such that $\angle AOC = \beta$. Let a be the radius of the circle. Then

$$OE = a\cos\alpha,$$
$$OF = a\cos\beta.$$

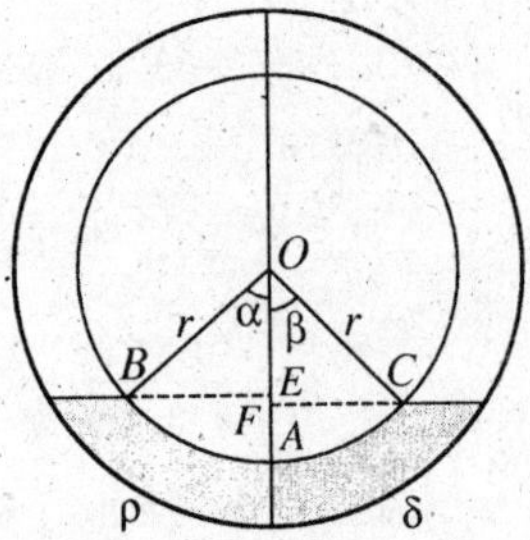

Fig. 3.25

Now pressure at A due to the liquid in AB

$$= \rho g AE = \rho g(OA - OE)$$
$$= \rho g(a - a\cos\alpha) = a\rho g(1 - \cos\alpha).$$

Also pressure at A due to the liquid in AC

$$= \delta g AF = \delta g(OA - OF)$$
$$= \delta g(a - a\cos\beta) = a\delta g(1 - \cos\beta).$$

As A is the lowest point, so equating the two pressures at A we get

$$a\rho g(1 - \cos\alpha) = a\delta g(1 - \cos\beta) \Rightarrow \rho 2\sin^2\frac{\alpha}{2} = \delta 2\sin^2\frac{\beta}{2}$$
$$\Rightarrow \frac{\rho}{\delta}\sin^2\frac{\alpha}{2} = \sin^2\frac{\beta}{2}$$
$$\Rightarrow \sqrt{\frac{\rho}{\delta}}\sin\frac{\alpha}{2} = \sin\frac{\beta}{2} \quad \text{Proved.}$$

▶ **Ex. 12.** A fine tube ABC of uniform bore, having the parts AB, BC of straight and perpendicular to one another, is held in a vertical plane and contains several liquids of different densities. If the tube be turned about the point B in its own plane and the liquids be allowed to settle down without spilling, prove that the weight of the liquid which has passed from one branch to the other bears to the weight of the whole the ratio $(\tan\alpha - \tan\beta) : 2$, where α, β are the inclinations to the vertical of the bisector of the angle ABC in the two positions.

Solution: Let BD be the bisector of $\angle ABC$.

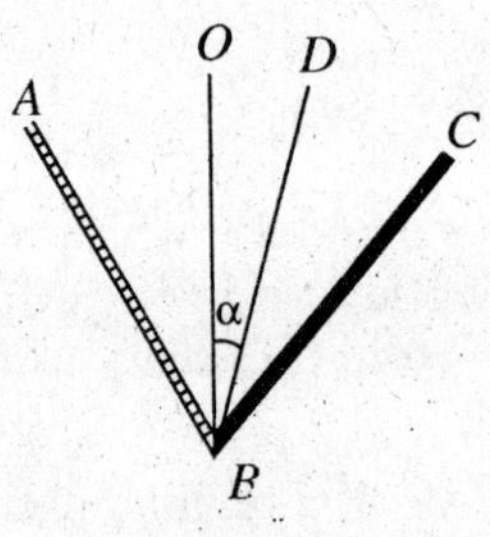

Fig. 3.26

Let $\angle OBD = \alpha$

$$\therefore \quad \angle ABO = \frac{\pi}{4} - \alpha$$
$$\text{and} \quad \angle CBO = \frac{\pi}{4} + \alpha.$$

Let W_1, W_2 be the weights of liquids in branches AB, BC, respectively. Then by equating pressures at B due to liquids in portion AB, BC, we have

$$W_1 \cos\left(\frac{\pi}{4} - \alpha\right) = W_2 \sin\left(\frac{\pi}{4} - \alpha\right)$$

or, $$\frac{W}{W_2} = \tan\left(\frac{\pi}{4} - \alpha\right) = \frac{1 - \tan\alpha}{1 + \tan\alpha}. \tag{1}$$

When the tube is turned into second position let W be the weight of the liquid that flows from AB to BC. Then the weights of the liquids in the two branches AB, BC are $W_1 - W$ and $W_2 + W$ respectively.

Equating the pressure for both sides at B, we have

$$(W_1 - W)\cos\left(\frac{\pi}{4} - \beta\right) = (W_2 - W)\sin\left(\frac{\pi}{4} - \beta\right)$$

or, $$\frac{W_1 - W}{W_2 + W} = \tan\left(\frac{\pi}{4} - \beta\right) = \frac{1 - \tan\beta}{1 + \tan\beta}. \tag{2}$$

From (1) by Componendo and Dividendo

$$\frac{W_2 - W_1}{W_2 + W_1} = \frac{1 + \tan\alpha - (1 - \tan\pi)}{1 + \tan\alpha + 1 - \tan\alpha}$$

or, $$\frac{W_2 - W_1}{W_2 + W_1} = \frac{2\tan\alpha}{2} = \tan\alpha. \tag{3}$$

Again, from (2) by Componendo and Dividendo

$$\frac{W_2 + W - (W_1 - W)}{W_2 + W + W_1 - W} = \frac{1 + \tan\beta - (1 - \tan\beta)}{1 + \tan\beta + 1 - \tan\beta}$$

or, $$\frac{W_2 - W_1 + 2W}{W_2 + W_1} = \frac{2\tan\beta}{2} = \tan\beta. \tag{4}$$

Subtracting (3) from (4), we get

$$\tan\beta - \tan\alpha = \frac{W_2 - W_1 + 2W - (W_2 - W_1)}{W_2 + W_1}$$

$$\tan\beta - \tan\alpha = \frac{2W}{W_1 + W_2}. \tag{5}$$

If the tube is so turned that liquid flow from BC to AB, we would get

$$\tan\alpha - \tan\beta = \frac{2W}{W_1 + W_2}. \tag{6}$$

Writing (5) and (6) together we get

$$\left|\frac{\tan\alpha - \tan\beta}{2}\right| = \frac{W}{W_1 + W_2} \quad \text{Proved.}$$

▶ **Ex. 13.** Equal volumes of two liquids of densities ρ and 3ρ, which do not mix together just fill a cone which is held with its axis vertical and vertex uppermost. Prove that the pressure at any point of the base is $3 - \sqrt[3]{4}$ times the pressure at the same point when the cone is filled with lighter liquid. [Bhag. 72H]

Solution: Let ABC be the cone of height H and semi-vertical angle α and liquid of density ρ, stand up to height h. So that liquid of density 3ρ stands up to height $H - h$ in the lower portion.

Since the volumes of the liquids in the two parts are equal. So the volume of the upper liquid is equal to half the volume of whole cone.

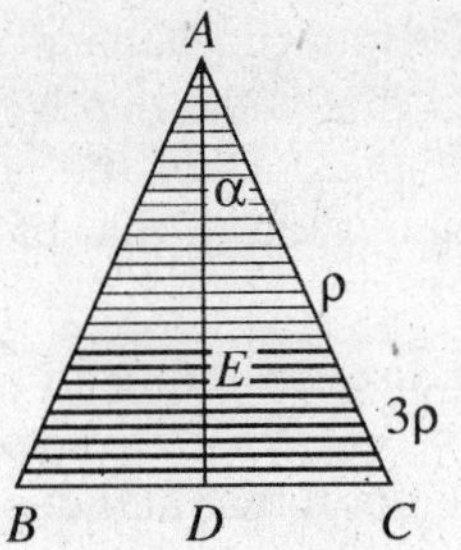

Fig. 3.27

$$\therefore \quad \frac{1}{3}\pi h^3 \tan^2\alpha = \frac{1}{2}\left[\frac{1}{3}\pi H^3 \tan^2\alpha\right]$$

or, $2h^3 = H^3$ or, $\frac{h}{H} = \frac{1}{\sqrt[3]{2}}$. (1)

Now, when the cone is filled completely with the lighter liquid, the pressure p' at any point on the base $= \rho g H$

and, when the cone is filled with equal volumes of two liquids, the pressure p at any point on the base

$$= \rho g h + g \cdot 3\rho(H - h) = \rho g(3H - 2h).$$

$\therefore$ Ratio of the two pressures

$$= \frac{p}{p'} = \frac{\rho g(3H - 2h)}{\rho g H}$$

$$= 3 - 2\frac{h}{H} = 3 - 2\frac{1}{2^{1/3}} \quad \text{[From (1)]}$$

$$= 3 - 2^{2/3} = 3 - (2^2)^{1/3}$$

$$= 3 - 4^{1/3} \quad \text{Proved.}$$

Exercises 3(II)

1. A cylindrical pipe which is filled with water opens into another pipe, the diameter of which is three times its own diameter. If a force of 20 lbs weight be applied to the water in the smaller pipe, find the force on the open end of the larger pipe, which is necessary to keep the water at rest.

[*Ans.* 180 lbs wt]

2. The pressure in the water pipe at the basement of a building is 35 lbs wt. to the square inch; whereas at the third floor it is only 19.375 lbs wt to the square inch. Find the height of the third floor. (Take a cubic foot of water to weight 62.5 lbs.) [*Ans.* 36 ft]

3. The pressure at the bottom of a

well is four times that at a depth of 2 ft. What is the depth of the well if the pressure of the atmosphere be equivalent to that of 30 feet of water.

Find also the depth of a well, the pressure at the bottom of which is three times that at a depth of 3 ft.

[*Ans.* 98 ft, 69 ft]

4. If a parallelogram be immersed in any manner in a homogeneous liquid, prove that the sum of the pressures at the extremities of each diagonal is the same. **[R.U. 98H]**

5. If ρ and ρ' be the densities of two fluids ($\rho < \rho'$) and the lengths of the arms of a U-tube in which they meet be m and n inches, respectively; prove that in order that the tube may be completely filled, the height of the column of the lighter liquid above the horizontal plane in which they meet must be $\frac{\rho'(m-n)}{\rho'-\rho}$ inches. **[M.U. 70A, 74A]**

6. A vessel contains n liquids, which do not mix, of densities $\rho, 2\rho, \ldots, n\rho$ beginning from the top. If the thickness of the layers be $h, 2h, \ldots, nh$, calculate the pressure at the bottom. **[Bhag. 71A]**

[*Ans.* $\frac{n(n+1)(2n+1).\rho g h}{\sigma}$]

7. The pressure of two points P and Q in a homogeneous liquid are p and p'. Prove that the pressure at the point R which divides PQ in the ratio $m : n$ is

$$\frac{mp' + np}{m+n}.$$

8. A triangle is immersed in a homogeneous liquid. Show that the sum of pressures at the vertices is three times the pressure at the centroid of the triangle.

9. If there be n liquids arranged in strata of equal thickness the density of the uppermost being ρ, that of the next 2ρ and so on, that of the lowest being $n\rho$, find the pressure at the lowest point of the lowest stratum.

[*Ans.* $\frac{(n+1)}{2}\rho g h$, where h is the thickness of each stratum]

10. A small uniform tube is bent into the form of a circle whose plane is vertical. Equal quantities of fluids, whose densities are ρ and δ, fill half of the tube, show that the radius passing through the common surface makes with the vertical an angle $\tan^{-1}\frac{\rho-\delta}{\rho+\delta}$.

[R.U. 81H, 83H, 92H, M.U. 87H]

11. A circular tube of uniform thin bore is half filled with equal volumes of three liquid of specific gravities 3, 4, 6 and is kept with its plane vertical. Prove that the diameter joining the two free surfaces makes an angle θ with the vertical such that $\tan\theta = \frac{19}{\sqrt{3}}$.

[R.U. 99H]

12. A circular tube of line uniform bore is half filled with equal volumes of four liquids which do not mix and whose densities are as $1 : 4 : 8 : 7$ and is held with its plane vertical. Show that the diameter joining the free surfaces makes an angle $\tan^{-1} 2$ with the vertical.

13. In a uniform circular tube two liquids are placed so as to subtend 90° each at the centre. If the diameter joining the two free surfaces be inclined at 60° to the vertical, prove that the densities of the liquids are

$$\sqrt{3}+1 : \sqrt{3}-1.$$

14. Two liquids of densities ρ and σ occupy areas subtending angles 2α and 2β at the centre of a fine uniform vertical circular tube. Prove that the radius to the surface of separation makes with the vertical an angle θ given by

$$\rho \sin\alpha \sin(\alpha \pm \theta) = \sigma \sin\beta \sin(\beta \pm \theta).$$

15. Three liquids of densities ρ_1, ρ_2, ρ_3 ($\rho_1 > \rho_2 > \rho_3$) fill a circular vertical tube and subtend at the centre angles $2\alpha, 2\beta, 2\gamma$. If the radius through the point on the circumference mid away between the ends of the lightest liquid makes with the vertical an angle θ, prove that

$$\frac{\rho_2 - \rho_3}{\rho_1 - \rho_3} = \frac{\sin\alpha \sin(\beta - \theta)}{\sin\beta \sin(\alpha + \theta)}.$$

16. If three liquids which do not mix and whose densities are ρ_1, ρ_2, ρ_3 fill a circular tube in a vertical plane and of α, β, γ are the angles which the radius to the common surfaces make with the vertical diameter measured in the same direction, prove that

$$\rho_1(\cos\beta - \cos\gamma) + \rho_2(\cos\gamma - \cos\alpha) + \rho_3(\cos\alpha - \cos\beta) = 0.$$

[M.U. 68A]

17. A fine uniform cycloidal tube is held with its axis vertical and vertex downwards and equal weights of two liquids occupy lengths a and b. Prove that the heights of the free surfaces of the liquids above the vertex are in the ratio

$$(3a + b)^2 : (3b + a)^2.$$

[R.U. 65H, 70H, 94H, 2001H, Bih. 76H, Pat. 81H]

18. A uniform tube in the form of an equilateral triangle is held in a vertical plane and completely filled with equal volumes of three liquids which do not mix and whose densities are in arithmetical progression.

(a) Prove that the straight line joining the surfaces of separation of the liquid of mean density is vertical.

(b) Also show that if the side that contains portions of the heaviest and lightest liquids make an angle θ with the vertical, the surface of separation divides the side in the ratio

$$\cos\left(\frac{\pi}{\sigma} - \theta\right) : \cos\left(\frac{\pi}{\sigma} + \theta\right).$$

[Bih. 80H]

19. A fine tube bent into the form of an ellipse is held with its major axis vertical, and is filled with three fluids whose densities are ρ_1, ρ_2 and ρ_3, taken in order round the elliptic tube. If r_1, r_2 and r_3 be the distances of the points of separation from either focus, prove that

$$r_1(\rho_1 - \rho_2) + r_2(\rho_2 - \rho_3) + r_3(\rho_3 - \rho_1) = 0.$$

[Bih. 67H, R.U. 66H, 69H, 73H, Bhag. 65H, M.U. 99H]

20. A fine parabolic tube is held with its axis vertical and vertex downwards and is filled with three liquids whose densities are ρ_1, ρ_2 and ρ_3 beginning from the top on one side. It r_1, r_2, r_3, r_4, be the

distances of the boundaries from the focus, prove that

$$\rho_1(r_1 - r_2) + \rho_2(r_2 - r_3) + \rho_3(r_3 - r_4) = 0.$$

21. A fine tube of uniform bore is bent into the form of a square and filled with equal volumes of three liquid of densities

$$\rho_1, \rho_2, \rho_3 \quad (\rho_1 < \rho_2 < \rho_3).$$

If the tube is placed with one side vertical, show that a side of the square will be filled with liquid of the first, third or second kind only according as

$$\rho_2 > \tfrac{1}{3}(2\rho_3 + \rho_1),$$
or, $< \tfrac{1}{3}(\rho_3 + 2\rho_1)$,
or, lies between these values.

[Pat. 61H]

22. A thin uniform cycloidal tube is held with its axis vertical and vertex downwards and contains liquids of densities ρ and σ in such a proportion that their surface of separation is at the vertex, prove that the weights of the liquids are in the ratio $\sqrt{\rho} : \sqrt{\sigma}$. [Bhag. 74H]

23. A uniform cycloidal tube contains equal volumes of four liquids of densities 1, 2, 3, 4 arranged in order given above. Show that if the axis of symmetry be inclined at an angle $\cot^{-1}\pi$ to the vertical and the tube be completely filled, no fluid will run out.

[Pat. 81H, R.U. 65H, 70H, Bih. 76H, 60H, 65H, 68H]

24. Prove that the pressure at any point within a liquid whose density varies as the depth is proportional to the square of the depth.

[L.N.M.U. 96H]

25. A semi-circular tube has its bounding diameter horizontal and contains equal volumes of n fluids of densities successively equal to ρ, 2ρ, 3ρ, ..., arranged in this order. If each fluid subtends an angle 2α at the centre and the tube just holds them all, show that

$$\tan n\alpha = (2n + 1)\tan\alpha.$$

[Bhag. 70H, M.U. 75H]

26. A fine parabolic tube is held with its axis vertical and vertex downwards. It contains liquids of successive densities $\rho_1, \rho_2, \ldots, \rho_n$ beginning from the top on one side and ending with the last liquid on the other. If $r_0, r_1, r_2, \ldots, r_n$ be the focal distances of the boundaries, prove that

$$\rho_1(r_0 - r_1) + \rho_2(r_1 - r_2) + \cdots + \rho_n(r_{n-1} - r_n) = 0.$$

27. Two vertical tubes of equal uniform bore are connected at their bases by a horizontal tube of small bore, mercury is first poured into them, and then water is poured down one, until there is equilibrium with the mercury surface at a height h above the common surface. Another communication between the tubes is now opened by means of a fine horizontal tube at a height c above the common surface, prove that the length of the column of water which will flow from one tube to other is the smaller of σc and $\frac{1}{2}\sigma h$; σ being the specific gravity of mercury.

28. A fine bent tube, on a vertical plane, has its branches AB, BC inclined to the verticals at angles α, β on opposite sides, and contains two fluids which fill lengths a, b of the tube, the fluids meeting in the branch AB at a distance c from B prove that they will meet at B, if the tube is turned in its own plane through an angle

$$\cot^{-1}\left\{\frac{b}{c}\sin\frac{\alpha+\beta}{2}\sec\frac{\alpha-\beta}{2}\sec\alpha-\tan\alpha\right\}.$$

29. A closed tube in the form of an ellipse with its major axis vertical is filled with three different liquids of densities ρ_1, ρ_2, ρ_3, respectively. If the distances of the surfaces of separation from either focus be r_1, r_2, r_3, respectively, prove that

$$r_1(\rho_2-\rho_3)+r_2(\rho_3-\rho_1)+r_3(\rho_1-\rho_2)=0.$$

[R.U. 66H, M.U. 80H, 86H, Pat. 79H, 82H, L.N.M.U. 95H]

CHAPTER 4

Thrust on Plane Surfaces

4.1 Resultant Thrust and Whole Pressure

We know that when a surface is in contact with a liquid, the pressure on every element of it is along the normal to the surface. Now, if we consider the pressures on all the elements of the surface, in contact with the fluid, and add up arithmetically we get the whole pressure or thrust of the fluid on the surface. Thus, *the whole pressure of a fluid on any surface in contact with it is the sum of the normal pressures exerted by the fluid on all the elements of the surface.*

In the case of a plane area in contact with a liquid, the pressures at various portions of the area will constitute a system of parallel forces, which can be reduced to a single force. This single force is called the *resultant thrust* on the plane area. Since this resultant thrust is the arithmetical sum of all the pressures, it is also known as *whole pressure* or *total thrust.*

However, in the case of a curved surface, the pressures on different elements will be in different directions. Hence, when they are added up; to get the whole pressure, we do not get the resultant thrust. The whole pressure on a curved surface has no physical meaning whatsoever. Resultant thrust on a curved surface will be discussed later on.

4.2 Whole Pressure on a Plane Surface

Theorem 4.2.1 *The thrust of a heavy homogeneous liquid on a plane area is measured by the product of the area and the pressure of its centre of gravity.*

Proof

Case I. *When atmospheric pressure is neglected.*

Let S be a plane area immersed in a liquid of density ρ. Let P be a point at a depth z_1 in the plane area The pressure at P is therefore $\rho g z$. Let ds_1 be an elementary area about the point P. The thrust on this element $= \rho g z_1, ds_1$. Now we divide the whole area S into an infinite number of elementary areas ds_1, ds_2, whose depths below the free surface are $z_2, z_3, \ldots$, respectively.

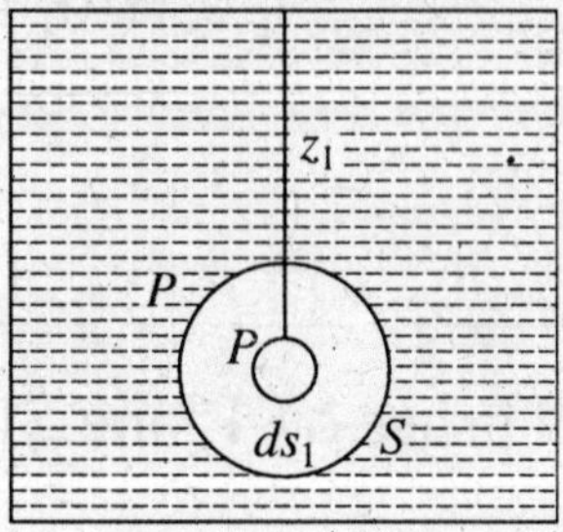

Fig. 4.1

The total thrust on the plane area

$$= \rho g z_1 ds_1 + \rho g z_2 ds_2 + \cdots$$
$$= \rho g(z_1 ds_1 + z_2 ds_2 + \cdots) \quad (4.1)$$

If $\bar{z}$ be the depth of the centre of gravity of the plane area S below the free surface, then

$$\bar{z} = \frac{z_1 ds_1 + z_2 ds_2 + \cdots}{ds_1 + ds_2 + \cdots}$$

or, $\bar{z}(ds_1 + ds_2 + \cdots)$

$$= z_1 ds_1 + z_2 ds_2 + \cdots \quad (4.2)$$

Hence, from (4.1), we get the total thrust on the plane area

$$= \rho g \bar{z}(ds_1 + ds_2 + \cdots)$$
$$= \rho g \bar{z} S \quad (4.3)$$

= pressure at the centre of gravity × area of the plane surface.

Equation (4.3) also states that *the whole pressure on a plane surface in contact with a homogeneous liquid is the weight of a column of the liquid whose cross-section is the same as the area of the plane surface and whose height is equal to the depth of the centre of gravity of the plane below the free surface.*

Case II. *When atmospheric pressure is taken into consideration.*

In this case, thrust on $ds_1 = (\pi + \rho g z_1) ds_1$, where π is the atmospheric pressure. Hence, the total thrust on the plane area

$$= (\pi + \rho g z_1) ds_1 + (\pi + \rho g z_2) ds_2 + \cdots$$
$$= \Sigma(\pi + \rho g z_r) ds_r = \pi \Sigma ds_r + \rho g \Sigma z_r ds_r$$
$$= \pi S + \rho g \bar{z} \Sigma ds_r \quad \text{[From 2]}$$
$$= \pi S + \rho g \bar{z} S = S(\pi + \rho g \bar{z}). \quad (4.4)$$

Let $\pi = \rho g h_1$, where h_1 is the height of the effective surface above the free surface.

Then from (4.4), the total thrust

$$= S(\rho g h_1 + \rho g \bar{z}) = \rho g(h_1 + \bar{z})S = \rho g \bar{z}_1 S, \quad (4.5)$$

where $\bar{z}_1 = h_1 + \bar{z}$ = depth of C.G. of the plane area below the effective surface.

4.3 Whole Pressure on a Plane Surface below the Layers of Different Liquids

Case I. *Plane surface in contact with the lowest liquid.*

Let us consider n liquids, in equilibrium of densities

$$\rho_1, \rho_2, \rho_3, \ldots; \rho_n$$
$$(\rho_1 < \rho_2 < \rho_3 < \cdots < \rho_n)$$

and of thickness $h_1, h_2, \ldots, h_n$, respectively.

Let a plane surface of area S be immersed in the liquid of density ρ_n.

We divide the area S into an infinite number of small elementary areas ds_1, $ds_2, \ldots$. Let $z_1, z_2, \ldots$ be the depths of areas $ds_1, ds_2, \ldots$ below lowest surface of separation.

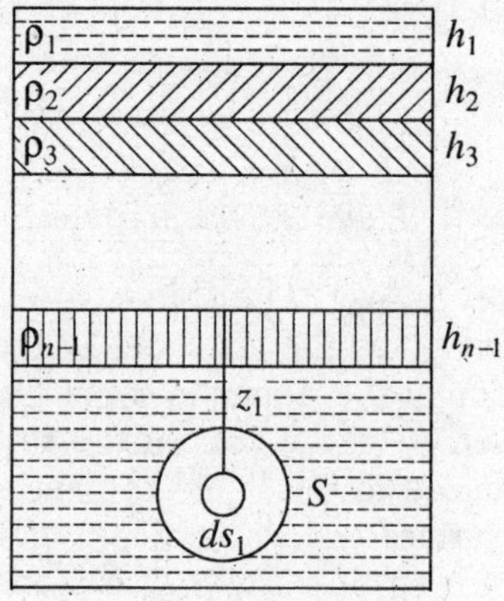

Fig. 4.2

Then, the pressure on the element

$$\begin{aligned} ds_1 &= (g\rho_1 h_1 + g\rho_2 h_2 + \cdots + g\rho_{n-1}h_{n-1} + g\rho_n z_r)\delta_r \\ &= \left(g\sum_{i=1}^{n-1} \rho_i h_i + g\rho_n z_r\right) ds_r. \end{aligned}$$

$\therefore$ Whole pressure on the surface of area S

$$= P = g\sum_{i=1}^{n-1} \rho_i h_i \Sigma ds_r + g\rho_n \Sigma z_r ds_r. \tag{4.6}$$

If $\bar{z}$ be the depth of the centre of gravity of the plane area S below the free surface, then

$$\bar{z}_1 = \frac{\bar{z}_1 ds_1 + z_2 ds_2 + \cdots}{ds_1 + ds_2 + \cdots},$$
$$\text{or,} \quad \bar{z}(ds_1 + ds_2 + \cdots) = z_1 ds_1 + z_2 ds_2 + \cdots,$$
$$\text{or,} \quad z\Sigma ds_r = \Sigma z_r ds_r.$$

Hence, from (4.6), we have

$$\begin{aligned} P &= g\sum_{i=1}^{n-1} \rho_1 h_i \Sigma ds_r + g\rho_n \bar{z}\Sigma ds_r \,. \\ &= g\sum_{i=1}^{n-1} \rho_i h_i S + g\rho_n \bar{z} S \quad (S = \Sigma ds_r) \\ &= g\left(\sum_{i=1}^{n-1} \rho_i h_i + \rho_n \bar{z}\right) S. \end{aligned} \tag{4.7}$$

If the atmospheric pressure π is taken into consideration, then (4.7) becomes

$$= \pi S + g\left(\sum_{i=1}^{n-1} \rho_i h_i + \rho_n \bar{z}\right) S$$

$$= \left(\pi + g\sum_{i=1}^{n-1} \rho_i h_i + g\rho_n \bar{z}\right) S.$$

Case II. *Plane surface in contact with more than one liquid.*

Let $\rho_1, \rho_2, \ldots, \rho_n$ $(\rho_1 < \rho_2, \ldots, \rho_n)$ be the densities of n liquids.

Let s_1 be the area and $\bar{z}_1$ the depth of the C.G. from the free surface of the portion of the lamina in contact with the n liquids from the above.

Let s_2 be the area of the surface and $\bar{z}_2$ the depth of its C.G., if the portion immersed in the liquid of density ρ_1 is removed. Similarly, let s_r be the area of the surface and $\bar{z}_r$ the depth of its C.G. of the portion immersed in the liquids of densities $\rho_1, \rho_2, \ldots, \rho_{r-1}$ is removed.

The thrust on the surface will remain the same even if we consider that one liquid of density ρ_1 is in contact with the whole surface s_1, another liquid of density $(\rho_2 - \rho_1)$ is in contact with the surface s_2 and soon. Therefore,

Thrust due to first liquid $= g\rho_1\bar{z}_1 s_1$

Thrust due to second liquid $= g(\rho_2 - \rho_1)\bar{z}_2 s_2$

Thrust due to rth liquid $= g(\rho_r - \rho_{r-1})\bar{z}_r s_r$ and so on.

Adding, we get the whole pressure on the surface S.

$$P = g[\rho_1\bar{z}_1 s_1 + (\rho_2 - \rho_1)\bar{z}_2 s_2 + \cdots + (\rho_n - \rho_{n-1})\bar{z}_n s_n]$$

$$= g\left[\rho_1\bar{z}_1 s_1 + \sum_{r=2}^{n} (\rho_r - \rho_{r-1})\bar{z}_r s_r\right].$$

4.4 Whole Pressure on a Horizontal Surface

The whole pressure on a horizontal surface in a homogeneous liquid is equal to the product of the area of the surface and the pressure at any point of the surface; i.e., $P = Ap$, where A is the area of a plane surface immersed horizontally in a homogeneous liquid and p is the pressure at any point of the surface (being equal for each point of this area).

It is clear that whole pressure or thrust on a horizontal surface does not depend on the shape of the vessel containing the liquid but only on the depth of the liquid and the area of the horizontal surface.

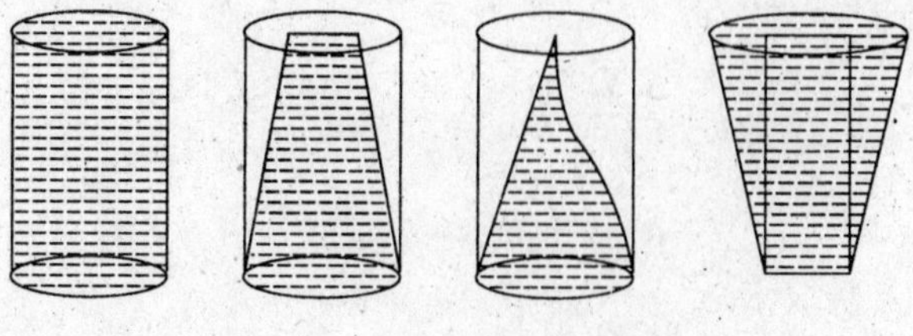

Fig. 4.3

In the above figure four vessels of different shapes but of equal heights and of equal bases are shown. These vessels contain different amounts of the same liquid. But the whole pressure on the base of each vessel will be the same, because whole pressure $= g\rho hA = pA$, where ρ is the density of the liquid, h the depth of the C.G. of the area A is the area of the base. Depth of the C.G. of the area is same for all vessels.

Illustrative Examples

▶ **Ex. 1.** In the vertical side of a water tank there is a square plate whose upper edge is horizontal and at a depth of 10 ft below the surface of the water. The length of the plate is 1.5 ft. Find the resultant pressure on the plate, taking the weight of 1 cu.ft of water to be 62.5 lb.

Solution: Area of the plate $= 1.5 \times 1.5$ sq.ft $= (1.5)^2$, i.e., $S = (1.5)^2 = 2.25$ sq.ft.

The depth of the C.G. of the plate below the free surface

$$= \left(10 + \frac{1}{2} \times 1.5\right) \text{ ft} = (10 + .75) \text{ ft} = 10.75 \text{ ft, i.e., } \bar{z} = 10.75 \text{ ft.}$$

The weight of 1 cubic foot of water = 62.5 ft. Hence, $W = 62.5$ lbs.

Hence, the required resultant pressure on the plate

$$= W\bar{z}S = 62.5 \times 10.75 \times 2.25 = 1511.72 \text{ lbs. wt. nearly.}$$

▶ **Ex. 2.** A square is placed in a liquid with one side in the surface. Show how to draw a horizontal line, in the surface dividing it into two portions, the thrusts on which are the same.

Solution: Let ABCD be a square with side AB in the surface. Let a be the length of each side of the square.

Let a line EF at a depth x below AB divide the square in two portions on which pressures are equal.

Let G_1 and G_2 be the centres of gravity of the portions ABCD and ABEF.

Then the depth of G_1 and G_2 below the free surface are $\frac{a}{2}$ and $\frac{x}{2}$, respectively.

Also the areas of the square ABCD and rectangle ABEF are a^2 and ax, respectively.

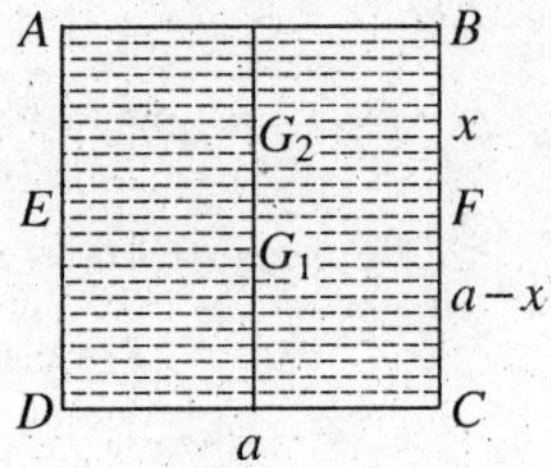

Fig. 4.4

Then

$$P_1 = \text{Thrust on the square ABCD} = \rho g \cdot \frac{a}{2} \cdot a^2$$

$$\text{and } P_2 = \text{Thrust on the rectangle ABEF} = \rho g \cdot \frac{x}{2} \cdot ax,$$

where ρ is the density of the liquid.

But in order to have the two thrusts to be equal, we must have

$$P_1 = 2P_2, \quad \text{or,} \quad \rho g \cdot \frac{a}{2} \cdot a^2 = 2 \cdot \rho g \cdot \frac{x}{2} \cdot ax$$

$$\text{or,} \quad x^2 = \frac{a^2}{2} \quad \text{or,} \quad x = \frac{x}{\sqrt{2}}.$$

Hence, the dividing line must be drawn at a depth

$$= \frac{1}{\sqrt{2}} \cdot \text{(side of square) below the free surface.}$$

▶ **Ex. 3.** A cone full of water is placed on its side on a horizontal table. Show that the thrust on the base is $3W \sin\alpha$, where W is the weight of the contained fluid and 2α is the vertical angle of cone.

[Pat. 72A, 74A, 68H, M.U. 63A, R.U. 65A, 71A, 68H, 97H, Bhag. 69A]

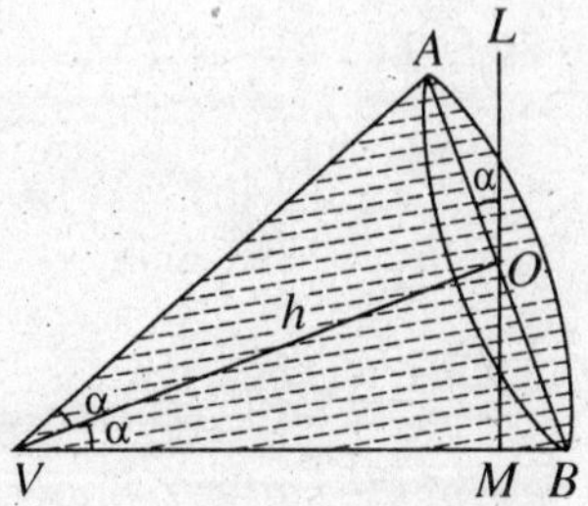

Fig. 4.5

Solution: Let a cone (V, AB) be placed with its generator VB horizontal. Let r and h be its radius and height, respectively.

Let LM be the vertical through O, the centre of base.

Pressure on the base of the cone

$$= \rho g \pi r^2 OL. \tag{1}$$

In

$$\triangle OAL \cos\alpha = \frac{OL}{OA} \quad \left(\because \text{ In } \triangle OAV \tan\alpha = \frac{OA}{OV} = \frac{r}{h}, \; r = h\tan\alpha\right)$$

$$OL = OA\cos\alpha.$$

Hence, from (1) pressure on the base of the cone

$$= \rho g \pi r^2 OA\cos\alpha = \rho g \pi r^2 r \cos\alpha \quad (\because OA = r)$$

$$= \rho g \pi r^2 h \tan\alpha \cos\alpha = \rho g \pi r^2 h \sin\alpha$$

$$= \rho g \frac{1}{3} \pi r^2 h 3 \sin\alpha = \text{(Weight of the liquid) } 3\sin\alpha$$

$$= W \cdot 3\sin\alpha \quad \left[\because W = V\rho g = \frac{1}{3}\pi r^2 h g\right].$$

▶ **Ex. 4.** The lighter of two liquids, whose specific gravities are 2 : 3 rests on the heavier, to a depth of four inches. A square is immersed in a vertical position with one side in the upper surface, determine the side of the square in order that the thrusts on the portions in the two liquids may be equal.

Solution: Let ABCD be a square immersed in liquid with side AB in the surface. Let AB = a be the length of each side of the square. The portion AEFB being in the upper liquid.

Let G_1 and G_2 be the centres of gravity of the portions ABFE and EFCD.

Let densities of the two liquids be 2ρ and 3ρ.

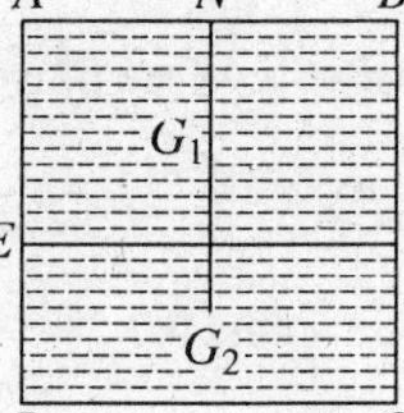

Fig. 4.6

Then the thrust P_1 on the portion ABFE $= 2\rho g : \frac{4}{2} \cdot 4a = 16\rho g a$

Also the thrust P_2 on the portion EFCD

$$= \left[3\rho g \frac{(a-4)}{2} + 2\rho g 4\right] a(a-4)$$

$$= \frac{\rho g}{2}[3a - 12 + 16]a(a-4)$$

$$= \frac{\rho g}{2}(3a+4)a(a-4).$$

Given $P_1 = P_2$.

Hence,

$$16\rho g a = \frac{\rho g}{2}(3a+4)a(a-4), \quad \text{or,} \quad 32 = (3a+4)(a-4)$$

$$\text{or,} \quad 32 = 3a^2 - 8a - 16$$

$$\text{or,} \quad 3a^2 - 8a - 48 = 0.$$

$$\therefore \quad a = \frac{8 \pm \sqrt{64 + 43.48}}{2.3} = \frac{8 \pm \sqrt{64+576}}{6} = \frac{8 \pm \sqrt{640}}{6}$$

$$a = \frac{8 \pm 8\sqrt{10}}{6} = \frac{4(1 \pm \sqrt{10})}{3}.$$

But $a \not< 0$. Hence, $a = \frac{4(1+\sqrt{10})}{3}$. **Ans.**

▶ **Ex. 5.** A hollow weightless hemisphere filled with a liquid, is suspended freely from a point in the rim of its base. Show that the thrust on the base is to the weight of the contained liquid as $12 : \sqrt{73}$.

[Pat. 69H, 73A, Bih. 71A, R.U. 66A, M.U. 64A, 72H, P.U. 95H, L.N.N.U. 96H]

Solution: Let ABC be the hemisphere which is suspended from a point A of its plane base. AB is a diameter of this plane base. Let G be the centre of gravity of the hemisphere and O is the centre of base of the hemisphere. Then OG $= \frac{3r}{8}$, where r is the radius of the hemisphere. For equilibrium AG must be vertical.

Let $\angle GAO = \theta = \angle AOL$.

In

$$\triangle OGA \tan\theta = \frac{OG}{OA} = \frac{3r}{8r} = \frac{3}{8}.$$

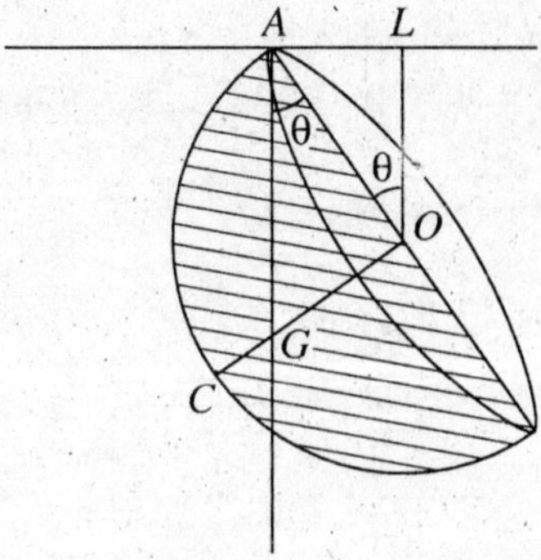

Fig. 4.7

$$\therefore \quad \cos\theta = \frac{1}{\sec\theta} = \frac{1}{\sqrt{1+\tan^2\theta}} = \frac{1}{\sqrt{1+\frac{9}{64}}} = \frac{8}{\sqrt{73}}.$$

$$\begin{aligned} OL &= \text{depth of O below A} \\ &= AO\cos\theta \\ &= \frac{8r}{\sqrt{73}}. \qquad (1) \end{aligned}$$

Then

$$\begin{aligned} P &= \text{Thrust on the base} = \text{Area of the base} \times \text{Pressure at its C.G.} \\ &= \pi r^2 \times (\text{depth of O below A}) \times \rho g = \pi r^2 \times OL \times \rho g \\ &= \pi r^2 \rho g \frac{8r}{\sqrt{73}} \quad \text{From (1).} \end{aligned}$$

Also weight W of the liquid contained in the hemisphere

$$W = \frac{1}{2}\left(\frac{4}{3}\pi r^3 \rho g\right).$$

Hence, required ratio $= \dfrac{P}{W} = \dfrac{\pi r^2 \rho g 8r \cdot 3}{\sqrt{73} \cdot 2\pi r^3 \rho g} = \dfrac{12}{\sqrt{73}}$ **Proved.**

▶ **Ex. 6.** A rectangular area is immersed in a heavy homogeneous liquid with two sides horizontal and is divided by horizontal lines into strips on which the total thrusts are equal. If a, b, c are the breadths of three consecutive strips, prove that

$$a(a+b)(b-c) = c(b+c)(a-b).$$

[M.U. 76H, 81H, Bih. 63H, Pat. 75A, 74, 79H, R.U. 79H, Cal. 83H, 91H, Bhag. 66H, R.U. 91H]

Solution: Let the rectangle ABGH be immersed in a liquid whose two sides AB and GH are horizontal.

Let AC, DF and EG be the three consecutive strips of breadths a, b, c, respectively and length x (*say*).

Let the upper edge of the strip ABCD be at a distance d from LM, where the free surface of the liquid meets with the plane of the lamina produced.

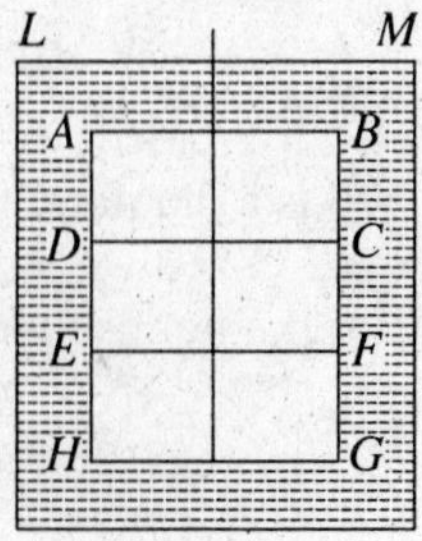

Fig. 4.8

Then thrust on the strip ABCD

$$= \rho g \times \text{(depth of the C.G. of the strip)} \times \text{(area of the strip)}$$
$$= \rho g a x \left(d + \frac{a}{2}\right).$$

Similarly, the thrust on the middle strip = $\rho g c x \left(d + a + \frac{b}{2}\right)$.

Similarly the thrust on the lower strip = $\rho g k x \left(d + a + b + \frac{c}{2}\right)$

From question these thrusts are equal.

$$\rho g a x \left(d + \frac{a}{2}\right) = \rho g b x \left(d + a + \frac{b}{2}\right) = \rho g c x \left(d + a + b + \frac{c}{2}\right)$$

or, $a(2d + a) = b(2d + 2a + b) = c(2d + 2a + 2b + c)$

or, $\dfrac{2d + a}{bc} = \dfrac{2d + 2a + b}{ac} = \dfrac{2d + 2a + 2b + c}{ab}$

or, $\dfrac{2d + 2a + b - 2d - a}{ac - bc} = \dfrac{2d + 2a + 2b + c - 2d - 2a - b}{ab - ac}$,

or, $\dfrac{a + b}{c(a - b)} = \dfrac{b + c}{a(b - c)}$

or, $a(a + b)(b - c) = c(b + c)(a - b)$ **Proved.**

► **Ex. 7.** The lighter of two liquids of density ρ rests on the heavier of density σ to a depth $\frac{h}{2}$. A trapezium is immersed vertically with one parallel side of length a in the surface and the opposite of length b at a depth h, prove that the thrust on the trapezium is $\frac{hg}{48}[(7a + 11b)\rho + (a + 5b)6]$.

Solution: Let ABCD be the trapezium completely immersed in the liquid with side AB in the surface. Join BD.

Let P_1 and P_2 be the thrusts on the triangles ABD and BCD.

The area of the $\Delta ABD = \frac{1}{2}ah$.

The depth of its centre of gravity below the free surface = $\frac{1}{3}h$.

$$P_1 = \rho g \frac{h}{3} \cdot \frac{1}{2} a h = \frac{\rho g a h^2}{\sigma}.$$

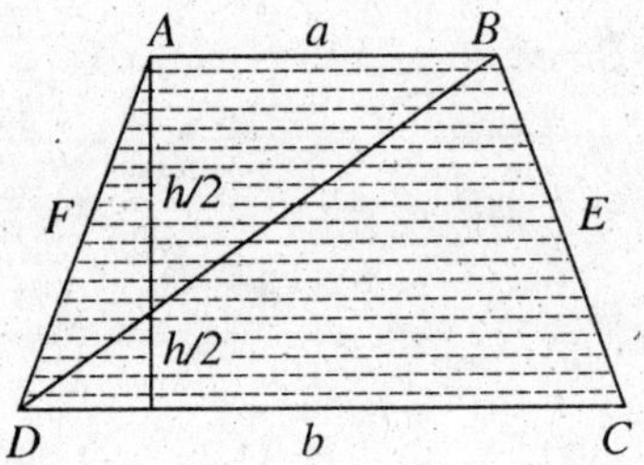

Fig. 4.9

The area of $\Delta BDC = \frac{1}{2}bh$.

The depth of its centre of gravity below the free surface = $\frac{2}{3}h$.

$$P_2 = \rho g \cdot \frac{2}{3} \cdot h \cdot \frac{1}{2} b h = \frac{2\rho g h^2}{\sigma} b.$$

Hence, the thrust on ABCD due to the lighter fluid only = $\frac{\rho g h^2}{\sigma}(2b + a)$.

Also the thrust on ECDF due to fluid of density $\sigma - \rho$ is

$$= \frac{(\sigma-\rho)gh^2}{\sigma.4}\left(\frac{a+b}{2}+2b\right).$$

Hence, the total thrust on ABCD due to the two liquids

$$= \frac{\rho g h^2}{\sigma}(a+2b) + \frac{g(\sigma-\rho)}{\sigma}\frac{h^2}{4}\left(\frac{a+b}{2}+2b\right)$$
$$= \frac{gh^2}{48}[\rho(4a+8b)+(\sigma-\rho)(a+5b)]$$
$$= \frac{gh^2}{48}[\rho(8a+6b-a-5b)+\sigma(a+5b)]$$
$$= \frac{gh^2}{48}[\rho(7a+11b)+\sigma(a+5b)] \quad \textbf{Proved.}$$

► **Ex. 8.** A parallelogram is immersed in homogeneous fluid of density ρ, open to the atmospheric pressure II, with the side AB in the surface. E is a point on AB such that $AE = \frac{2}{3}AB$. A straight line joins E to a point F in CD. Show that if the thrusts on AEFD and EBCF are equal, DF is given by $6DF(3\pi + 2g\rho b) = AB(6\pi + 5gab)$, where b is the depth of CD.

Solution: From A, we draw AK perpendicular to CD, then AK = b from question.

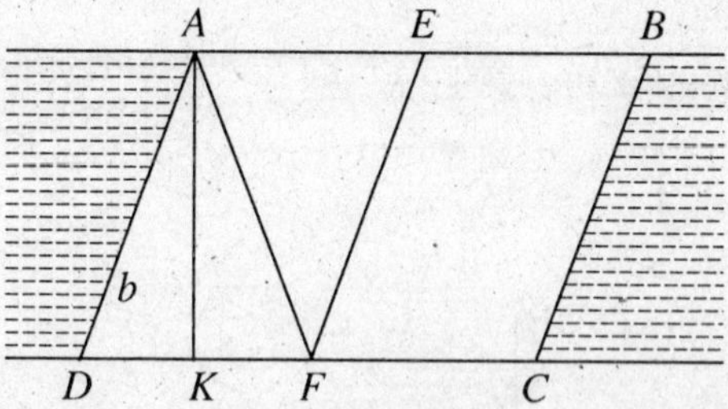

Fig. 4.10

Depth of centre of gravity of the parallelogram ABCD, from the free surface is $\frac{b}{2}$.

Let $\bar{z}$ be the depth of the C.G. of the area AEFD.

Since AEFD = ΔAFD + ΔAFE

$$\therefore \quad \bar{z} = \frac{W_1Z_1 + W_2Z_2}{W_1+W_2}, \qquad (1)$$

where Z_1 and Z_2 are the depths of centre of gravities of ΔAFD and ΔAFE. Also W_1, W_2 are the weights of the ΔAFD and ΔAFE.

$$Z_1 = \frac{2}{3}AK = \frac{2}{3}b \qquad \Bigg| \qquad Z_2 = \frac{1}{3}AK = \frac{1}{3}b$$

$$W_1 = \rho g \cdot \text{area of } \Delta AFD = \rho g\frac{1}{2}DF \cdot AK = \frac{\rho g}{2}DFb$$
$$W_2 = \rho g \cdot \text{area of } \Delta AFE = \rho g\frac{1}{2}AE \cdot b = \frac{\rho g}{2}AEb.$$

Hence, from (1)

$$\bar{z} = \frac{\frac{2}{3}b\frac{\rho g}{2}DFb + \frac{\rho g}{2}AEb\frac{1}{3}b}{\frac{\rho g}{2}DFb + \frac{\rho g}{2}AEb} = \frac{\frac{\rho g b^2}{2}\left[\frac{2}{3}DF + \frac{1}{3}AE\right]}{\frac{\rho g b}{2}(DF + AE)}$$

$$= \frac{\frac{b}{3}[2DF + AE]}{(DF + AE)} = \frac{\frac{b}{3}(2DF + \frac{2}{3}AB)}{DF + \frac{2}{3}AB} \quad \left[AE = \frac{2}{3}AB\right]$$

$$= \frac{2b(3DF + AB)}{3(3DF + 2AB)}.$$

Now, the thrust on ABCD

$$\left(\pi + \rho g\frac{b}{2}\right) ABb.$$

Also, the thrust on AEFD

$$= (\pi + g\rho\bar{z})\frac{1}{2}(AE + DF)b$$

$$= (\pi + g\rho\bar{z})\frac{1}{2}\left(\frac{2}{3}AB + DF\right) b$$

$$= \frac{b}{6}(\pi + g\rho\bar{z})(2AB + 3DF).$$

Again, thrust on ABCD = 2(thrust on AFED).

$$\left(\pi + \rho g\frac{b}{2}\right) ABp = 2\cdot\frac{b}{6}(\pi + g\rho\bar{z})(3DF + 2AB)$$

or, $3(2\pi + \rho g b)AB = 2(\pi + g\rho\bar{z})(3DF + 2AB)$

or, $3(2\pi + \rho g b)AB = 2\left[\pi + \frac{g\rho 2b(3DF + AB)}{3(3DF + 2AB)}\right](3DF + 2AB)$

or, $9(2\pi + \rho g b)AB = 2[3\pi(3DF + 2AB) + \rho g 2b(3DF + AB)]$

or, $AB[18\pi + 9\rho g b - 12\pi - 4b\rho g] = 6DF(3\pi + 2b\rho g)$

or, $AB(6\pi + 5b\rho g) = 6DF(3\pi + 2g\rho b)$ **Proved.**

▶ **Ex. 9.** A vessel contains n different liquids resting in horizontal layers and of densities $\rho_1, \rho_2, \ldots, \rho_n$ starting from the highest fluid. A triangle is held with its base in the upper surface of the highest liquid, and with its vertex in the nth liquid. Prove that if Δ be the area of the triangle and $h_1, h_2, \ldots, h_n$ be the depths of the vertex below the upper surfaces of the 1st, 2nd, nth liquids, respectively; the thrust on the triangle is

$$\frac{1}{3}\frac{g\Delta}{h_1^2}\{\rho_1(h_1^3 - h_2^3) + \rho_2(h_2^3 - h_3^3) + \cdots + \rho_n h_n^3\}.$$

Solution: Let AB_1C_1 be the triangle immersed in n liquids with its base B_1C_1 in the upper surface of the liquid of density ρ_1 and the vertex A of the triangle be in the liquid of density ρ_n. Given that $AD_1 = h_1$, $AD_2 = h_2, \ldots$.

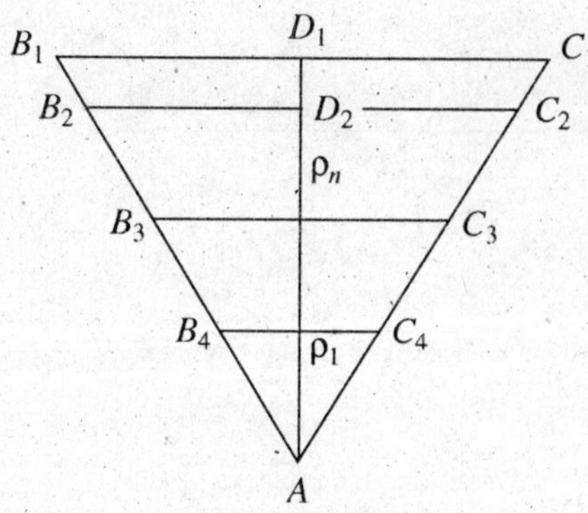

Fig. 4.11

Also $\Delta = $ area of $\Delta AB_1C_1 = \frac{1}{2}B_1C_1 \cdot h_1$

$$B_1C_1 = \frac{2\Delta}{h_1}.$$

From the similar triangles AB_1C_1 and AB_2C_2, we have

$$\frac{B_2C_2}{B_1C_1} = \frac{h_2}{h_1}.$$

$$B_2C_2 = \frac{B_1C_1h_2}{h_1} = \frac{2\Delta h_2}{h_1^2}.$$

Similarly

$$B_3C_3 = \frac{2\Delta h_3}{h_1^2} \text{ and so on.}$$

Hence, areas of triangles $AB_1C_1, AB_2C_2, \ldots, AB_nC_n$ are

$$\Delta, \frac{\Delta h_2^2}{h_1^2}, \frac{\Delta h_3^2}{h_1^2}, \ldots, \frac{\Delta h_n^2}{h_1^2}, \text{ respectively.}$$

Now the thrust on the whole

ΔAB_1C_1 = The thrust on ΔAB_1C_1 in liquid of density ρ_1 with B_1C_1 in free surface
+ The thrust on ΔAB_2C_2 in liquid of density $(\rho_2 - \rho_1)$ with B_2C_2 in free surface
+ The thrust on ΔAB_3C_3 in liquid of density $(\rho_3 - \rho_2)$ with B_3C_3 in free surface
+ ....
+ The thrust on ΔAB_nC_n in liquid of density $(\rho_n - \rho_{n-1})$ with B_nC_n in free surface

$$= \rho_1 g\frac{h_1}{3}\Delta + g(\rho_2 - \rho_1)\frac{h_2}{3}\frac{\Delta h_2^2}{h_1^2} + g(\rho_3 - \rho_2)\frac{h_3}{3}\frac{\Delta h_3^2}{h_1^2} + \cdots$$
$$+ g(\rho_n - \rho_{n-1})\frac{h_n}{3}\frac{\Delta h_n^2}{h_1^2}$$
$$= \frac{g\Delta}{3h_1^2}[\rho_1 h_1^3 + (\rho_2 - \rho_1)h_2^3 + (\rho_3 - \rho_2)h_3^3 + \cdots + (\rho_n - \rho_{n-1})h_n^3]$$
$$= \frac{g\Delta}{3h_1^2}[\rho_1(h_1^3 - h_2^3) + \rho_2(h_2^3 - h_3^3) + \cdots + \rho_n h_n^3] \quad \textbf{Proved.}$$

▶ **Ex. 10.** An ellipse is placed with its major axis on the surface of the water and its plane vertical, a circle—the auxiliary circle is described on the major axis as diameter. Find the thrust of water on the portion of the area enclosed between the auxiliary circle and the ellipse.

Solution: Let the equation of the ellipse whose diameter is AB be

$$\frac{x^2}{a^2}+\frac{y^2}{b^2}=1.$$

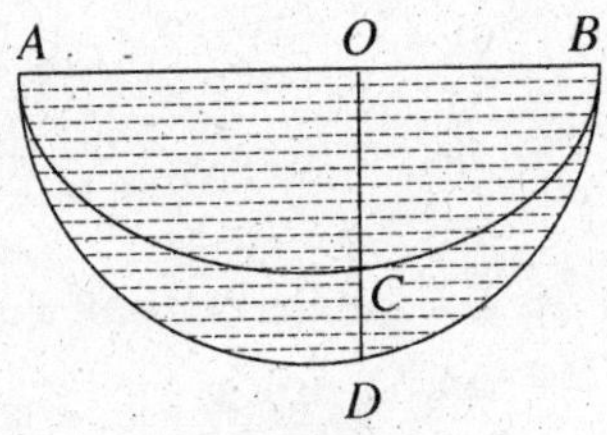

Fig. 4.12

Then the equation of the auxiliary circle is $x^2+y^2=a^2$.

The depth of the C.G. of the semi-ellipse ABC is $\frac{4b}{3\pi}$.

Hence, the thrust on the semi-ellipse ABC

$$=\frac{1}{2}\pi ab\frac{4b}{3\pi}\rho g=\frac{2ab^2}{3}\rho g.$$

But ρ = density of water $=1=\frac{2ab^2}{3}g$.

Similarly, the thrust on the semi-circle ABD

$$=\frac{1}{2}\pi a^2\frac{4a}{3\pi}g=\frac{2a^3g}{3}.$$

Hence, the thrust on the portion ADBC $=\frac{2}{3}ag(a^2-b^2)$.

▶ **Ex. 11.** A trapezoidal plane having its parallel sides of length x and y $(x>y)$ at a distance z apart, is immersed vertically in water into n side uppermost (horizontal) at a depth d below the water surface. Find the total thrust on the surface. [I.A.S. 2000]

Solution: Let ABCD be the trapezium, whose side $AB=x$ and $CD=y$. the side AB is at a depth d below the free surface. E and F are the mid points of the sides AB and CD. Also $EF=z$.

Then

$$GL=GE+EL=GE+d. \quad (1)$$

Also

$$\frac{GE}{GF}=\frac{x+\frac{y}{2}}{y+\frac{x}{2}}=\frac{2x+y}{2y+x} \Rightarrow \frac{GF}{GE}=\frac{2y+x}{2x+y}$$

$$\Rightarrow \frac{GF}{GE}+1=\frac{3x+3y}{2x+y}$$

$$\Rightarrow \frac{EF}{GE}=\frac{3x+3y}{2x+y}$$

$$\Rightarrow \frac{z}{GE}=\frac{3x+3y}{2x+y}$$

$$\Rightarrow GE=\frac{(2x+y)z}{3x+3y}.$$

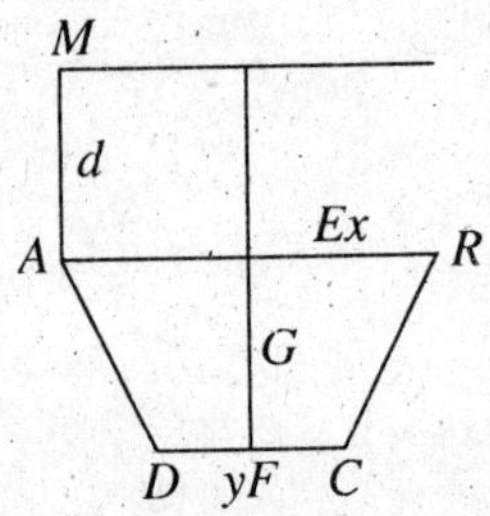

Fig. 4.13

So from (1)

$$GL = \frac{(2x+y)z}{3(x+y)} + d.$$

Also area of the trapezium $= \frac{1}{2}(x+y)z$.

Hence the required total thrust

$$= W\bar{z}\cdot S = W\left[\frac{(2x+y)z}{3(x+y)} + d\right]\cdot\frac{1}{2}(x+y)z$$
$$= \frac{W[(2x+y)z + 3d(x+y)]z}{6},$$

where W is the weight per unit volume of the liquid.

▶ **Ex. 12.** A semicircular lamina is immersed in a liquid with the diameter in the surface and its plane vertical. Show how to divide it into three sectors the thrust on each of the which is the same. **[P.U. 71A, M.U. 71A, 72A, 92H, L.N.M.U. 88H, 93H, 76A]**

Solution: Let AOB be the diameter of the semicircle ACDB in the surface of the liquid. Let AOC, COD and DOB be three sectors on which thrusts are equal.

Let $\angle AOC = \theta$.

The C.G. of this sector lies on the bisector of $\angle AOC$ at a distance $\frac{4r}{3\theta}\sin\frac{\theta}{2}$ from O.

$\therefore$ depth of the C.G. below O

$$= \frac{4r}{3\theta}\sin\frac{\theta}{2}\times\sin\frac{\theta}{2} = \frac{4r}{3\theta}\sin^2\frac{\theta}{2}.$$

Area of this sector $= \frac{1}{2}r^2\theta$.

$\therefore$ the thrust on this sector

$$= \frac{1}{2}r^2\theta\times\frac{4r}{3\theta}\sin^2\frac{\theta}{2}\rho g = \frac{2r^3}{3\theta}\sin^2\frac{\theta}{2}\rho g,$$

where ρ is the density of the liquid.

Thrust on semicircle

$$= \frac{\pi r^2}{2}\times\frac{4r}{3\pi}\rho g = \frac{2r^3}{3}\rho g.$$

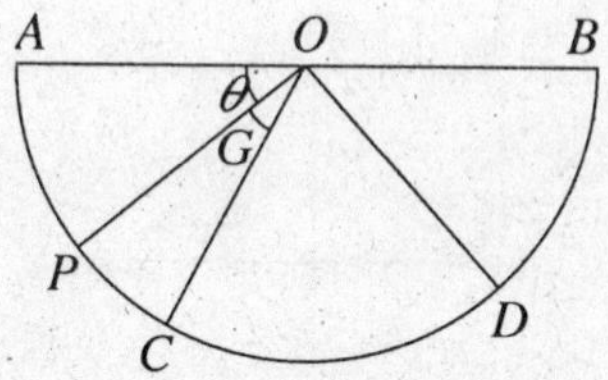

Fig. 4.14

But the thrust on sector AOC $= \frac{1}{3}\times$ thrust on semicircle.

$$\Rightarrow \frac{2r^3}{3}\sin^2\frac{\theta}{2}\rho g = \frac{1}{3}\times\frac{2r^3}{3}\rho g$$
$$\Rightarrow \sin^2\frac{\theta}{2} = \frac{1}{3} \Rightarrow \frac{1-\cos\theta}{2} = \frac{1}{3}$$
$$\Rightarrow 1-\cos\theta = \frac{2}{3} \Rightarrow \frac{1}{3} = \cos\theta.$$

Thus, radii which make angles with horizon, whose cosines are $\frac{1}{3}$ divide the semicircle into three sectors, the thrusts on which are equal.

Now $r\cos\theta = \frac{r}{3}$.

Hence the radii passing through the points on the circumference where it is cut by perpendiculars to the diameter at distances equal to one-third the radius from the centre divide the semicircle into three sectors the thrusts on which are equal.

▶ **Ex. 13.** A hollow cone whose axis is vertical and base downwards is filled with equal volumes of two liquids whose densities are in the ratio 3 : 1, show that the thrust on the base is $(3-3\sqrt{4})$ times as it is when the vessel is filled with the lighter fluid. **[P.U. 65, R.U. 70A, 86H, 92A, 95H]**

Solution: Let h be the height and α the semivertical angle of the cone.

Let the common surface CD be at a depth x below V. Since the volumes are equal

$$\therefore\ \frac{1}{3}\pi x^3 \tan^2\alpha = \frac{1}{2}\frac{1}{3}\pi h^3 \tan^2\alpha \Rightarrow x = \left(\frac{1}{2}\right)^{1/3} h. \qquad (1)$$

Let the densities of two fluids be 3ρ and ρ.

$\therefore$ pressure at a point on the base when filled with the two liquids of equal volumes.

$$= x\rho g + (h-x)3\rho g = (3h-2x)\rho g$$

$$= \left[3h - 2\left(\frac{1}{2}\right)^{1/3} h\right] \rho g \text{ [From (1)]}$$

$$= \left[3 - \sqrt[3]{4}\right] h\rho g.$$

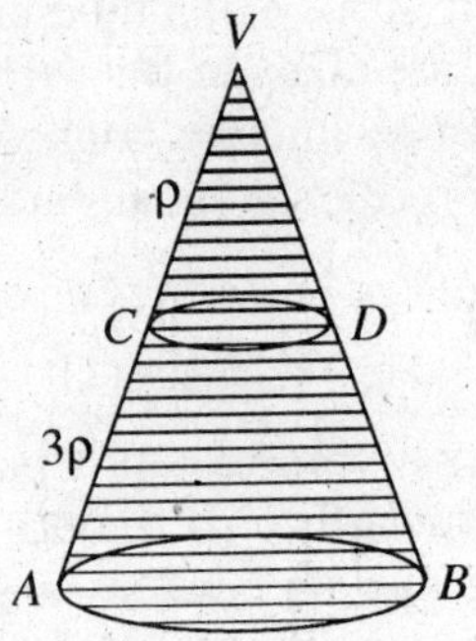

Fig. 4.15

So the thrust on the base

$$= \text{pressure at its C.G.} \times \text{area}$$
$$= \left(3 - \sqrt[3]{4}\right) h\rho g(\pi h^2 \tan^2\alpha)$$
$$= \left(3 - \sqrt[3]{4}\right) \pi h^3 \rho g \tan^2\alpha$$
$$= 17 \text{ (say)} \qquad (2)$$

Again thrust when only lighter liquid is filled in the cone

$$= \pi h^2 \tan^2 \alpha\rho g h = P_2 \text{ (say)} \qquad (3)$$

Now from (2) and (3), we get

$$\frac{P_1}{P_2} = \left(3 - \sqrt[3]{4}\right) \Rightarrow P_1 = \left(3 - \sqrt[3]{4}\right) P_2 \text{ Proved.}$$

Exercises 4(III)

1. A rectangle ABCD is immersed in water with the side AB in the surface. Draw a straight line through A which shall divide the lamina into two parts, the thrusts on which are equal.

   [*Ans.* DE $= \frac{3}{4}$DC, where E is a point, on CD]

   [Pat. 79A, 74A, M.U. 67A]

2. The side AD of a triangle ABD is in the surface of a fluid and points C, E are taken in BD such that the pressures on the triangles CAD, CAE, ABE are equal, find the ratio DC : CE : EB.

   [*Ans.* $1 : \sqrt{2}-1 : \sqrt{3}-\sqrt{2}$]

3. A heavy conical cup is placed vertex upwards on a smooth horizontal plane and water is gradually poured in through a hole at the top. The weight of the cup is $\frac{5}{8}$ of the weight of the water which would just fill of. Prove that the cup will be on the point of rising from the plane when the water, has reached half the height of the cup.

4. A solid triangular prism, the faces of which include angle α, β, γ is placed in any position entirely within a liquid. If P, Q, R be the thrusts on the three faces, respectively opposite to angles α, β, γ, prove that

$$P \operatorname{cosec} \alpha + Q \operatorname{cosec} \beta + R \operatorname{cosec} \gamma = 0$$

   is in variable so long as the depth of the centre of gravity of the prism is unchanged.

5. A triangle ABC is immersed in a liquid, its plane being vertical and the side AB in the surface. If O be the centre of the circumscribing circle of $\triangle ABC$, prove that

$$\frac{\text{Thrust on } \triangle OCA}{\text{Thrust on } \triangle OCB} = \frac{\sin 2B}{\sin 2A}.$$

6. A triangle is immersed in a homogeneous liquid with one side in the surface, show how to draw a horizontal line dividing it into two portions on which thrusts are equal.

   [*Ans.* At half the depth of the vertex]

7. ABC is a triangular lamina with the side AB in the surface of a heavy homogeneous liquid. A point D is taken in AC such that the thrusts on the areas ABD, DBC are equal. Show that $AD : AC = 1 : \sqrt{2}$. **[R.U. 76H]**

8. A triangle ABC is immersed in a fluid, with one side BC in the surface. Find a point O within the triangle such that the thrusts on the triangles OAB, OBC, OCA may be equal.

   [*Ans.* O lies on the median AD such that $AD = \sqrt{3}\,OD$]

9. Calculate the thrust on a rectangular plate whose sides are a and b, the side a being horizontal and at a depth c below the surface and the plane of the rectangle being inclined at an angle θ to the vertical.

   [*Ans.* $\left(c + \frac{b}{2}\cos\theta\right) ab\rho g$]

10. A rectangular plate is immersed vertically in a liquid with one side in the surface. Divide it into two

parts by a line parallel to the surface such that the pressure on the upper part is double the pressure on the lower part. **[Bih. 66H]**

[*Ans.* The line divides the vertical side in the ratio $\sqrt{2} : \sqrt{3}$]

11. Divide a square lamina immersed vertically in a fluid with one side in the surface by a straight line parallel to a diagonal so that the thrusts on the two parts may be equal. **[M.U. 66A, Mith. 82]**

[*Ans.* The line cuts the vertical side of length a at a distance x given by the equation $2x^3 - 6a^2x + a^3 = 0$]

12. A semi-circular area is immersed in a liquid vertically with the diameter in the surface. If the line through an extremity of the diameter dividing the semi-circle into two parts the thrusts on which are equal makes with the horizontal an angle θ, prove that

$$2\cos^4\theta = 1.$$

13. Find the whole pressure on a plane area, in the form of a trapezium, immersed vertically in a heavy homogeneous liquid with one of the parallel sides, length a, in the surface and the other, of length b, at a depth h below the surface.

[*Ans.* $\frac{\rho g h^2}{6}(a + 2b)$]

14. A hollow weightless cone of semi-vertical angle α and of height h, is filled with liquid and freely hung from a point on the rim of its base, prove that the thrust of the water on base is

$$\frac{4\pi h^3 \tan^3\alpha \sin\alpha}{\sqrt{1 + 15\sin^2\alpha}}\rho g.$$

Also show that this thrust bears to the weight of water in the cone a ratio of

$$12\sin^2\alpha : \cos\alpha\sqrt{1 + 15\sin^2\alpha}.$$

[Bih. 67A, Bhag. 66A, Pat. 78H, M.U. 82H.]

15. A hollow weightless cylinder filled with water is suspended freely from a point on the rim of its plane end. If the height of the cylinder be equal to the diameter of the plane circular ends, show that the thrusts on the ends are in the ratio $3 : 1$.

16. The lighter of two liquids of density ρ rests on the heavier of density σ, to a depth b. If the square of side a is immersed in a vertical position with one side in the surface of the upper liquid and the thrusts on the two portions of the square in contact with the two liquids be equal, prove that

$$b\rho(3b - 2a) = \sigma(a - b)^2.$$

[Bhag. 62A, 65A, R.U. 72A, 72H, 82H, 96H]

17. A liquid whose specific weight is w' rests to a depth a upon another liquid, with which it does not mix, whose specific weight is w. A square of side $b(> a)$ is immersed in the two liquids, its upper edge being in the surface of the upper liquid and its plane being vertical. Find the thrust on the square.

[*Ans.* $\frac{bw}{2}(b - a)^2 + \frac{abw'}{2}(2b - a)$]

18. A cubic vessel is filled with two liquids of given densities ρ and σ, the volumes of each being the same, find the pressure on the base and on any side of the vessel.

[*Ans.* $\frac{a^3}{2}(\rho + \sigma)g$, $\frac{a^3}{8}(3\rho + \sigma)\, g (\rho < \sigma)$]

19. ABC is a triangle immersed vertically in water with C in the surface and the sides AC, BC equally inclined to the surface, prove that the vertical through C divides the triangle into two parts the thrusts on which are in the ratio

$$b^3 + 3ab^2 : a^3 + 3a^2b.$$

[Bhag. 67H, R.U. 95H]

20. ABC is a triangle immersed in a liquid with vertex A in the surface and the median vertical. Prove that the ratio of the thrusts on the two parts into which the triangle is divided by the median is

$$\frac{5b^2 + 3c^2 - 2a^2}{3b^2 + 5c^2 - 2a^2}.$$ [Bih. 69H]

21. A triangle ABC is immersed vertically in two liquids with its base BC in the surface. If the densities of the upper and lower liquids be σ and ρ, respectively and the depth of the surface of separation be half the depth of the vertex, show that the ratio of the thrust on the portions in the two liquids is $4\sigma : 3\sigma + \rho$.

[P.U. 79H, R.U. 71H, 2000H]

22. The centre board of a yacht is in the form of a trapezoid in which the two parallel sides are 3 and 5 ft, respectively and the side perpendicular to these two is 4 ft in length. Assuming that the last named side is parallel to the surface of the water, at a depth of 1 foot and that the parallel sides are vertical; find the total pressure on the board. [*Ans.* $3041\frac{2}{3}$ lbs]

23. The sides of a rectangle are in the ratio $\pi : 4$ and semi-circles are described on the longer side as diameters, prove that if the rectangle is immersed in water, with one of the shorter sides in the surface, the thrusts on the two parts external to both semi-circles will together be equal to that on the part common to them.

24. A trapezoidal lamina ABCD is immersed in water with the parallel sides AB and CD horizontal. Prove that the locus of a point P on the area such that the thrust on PAD is n times the thrust on PBC is a straight line through the point of intersection of the non-parallel sides.

25. A parabolic area, terminated by a double ordinate of length twice the latus rectum, is immersed with the vertex downwards and the double ordinate in the free surface. Calculate the ratio of the thrusts on the portions into which it is divided by the latus rectum.

[*Ans.* 17 : 47]

26. A parabolic area bounded by the latus rectum is just immersed vertically in a heavy homogeneous liquid at rest with its vertex in the surface. Find the thrust on it.

[*Ans.* $\frac{8}{5}Wa^3$, $4a$ is the latus rectum of the parabola and W is the weight of the liquid per unit volume]

27. The inclinations of the axis of a submerged solid cylinder to the vertical in two different positions are complementary to each other. If p and p' be the differences between the pressures on the two ends in the two cases, prove that the weight of the displaced fluid is equal to $(p^2 + p'^2)^{1/2}$.

28. A hollow right cone with weight is filled with liquid and suspended by a point in the rim of its base. If the axis of the cone is inclined

to the vertical at an angle θ, prove that

$$\cot 2\theta = \cot \alpha - \tfrac{3}{4} \operatorname{cosec} \alpha,$$

where α is the vertical angle of the cone.

29. An ellipse is placed with its minor axis on the surface of water and its plane vertical, a circle is described on the minor axis as diameter. Find the pressure of water on the portion of the area enclosed between the ellipse and the circle.

[*Ans.* $\frac{2bg}{3}(a^2 - b^2)$, where $2a$ and $2b$ are the lengths of major and minor axes of the ellipse]

30. A hole in the side of a ship is closed by a circular door 6 feet in diameter hinged at the highest point and held in side against the water pressure by a fastening at its lowest point. If the highest and the lowest points of the door are at depths of 5 and 10 feet, show that the least force exerted by the fastening must be 3.2 tons.

31. Find the whole pressure on a triangle the depths of whose vertices are α, β, γ, the liquid being homogeneous and S is the area of the triangle. [*Ans.* $\frac{\alpha+\beta+\gamma}{3} g\rho S$]

[R.U. 97H]

32. A rectangular lamina is immersed vertically in a liquid with one side in the surface. Show how to draw a horizontal line dividing it into two parts such that the ratio of the thrusts on the upper and lower parts is 9 : 7.

[*Ans.* $BF = \frac{\sqrt{7}}{3} BC$, where ABCD is a rectangle and F is a point on BC]

33. A rectangular vessel contains three liquids which do not mix and whose specific gravities are 1.0, 1.2 and 1.6, the thicknesses of which are 4, 3 and 2 inches, respectively. Compare the total thrusts of the liquids on the different parts of a side of the vessel.

[*Ans.* 40 : 87 : 92]

34. A cylindrical tumbler, half filled with a liquid of density ρ is filled up with a liquid of density ρ' which does not mix with the former one. Show that the thrust on the base of the tumbler is to the whole pressure on its curved surface as $2r(\rho + \rho')$ to $h(\rho + 3\rho')$, where h is the height and r the radius of the base of the cylinder.

35. A triangle of altitude h is immersed in a liquid whose density at a depth Z is $\sigma + \lambda Z$, with the vertex in the surface and base horizontal. Prove that the thrust upon it is

$$a\left[\frac{\pi h}{2} + g\left(\frac{\sigma h^2}{3} + \frac{\lambda h^3}{8}\right)\right].$$

36. A parallelogram is immersed in water, one side being in the surface. Divide it into n parts by horizontal lines so that the thrust on these parts may be equal. Show that the depths of the dividing lines are proportional to the square root of the natural numbers.

37. A cubical box with vertical sides is filled with equal volumes of n different liquids which do not mix, the density of the uppermost being ρ that of the next 2ρ, ... and that of the lowest $n\rho$. Show that the thrust on the base is $(n + 1)$ times the thrust on that part of one of

the sides which is contact with the lowest liquid. Also find the thrust on a vertical side.

[*Ans.* $\frac{\rho g a^3(n+1)(2n+1)}{12n}$]

38. A parallelogram is immersed in a fluid with a diagonal vertical, an extremity of which is in the surface of the liquid. Through this point lines are drawn, dividing the parallelogram into three equal parts. Compare the pressures on three parts and if P_2 be the pressure on the middle part and P_1, P_2 those on the other two, prove that

$$16P_2 = 11(P_1 + P_2).$$

[B.U. 76H]

39. A square of side a is dipped vertically in two liquids of densities ρ and σ with the upper side in the free surface. The depth of the upper fluid of density ρ is $b(< a)$. Find the thrust on the square.

[Pat. 81H]

[*Ans.* $\frac{g\rho}{2}ab(2a-b) + \frac{1}{2}g\sigma a(a-b)^2$]

40. An isosceles triangle is immersed vertically in a fluid with its vertex in the surface and base horizontal. Determine how it must b divided by a line parallel to the base so that the pressure on upper and lower portions may be respectively in the ratio $m : n$.

[*Ans.* Depth on the line below the vertex = $\left(\frac{m}{m+n}\right)^{1/3} . h$, where h is the height of the triangle]

CHAPTER 5

Centre of Pressure

5.1 Introduction

If a plane area be immersed in liquid, the pressure at any point of it is perpendicular to the plane area and is proportional to the depth of the point. So, the pressures or thrusts on all points of a given plane area set of parallel forces which can be reduced to single resultant force. The point of the plane area at which the resultant force acts is called the *centre of pressure.*

Definition 5.1.1 *The point of a plane area at which the resultant thrust, exerted by the fluid in which it is immersed, acts, is called the* centre of pressure *(C.P.) of the area.*

5.2 To find the C.P. of a Plane Area immersed in a Liquid

Let A be the plane area making an angle θ with the horizontal. Let AB be its line of intersection with the free surface. Let dA be an elementary area at a depth h below the free surface. Then the thrust on elementary area $= \rho g h dA$, where ρ is the density of the liquid. The total thrust on the area being the resultant of elementary thrusts $= \Sigma \rho g h dA$.

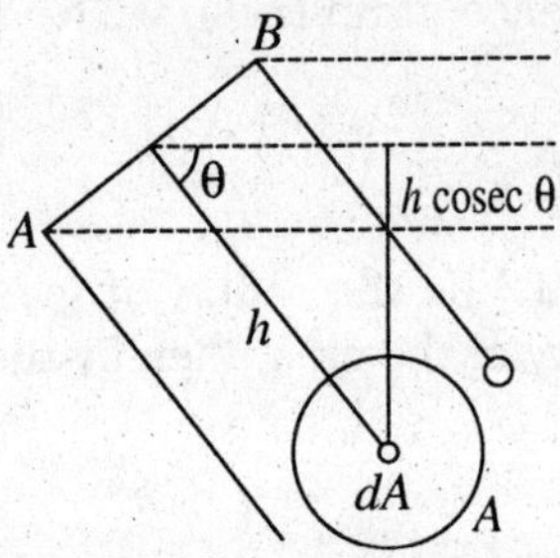

Fig. 5.1

If H be the depth of the C.P. of the whole area A, we have by taking moments about AB,

$$H \operatorname{cosec}\theta \times \Sigma\rho g h dA = \Sigma(\rho g dA \times h \operatorname{cosec}\theta)$$

$$H = \frac{\Sigma\rho g h^2 dA}{\Sigma\rho g h dA} = \frac{\Sigma h^2 dA}{\Sigma h dA}$$

$$H = \frac{\int h^2 dA}{\int h dA}, \tag{5.1}$$

where the integration is to be taken throughout the whole area.

If the area dA be expressed by $dx\,dy$, then from (5.1)

$$H = \frac{\iint h^2 dx\,dy}{\iint h dx\,dy}.$$

If the liquid is not homogeneous, then from (5.1)

$$H = \frac{\iint hp\,dx\,dy}{\iint p\,dx\,dy},$$

where p is the pressure at a depth h.

Equation (5.1) shows that *the position of the C.P. of a plane area is independent of the inclination of the plane.* [C.U. 90H]

Also from eqn. (5.1), we have

$$\int h^2 dA = \text{Moment of inertia of the area about AB}$$

$$= AK^2, \text{ where K is the radius of gyration of the area about AB.}$$

Also, $\int h dA$ = First moment of the area about AB = Ah.

Hence, from (5.1), we have

$$H = \frac{AK^2}{Ah} = \frac{K^2}{h}. \tag{5.2}$$

If K' be the radius of gyration about a parallel axis through G, the centre of gravity of the area, then by the theorem on parallel axes

$$K^2 = K'^2 + h^2.$$

Hence, from (5.2), we get

$$H = \frac{K'^2 + h^2}{h} = \frac{K'^2}{h} + h. \tag{5.3}$$

Thus, the depth of the C.P. below the C.G. is $\frac{K'^2}{h}$. [C.U. 94H, 03H]

K'^2 being constant we can say that *the vertical distance between the centre of pressure and the centre of gravity varies inversely as the depth of the centre of gravity, however, the lamina is moved without rotation.*

Differentiating (5.3), we get

$$\frac{d}{dt}(H-h) = \frac{-K'^2}{h^2}\frac{dh}{dt}.$$

This shows that the centre of pressure approaches the horizontal through the centre of gravity with a velocity which it inversely proportional to the square of the depth of its centre of gravity.

In case the fluid is heterogeneous with a slowly varying density, let p be the pressure at G, the centre of gravity when it is at a depth h. At considerable depths dimensions of the area are small compared with h. As such we may suppose the density over the area to be constant and equal to that at G, say ρ.

Then the thrust on a small area α, at a vertical distance z from G $= \alpha(p + g\rho z)$ and the vertical distance between the C.P. and C.G.

$$= \frac{\Sigma\alpha(p+g\rho z)z}{\Sigma\alpha(p+g\rho z)} = \frac{p\Sigma\alpha z + g\rho\Sigma\alpha z^2}{p\Sigma\alpha + g\rho\Sigma\alpha z}$$

$$= \frac{g\rho\Sigma\alpha z^2}{p\Sigma\alpha} \quad [\because \Sigma\alpha z = 0]$$

$$= \frac{g\rho}{p}\frac{AK'^2}{A} = \frac{gK'^2\rho}{p}$$

gK'^2 being constant, the quantity varies as $\frac{\rho}{p}$.

Hence, *in a heterogeneous fluid, with a slowly varying density, the vertical distance between the centre of gravity and the centre of pressure will at considerable depth, be proportional to the quotient of the density by the pressure at the centre of gravity.*

5.3 To find C.P. of an area composed of two portions whose C.P. are known

- If T_1 and T_2 be the thrust on the two parts of a given plane area, and Z_1 and Z_2 be the depths of their centres of pressure below the surface, then the depth $\bar{z}$ of the C.P. of the whole area below the free surface is given by

$$\bar{z} = \frac{T_1Z_1 + T_2Z_2}{T_1 + T_2}.$$

The generalised form of the above formula is

$$\bar{z} = \frac{T_1Z_1 + T_2Z_2 + T_3Z_3 + \cdots}{T_1 + T_2 + T_3 + \cdots}$$

To find when a portion of a plane lamina is cut off; to find the C.P. of the remainder.

- If T_1 and T_2 be the thrusts on the whole plane area and the portion of the area removed and Z_1 and Z_2 the depths of their centres of pressures below the free surface, then the depth $\bar{z}$ of the C.P. of the remaining portion is given by

$$\bar{z} = \frac{T_1Z_1 - T_2Z_2}{T_1 - T_2}.$$

5.4 The Depth of the Centre of Pressure of a Plane Area immersed in a Liquid is greater than the Depth of its Centre of Gravity

[C.U. 91H]

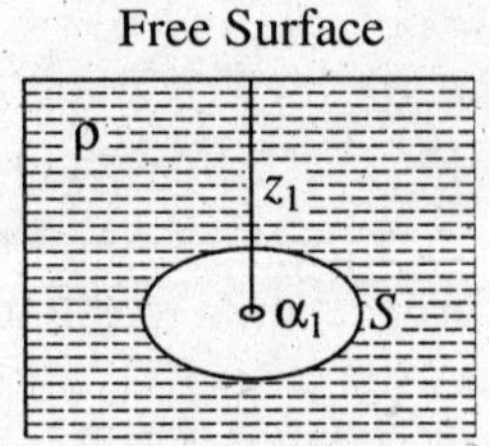

Fig. 5.2

Let the given plane area S be divided into small elementary areas $\alpha_1, \alpha_2, \alpha_3, \ldots$ whose depths are $z_1, z_2, z_3, \ldots,$ respectively, below the free surface. Let ρ be the density of the liquid.

Thrust on $\alpha_1, \alpha_2, \ldots, \alpha_n$ are $\rho g z_1 \alpha_1, \rho g z_2 \alpha_2, \ldots, \rho g z_n \alpha_n, \ldots$.

∴ The depth of C.P. below the free surface

$$= \frac{\rho g z_1 \alpha_1 \cdot z_1 + \rho g z_2 \alpha_2 \cdot z_2 + \cdots + \rho g z_n \alpha_n \cdot z_n + \cdots}{\rho g z_1 \alpha_1 + \rho g z_2 \alpha_2 + \cdots + \rho g z_n \alpha_n + \cdots}$$

$$= \frac{\rho g(\alpha_1 z_1^2 + \alpha_2 z_2^2 + \cdots + \alpha_n z_n^2 + \cdots)}{\rho g(\alpha_1 z_1 + \alpha_2 z_2 + \cdots + \alpha_n z_n + \cdots)}$$

$$= \frac{\alpha_1 z_1^2 + \alpha_2 z_2^2 + \cdots + \alpha_n z_n^2 + \cdots}{\alpha_1 z_1 + \alpha_2 z_2 + \cdots + \alpha_n z_n + \cdots}.$$

And the depth of C.G. below the free surface

$$= \frac{\rho g \alpha_1 z_1 + \rho g \alpha_2 z_2 + \cdots + \rho g \alpha_n z_n + \cdots}{\rho g \alpha_1 + \rho g \alpha_2 + \cdots + \rho g \alpha_n + \cdots}$$

$$= \frac{\rho g(\alpha_1 z_1 + \alpha_2 z_2 + \cdots + \alpha_n z_n + \cdots)}{\rho g(\alpha_1 + \alpha_2 + \cdots + \alpha_n + \cdots)}$$

$$= \frac{\alpha_1 z_1 + \alpha_2 z_2 + \cdots + \alpha_n z_n + \cdots}{\alpha_1 + \alpha_2 + \cdots \alpha_n + \cdots}$$

Now depth of C.P. − depth of C.G.

$$= \frac{\alpha_1 z_1^2 + \alpha_2 z_2^2 + \cdots + \alpha_n z_n^2 + \cdots}{\alpha_1 z_1 + \alpha_2 z_2 + \cdots + \alpha_n z_n + \cdots} - \frac{\alpha_1 z_1 + \alpha_2 z_2 + \cdots + \alpha_n z_n + \cdots}{\alpha_1 + \alpha_2 + \cdots + \alpha_n + \cdots}$$

$$= \frac{(\alpha_1 z_1 + \alpha_2 z_2 + \cdots + \alpha_n z_n^2 + \cdots)(\alpha_1 + \alpha_2 + \cdots + \alpha_n + \cdots) - (\alpha_1 z_1 + \alpha_2 z_2 + \cdots + \alpha_n z_n + \cdots)^2}{(\alpha_1 z_1 + \alpha_2 z_2 + \cdots + \alpha_n z_n + \cdots)(\alpha_1 + \alpha_2 + \cdots + \alpha_n + \cdots)}$$

$$= \frac{\alpha_1 \alpha_2 (z_1 - z_2)^2 + \alpha_1 \alpha_3 (z_n - z_3)^2 + \cdots}{(\alpha_1 z_1 + \alpha_2 z_2 + \cdots + \alpha_n z_n + \cdots)(\alpha_1 + \alpha_2 + \cdots + \alpha_n + \cdots)}$$

which is positive.

Hence, the depth of C.P. is greater than the depth of C.G.

Corollary 5.4.1 *If the plane area be horizontal, then* $z_1 = z_2 = \cdots = z_n = \cdots$

Hence, the depth of C.P. − the depth of C.G. = 0.
So, the depth of C.P. = the depth of C.G.

5.5 Geometrical Method for finding Centre of Pressure

Let ABCD be the plane area (not in a vertical position) immersed in a fluid and suppose that vertical lines through every point of the perimeter meet the surface of the fluid in the curve $A'B'C'D'$. Then the fluid enclosed in the cylinder $(ABCD, A'B'C'D')$ is called the *super-incumbent liquid.*

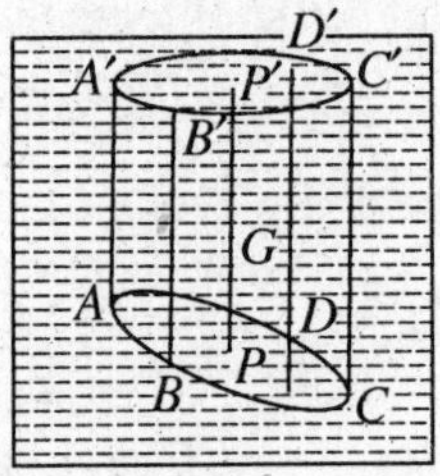

Fig. 5.3

This super-incumbent fluid is in equilibrium under the following forces:

- Its weight acting vertically downwards through its centre of gravity G.
- Resultant fluid thrust acting normally in area ABCD through P, the centre of pressure.
- The thrusts on the curved surface acting horizontally.

Resolving all these forces vertically we find that the upward vertical thrust on the plane area ABCD balances the weight of the super-incumbent liquid acting vertically downwards. This is possible only when G and P are in the same vertical line.

So, we conclude that *the vertical line through the centre of gravity of the super-incumbent liquid meets the plane area at its centre of pressure.*

5.6 Effect of Further Immersion

[I.A.S. 77, C.U. 95H]

If the plane area is immersed horizontally the C.P. and the C.G. coincide and hence further immersion in no way affects the relative position of the C.P.

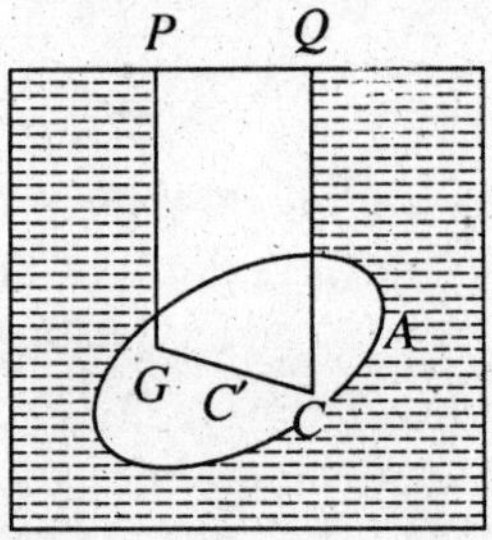

Fig. 5.4

When the plane is not horizontal let the centre of gravity G of a plane area A be at a depth h. The resultant thrust ρghA, on the area, acts at its C.P. C. If the area is lowered, without rotation by a further depth H, pressure at every point of the area is increased by ρgH. So, in the new position, we have two parallel forces ρghA and ρgHA acting at C and G, respectively. Their resultant, will act at C', the new centre of pressure on CG. such that

$$\rho ghACC' = \rho gHAGC' \quad \text{or,} \quad \frac{CC'}{GC'} = \frac{H}{h}.$$

Thus, the effect of lowering a plane area to a greater depth without rotation, is to displace the centre of pressure along a straight line on the area towards the centre of gravity.

Also the depth $\bar{Z}$ of C′ below the free surface is given by

$$\bar{Z} = \frac{\rho ghA(h' + H) + \rho gHA(h + H)}{\rho ghA + \rho GHA},$$

where h′ is the depth of the C.P. in the original position.

$$\bar{Z} = \frac{hh' + hH + Hh + H^2}{h + H} = \frac{hh' + 2Hh + H^2}{h + H}.$$

In the new position the difference of the depths of the C.P. and C.G.

$$= \frac{hh' + 2Hh + H^2}{h + H} - (h + H)$$

$$= \frac{hh' + 2hH + H^2 - (h^2 + 2Hh + H^2)}{h + H}$$

$$= \frac{hh' - h^2}{h + H}$$

h, h′ are known. Therefore, *the vertical distance between the C.G. and the C.P. varies inversely as the depth of the centroid.*

When $H \to \infty$, then the difference between, the depth of C.P. and C.G. is zero. So, we conclude that *if the area is lowered further by an infinite distance, C.P. and C.G. coincide.*

Combining together we see that if *a plane area is lowered vertically in a liquid, the C.P. approaches and ultimately coincides with the centre of gravity.* **[Mith 93H]**

Also we have

$$\bar{Z} - (h + H) = \frac{hh' - h^2}{h + H} \cdots \quad \therefore \frac{d}{dt}[\bar{z} - (h + H)] = (hh' - h^2)\frac{d}{dt}\left(\frac{1}{h + H}\right)$$

$$\text{or,} \quad \frac{d\bar{Z}}{dt} - \frac{d}{dt}(h + H) = -\frac{(hh' - h^2)}{(h + H)^2} \cdot \frac{d}{dt}(h + H).$$

Hence, *the centre of pressure approaches the horizontal through the centre of gravity with a velocity which is inversely proportional to the square of the depth of its centre of gravity.* **[C.U. 92H, 2000H]**

5.7 Centre of Pressure of a Parallelogram immersed in a Homogeneous Liquid with one side in the Surface **[M.U. 98H, 02H]**

Let ABCD be the parallelogram immersed in a liquid with side AB in the free surface.

Let AB = b and AD = a.

Without loss of generality we take the plane of the parallelogram vertical. Let E and F be the middle points of the sides AB and CD. Join EF. Let EF be inclined at an angle θ to the horizontal.

We divide the whole parallelogram into an infinite number of thin horizontal strips PQRS.

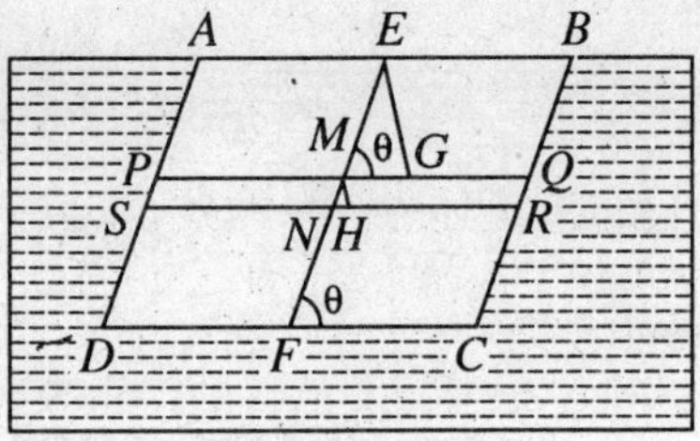

Fig. 5.5

The C.P. of all such strips will be at their middle points. Hence C.P. of the parallelogram will lie on EF.

Let EM = x, MN = dx.

Let ρ be the density of the liquid.

Then the depth of the strip PQRS from the free surface = EG = x sin θ.

Thickness of the strip = MH = dx sin θ.

Area of the strip PQRS = PQ dx sin θ = AB dx sin θ = a dx sin θ.

Thrust at any point of the strip = ρgx sin θ · a dx sin θ.

If $\bar{x}$ be the depth of C.P. of parallelogram ABCD from free surface AB, then

$$\bar{x} = \frac{\int_0^b x\rho g x \sin\theta\, a\, dx \sin\theta}{\int_0^b \rho g x \sin\theta\, a\, dx \sin\theta} = \frac{\rho g \sin^2\theta\, a \int_0^b x^2 dx}{\rho g \sin^2\theta\, a \int_0^b x\, dx}$$

$$= \frac{\left[\frac{x^3}{3}\right]_0^b}{\left[\frac{x^2}{2}\right]_0^b} = \frac{2}{3}\frac{b^3}{b^2} = \frac{2}{3}b = \frac{2}{3}EF(EF = AD = b).$$

Thus, the depth of the C.P. of a parallelogram with a side in the surface is $\frac{2}{3}$ times the depth of the side immersed.

Corollary 5.7.1 *In case of rectangle* $\theta = \frac{\pi}{2}$. *So,* $\bar{x} = \frac{2}{3}EF$. [L.N.M.U 96H]

Corollary 5.7.2 *As the inclination of the plane with the horizontal is* θ, *the vertical depth of the C.P.*

$$= \frac{2}{3} EF \sin\theta$$

$$= \frac{2}{3} \times \text{vertical depth of the lower horizontal side of the rectangle.}$$

5.8 C.P. of a Triangle immersed in a Liquid with Vertex in the Surface and Base Horizontal

[M.U. 84H, 87H, 99H, R.U. 92H, 03H]

Let the vertex A of the triangular area ABC immersed vertically in a liquid be in the surface and the side BC horizontal. Let the median AD makes an angle θ with the horizontal.

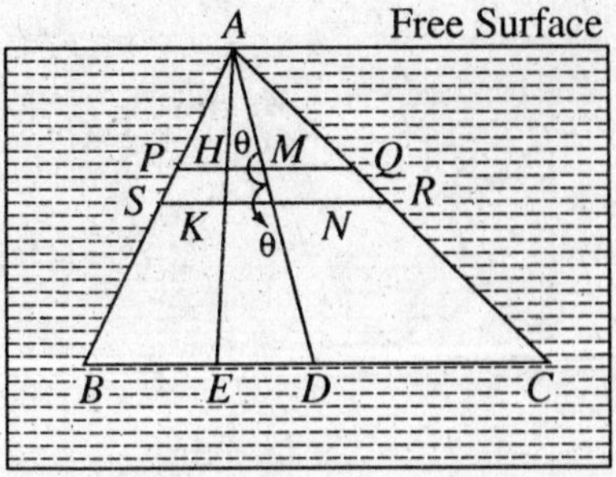

Fig. 5.6

We divide the triangle into an infinite number of horizontal strips PQRS. The C.P. of all such strips will act at their middle points which lie on AD. Hence, the C.P. of the triangle will lie on the median AD.

Let $BC = a$, $AD = h$, $AM = x$, $MN = dx$.

Now, the depth of the strip PQRS from the free surface $= AH = x\sin\theta$.

Thickness of the strip $= HK = dx\sin\theta$.
Now, $\triangle APQ$ and $\triangle ABC$ are similar.

$$\frac{PQ}{BC} = \frac{AM}{AD}, \quad \frac{PQ}{a} = \frac{x}{h}, \quad PQ = \frac{ax}{h}.$$

Area of the strip PQRS

$$= PQ\,dx\sin\theta = \frac{ax}{h}\,dx\sin\theta.$$

Thrust of any point of the strip

$$= \rho g x\sin\theta\frac{ax}{h}\,dx\sin\theta.$$

If $\bar{x}$ be the distance of C.P. from A along AD, then

$$\bar{x} = \frac{\int_0^h x\rho g x\sin\theta\frac{ax}{h}\sin\theta dx}{\int_0^h \rho g x\theta\frac{ax}{h}\sin\theta dx} = \frac{\frac{\rho g a\sin^2\theta}{h}\int_0^h x^3dx}{\frac{\rho g a\sin^2\theta}{h}\int_0^h x^2dx}$$

$$= \frac{h^4}{4}\cdot\frac{3}{h^3} = \frac{3}{4}h = \frac{3}{4}AD.$$

Hence, the depth of the C.P. is $\frac{3}{4}$ times the depth of the horizontal base.

5.9 C.P. of a Triangular Area immersed in a Liquid with One Side in the Surface

[T.M.B.U. 94H]

Let the side BC of the triangular area ABC immersed vertically in a liquid of density ρ. Let the median AD makes an angle θ with the horizontal.

Without loss of generality we take the plane of the triangle to be vertical.

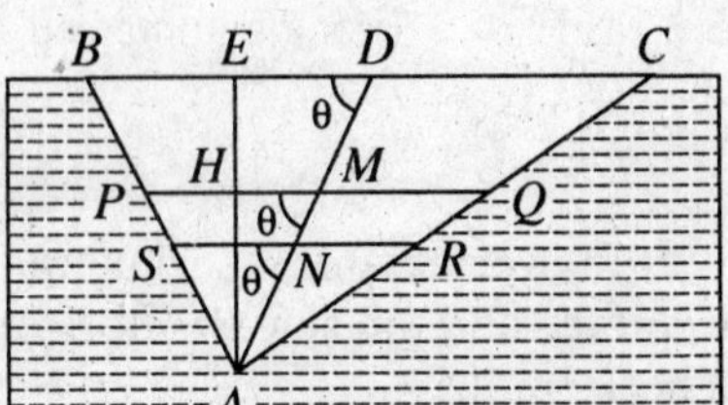

Fig. 5.7

We divide the triangular area into an infinite number of horizontal strips PQRS. From A, we draw AE perpendicular to BC.

The C.P. of all such strips will act at their middle points. Hence, the C.P. of the triangle will be on the median AD.

Let $BC = a$, $AD = h$, $MD = x$, $MN = dx$.

Then the thickness of the strip PQRS $= dx \sin\theta$.

The depth of the strip PQRS from the free surface, $BC = EH = x\sin\theta$.

Since, $\triangle ABC$ and $\triangle APQ$ are similar

$$\frac{BC}{PQ} = \frac{AD}{AM}, \quad \frac{a}{PQ} = \frac{h}{AM} = \frac{h}{AD - DM} = \frac{h}{h - x}.$$

Hence,

$$PQ = \frac{a(h-x)}{h}.$$

Area of the strip PQRS $= PQ\,dx \sin\theta = \dfrac{a(h-x)}{h}\,dx \sin\theta$.

Pressure at any point of the strip PQRS $= \rho g x \sin\theta$.

If $\bar{x}$ be the distance of the C.P. of the $\triangle ABC$ from D along DA, then

$$\bar{x} = \frac{\int x\rho g x \sin\theta \frac{a(h-x)}{h}\,dx \sin\theta}{\int \rho g x \sin\theta \frac{a(h-x)}{h}\,dx \sin\theta} = \frac{\rho g \sin^2\theta \frac{a}{h}\int_0^h x^2(h-x)\,dx}{\rho g \sin^2\theta \frac{a}{h}\int_0^h x(h-x)\,dx}$$

$$= \frac{\int_0^h (hx^2 - x^3)\,dx}{\int_0^h (xh - x^2)\,dx} = \frac{h\frac{h^3}{3} - \frac{h^4}{4}}{h\frac{h^2}{2} - \frac{h^3}{3}} = \frac{h^4}{12} \times \frac{6}{h^3} = \frac{h}{2}.$$

Thus, the depth of the C.P. is at half the depth of the vertex.

5.10 Centre of Pressure of a Vertical Circular Area of Radius a, totally immersed in a Homogeneous Liquid with its Centre at a depth h below the Free Surface

[M.U. 93H, 04H, R.U. 91H, 95H, 99H, T.M.B.U. 04H]

Let O be the centre of the circular area at a depth h below the free surface, i.e., $OM = h$.

We divide the circular area into a large number of thin strips of the type PQRS parallel to the y-axis.

The C.P. of all such strips will lie at their middle points. Hence, the C.P. of the circular area will lie on AB, where AB is the vertical diameter.

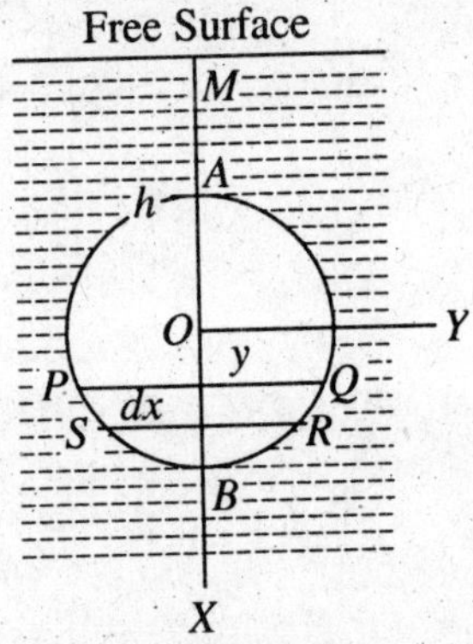

Fig. 5.8

Taking O as origin, OX the downward, vertical as x-axis and OY, the horizontal as y-axis.

Equation of circle is $x^2 + y^2 = a^2$. Let ρ be the density of the liquid.
Let thickness of the strip PQRS be δx and at a depth x below O.
Let coordinates of Q be (x, y). Now, area of the strip PQRS $= 2y\,dx$.
Pressure at any point of the strip $= \rho g(h + x)$.
Hence, if $\bar{x}$ be the depth of C.P. of the circular area, then

$$\bar{x} = \frac{\int xp\,ds}{\int p\,ds} = \frac{\int_{-a}^{a} x\rho g(h+x)2y\,dx}{\int_{-a}^{a} \rho g(h+x)2y\,dx}$$

$$= \frac{2\rho g}{2\rho g}\frac{\int_{-a}^{a} x(h+x)y\,dx}{2\rho g\int_{-a}^{a} (h+x)y\,dx} = \frac{\int_{-a}^{a} x(h+x)y\,dx}{\int_{-a}^{a} (h+x)y\,dx}.$$

Putting $x = a\cos\theta$, $y = a\sin\theta$. $\quad\therefore dx = -a\sin\theta\,d\theta$.
When $x = a$, $\theta = 0$, $x = -a$, $\theta = \pi$.

$$= \frac{\int_{\pi}^{0} a\cos\theta a\sin\theta(h + a\cos\theta)(-a\sin\theta)\,d\theta}{\int_{\pi}^{0} a\sin\theta(h + a\cos\theta)(-a\sin\theta)\,d\theta}$$

$$= \frac{a^3\int_0^{\pi} (h + a\cos\theta)\cos\theta\sin^2\theta\,d\theta}{a^2\int_0^{\pi} (h + a\cos\theta)\sin^2\theta\,d\theta}$$

$$\bar{x} = \frac{ah\int_0^{\pi}\cos\theta\sin^2\theta\,d\theta + a^2\int_0^{\pi}\cos^2\theta\sin^2\theta\,d\theta}{h\int_0^{\pi}\sin^2\theta\,d\theta + a\int_0^{\pi}\cos\theta\sin^2\theta\,d\theta}. \tag{5.4}$$

Let

$$I_1 = \int_0^{\pi}\cos\theta\sin^2\theta\,d\theta.$$

Let $t = \sin\theta$. $\therefore dt = \cos\theta\,d\theta$

$$I_1 = \int_0^{\pi} t^2 dt = \left[\frac{t^3}{3}\right]_0^{\pi} = \left[\frac{\sin^3\theta}{3}\right]_0^{\pi} = 0.$$

$$\int_0^{\pi}\sin^2\theta\cos^2\theta\,d\theta = \frac{1}{4}\int_0^{\pi} 4\sin^2\theta\cos^2\theta\,d\theta = \frac{1}{4}\int_0^{\pi}\sin^2 2\theta\,d\theta$$

$$= \frac{1}{8}\int_0^{\pi}(1 - \cos 4\theta)\,d\theta = \frac{1}{8}\left[\theta - \frac{\sin 4\theta}{4}\right]_0^{\pi} = \frac{\pi}{8}$$

$$\int_0^{\pi}\sin^2\theta\,d\theta = \int_0^{\pi}\frac{(1-\cos 2\theta)}{2}\,d\theta = \frac{1}{2}\left[\theta - \frac{\sin 2\theta}{2}\right]_0^{\pi} = \frac{\pi}{2}.$$

Thus, from (5.4), we have

$$\bar{x} = \frac{ah\cdot 0 + a^2\cdot \frac{\pi}{8}}{h\cdot \frac{\pi}{2} + a\cdot 0} = \frac{a^2}{4h}.$$

Therefore, the depth of C.P. below the free surface $= h + \frac{a^2}{4h}$.

Corollary 5.10.1 *When the circle is not vertical.*

[MU 96H, B.B.A.B.U 94H, T.M.B.U. 93H, 95H, 03H]

Let the plane of the circle be inclined to the surface at an angle θ and h be the vertical depth of the centre below free surface.

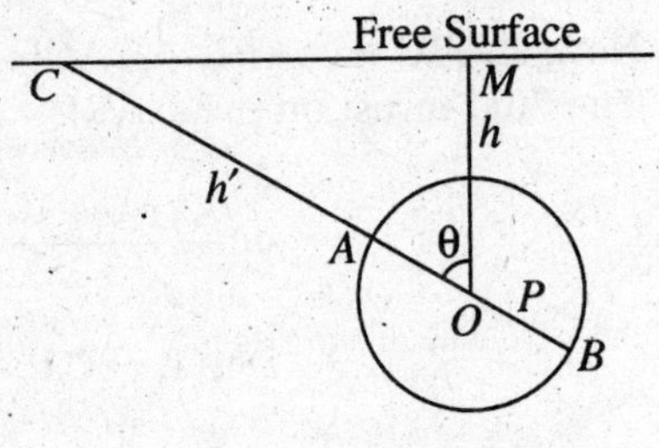

Fig. 5.9

We draw a line OAC in the plane of the circle cutting surface at C. If $OC = h'$, then $OP = \frac{a^2}{4h'}$.

In $\triangle OCM$

$$\cos\theta = \frac{OM}{OO'} = \frac{h}{h'}, h' = \frac{h}{\cos\theta}.$$

Hence,

$$OP = \frac{a^2\cos\theta}{4h}.$$

Corollary 5.10.2 *If the circular area is just immersed, then* $h = a$.

Hence,

$$\bar{x} = a + \frac{a^2}{4a} = a + \frac{a}{4} = \frac{5a}{4}.$$

Alliter. We know that

$$H = \frac{K'^2}{h} + h, \tag{5.5}$$

where H is the depth of C.P. of the plane area and h is the depth of C.G. of the plane area K' is the radius of gyration about a parallel axis through the C.G. G.

Here $h = h$, the depth of the centre of the circle which is also its C.G.

Also in *case* of circle $K'^2 = \frac{a^2}{4}$.

Hence, from (5.5)

$$H = \frac{a^2}{4h} + h.$$

5.11 The Depth of the C.P. of a Triangle in terms of the Depths of the Vertices

[B.B.A.B.U. 93H, 95H]

We shall first consider the case when one vertex is in the surface and the other two vertices are inside the fluid and then will deduce from it the depth of C.P. of the triangle whose all the three vertices are inside the fluid.

Case I. *When one vertex is in the surface and the two inside the fluid.*

[M.U. 94H, L.N.M.U. 94H, T.M.B.U. 93H]

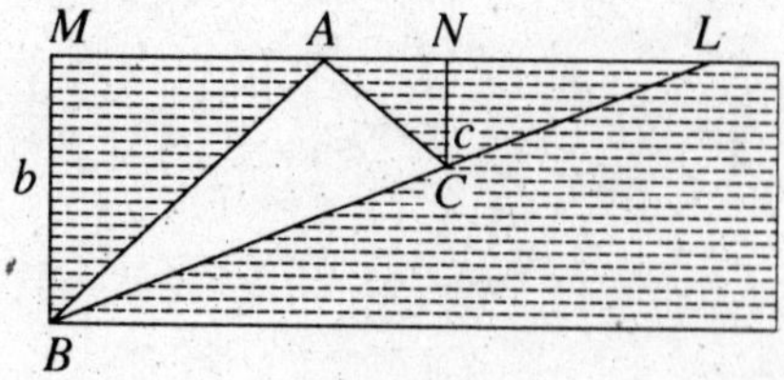

Fig. 5.10

Let ABC be the triangle immersed in a liquid of density ρ with vertex A in the free surface and the vertices B and C are inside the fluid at depths b and c from the free surface, respectively. Produce BC to meet the free surface in L.

Now, $\triangle ABC = \triangle ABL - \triangle ACL$.
The fluid thrust on the $\triangle ABL$

$$= \rho g \frac{(0+0+b)}{3}\left(\frac{1}{2}AL \cdot b\right) = \frac{\rho g b^2}{6} AL = T_1 \ (say)$$

$$Z_1 = \text{Depth of C.P. of } \triangle ABL = \frac{1}{2}b.$$

The fluid thrust on $\triangle ACL$

$$= \frac{\rho g(0+0+c)}{3}\left(\frac{1}{2}AL \cdot c\right) = \frac{\rho g c^2}{6} AL = T_2 \ (say)$$

$$Z_2 = \text{Depth of C.P. of } \triangle ACL = \frac{1}{2}c.$$

If $\bar{Z}$ be the depth of C.P. of $\triangle ABC$ below the free surface, then

$$\bar{Z} = \frac{T_1Z_1 - T_2Z_2}{T_1 - T_2} = \frac{\rho g \frac{b^2}{6} AL \cdot \frac{b}{2} - \rho g \frac{c^2}{6} AL \cdot \frac{c}{2}}{\rho g \frac{b^2}{6} AL - \rho g \frac{c^2}{6} AL}$$

$$= \frac{\rho g \frac{AL}{12}(b^3 - c^3)}{\rho g \frac{AL}{6}(b^2 - c^2)} = \frac{b^3 - c^3}{2(b^2 - c^2)}$$

$$= \frac{(b-c)(b^2 + bc + c^2)}{2(b-c)(b+c)} = \frac{b^2 + bc + c^2}{2(b+c)}.$$

[U.N.M.U. 89H, M.U. 92H, 2001H, T.M.B.U. 2002H, 98H, 96H, 04H, C.U. 97H, 02H]

Case II. *When all the three vertices are inside the fluid.* [M.U. 92H]

Let ABC be the triangle immersed in the fluid such that the depths of A, B and C from the free surface are a, b, c, respectively. Produce AC to meet the free surface in D. Join BD. Through A, B, C draw verticals to meet the free surface in L, M, N, respectively.

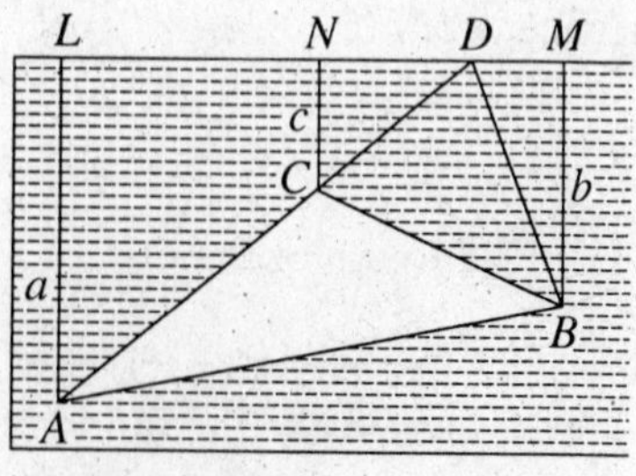

Fig. 5.11

Now, $\triangle ABC = \triangle ABD - \triangle ACD$.

Now,

$$\begin{aligned}
T_1 &= \text{Thrust on } \triangle ABD = \frac{\rho g(a+b+0)}{3}\triangle ABD \\
&= \frac{\rho g(a+b)}{3}\triangle ABD
\end{aligned}$$

$$Z_1 = \text{Depth of C.P. of } \triangle ABD = \frac{1}{2}\frac{a^2+ab+b^2}{a+b} \qquad \text{(By case I)}$$

$$\begin{aligned}
T_2 &= \text{Thrust on } \triangle BCD = \frac{\rho g(b+c+0)}{3}\triangle BCD \\
&= \frac{\rho g(b+c)}{3}\triangle BCD
\end{aligned}$$

$$Z_2 = \text{Depth of C.P. of } \triangle BCD = \frac{b^2+bc+c^2}{2(b+c)}.$$

If $\bar{z}$ be the depth of C.P. of $\triangle ABC$ below the free surface, then

$$\begin{aligned}
\bar{z} &= \frac{T_1Z_1 - T_2Z_2}{T_1 - T_2} \\
&= \frac{\frac{\rho g(a+b)}{3}\triangle\frac{ABD(a^2+ab+b^2)}{2(a+b)} - \frac{\rho g(b+c)}{3}\triangle\frac{BCD(b^2+bc+c^2)}{2(b+c)}}{\frac{\rho g(a+b)}{3}\triangle ABD - \frac{\rho g(b+c)}{3}\triangle BCD} \\
&= \frac{\frac{\rho g}{6}\triangle BCD\left[\frac{\triangle ABD}{\triangle BCD}(a^2+ab+b^2) - (b^2+bc+c^2)\right]}{\frac{\rho g}{3}\triangle BCD\left[(a+b)\frac{\triangle ABD}{\triangle BCD} - (b+c)\right]} \\
&= \frac{\frac{\triangle ABD}{\triangle BCD}(a^2+ab+b^2) - (b^2+bc+c^2)}{2\left[(a+b)\frac{\triangle ABD}{\triangle BCD} - (b+c)\right]}. \qquad (5.6)
\end{aligned}$$

But,

$$\begin{aligned}
\frac{\triangle ABD}{\triangle BCD} &= \frac{AD}{CD} \qquad \text{(Since } \triangle\text{s have the same height)} \\
&= \frac{AL}{CN} \qquad (\because \triangle ALD \text{ and } \triangle CND \text{ are similar}) \\
&= \frac{a}{c}.
\end{aligned}$$

Hence, from (5.6)

$$\begin{aligned}\bar{z} &= \frac{\frac{a}{c}(a^2+ab+b^2)-(b^2+bc+c^2)}{2\left[(a+b)\frac{a}{c}-(b+c)\right]}\\ &= \frac{a^3+a^2b+ab^2-c^3-b^2c-bc^2}{2(a^2+ab-bc-c^2)}\\ &= \frac{(a^3-c^3)+b(a^2-c^2)+b^2(a-c)}{2[(a^2-c^2)+b(a-c)]}\\ &= \frac{(a-c)(a^2+ac+c^2)+b(a-c)(a+c)+b^2(a-c)}{2[(a-c)(a+c)+b(a-c)]}\\ &= \frac{(a-c)(a^2+ac+c^2)+(ab+bc+b^2)}{2(a-c)(a+c+b)}\\ \bar{z} &= \frac{a^2+b^2+c^2+ab+bc+ca}{2(a+b+c)}. \end{aligned} \quad (5.7)$$

Corollary 5.11.1 *Equation (5.7) can be written as*

$$\bar{z} = \frac{k\left(\frac{a+b}{2}\right)\left(\frac{a+b}{2}\right)+k\left(\frac{b+c}{2}\right)\left(\frac{b+c}{2}\right)+k\left(\frac{c+a}{2}\right)\left(\frac{c+a}{2}\right)}{k\left(\frac{a+b}{2}\right)+k\left(\frac{b+c}{2}\right)+k\left(\frac{c+a}{2}\right)}. \quad (5.8)$$

Now, since a, b and c are the depths of the vertices A, B and C, respectively, therefore

$$\frac{a+b}{2}, \frac{b+c}{2}, \frac{c+a}{2}$$

are the depths of the middle points of the sides of the triangle. Thus, the eqn. (5.8) represents the depth of the point through which resultant of three parallel forces

$$k\frac{(a+b)}{2}, k\frac{(b+c)}{2}, k\frac{(c+a)}{2}$$

acting at the middle points of the sides and proportional to the corresponding depths.

In other words, we can say that *the C.P. of a triangular lamina, immersed in a homogeneous liquid, coincides with the centre of three parallel forces (or masses) acting at middle points of the sides and of magnitudes proportional respectively to the depths of their middle points.* **[R.U. 2000H]**

Corollary 5.11.2 *Equation (5.7) can also be written as*

$$\bar{z} = \frac{k(2a+b+c)a+k(a+2b+c)b+k(a+b+2c)}{k(2a+b+c)+k(a+2b+c)+k(a+b+2c)}.$$

It shows that the *C.P. of a triangle ABC whose vertices A, B and C are at depths a, b and c below the free surface, coincides with the parallel forces (or masses) acting at A, B and C and proportional to $2a+b+c$, $a+2b+c$, $a+b+2c$, respectively.*

5.12 C.P. of an Ellipse when the Major Axis is along a Line of Greatest Slope

Let us first take a special case. Let the elliptic plane AGP be the projection of a vertical circle A′G′P′ on a plane passing through LM, the line of intersection of the plane of the circle with the effective surface. Let us imagine a cylinder bounded at one end by the circular section and at the other by the elliptic plane.

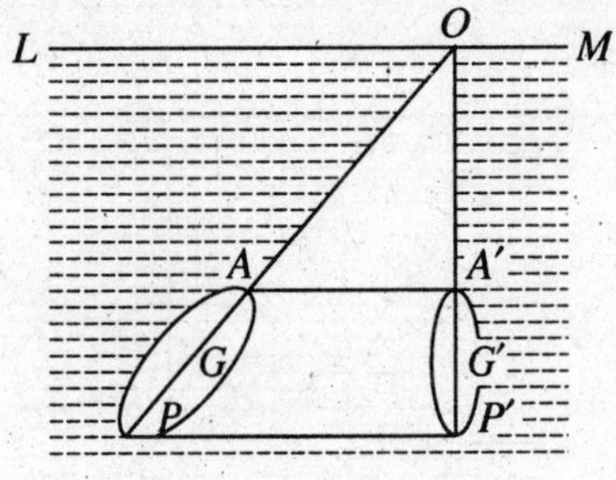

Fig. 5.12

Forces acting on the cylinder are:

- weight of the liquid contained in it,
- thrust on the curved surface,
- thrust on the circular section,
- thrust on the elliptic sections.

Resolving these forces horizontally, we see that the thrust on the circle must balance the horizontal component of the thrust on the ellipse. Therefore, if the horizontal line through P′, the C.P. of the circle meets the ellipse in P, the thrust on the ellipse must act through it. Thus, the points A′, G′, P′ are the orthogonal projections of A, G, P.

$$\frac{GP}{AG} = \frac{G'P'}{A'G'}. \tag{5.9}$$

If A, G, P and A′G′P′ meet in O, O must be in the effective surface. Then G′P′, the distance between C.P. and C.G. of the circle

$$= \frac{1}{4}\frac{(A'G')^2}{OG'}.$$

Hence, from (5.9), we have

$$\frac{GP}{AG} = \frac{1}{4}\frac{(A'G')^2}{OG'}\cdot\frac{1}{A'G'} = \frac{A'G'}{4OG'}$$

$$\Rightarrow \frac{GP}{AG} = \frac{1}{4}\frac{AG}{OG}$$

$$\Rightarrow GP = \frac{1}{4}\frac{AG^2}{OG}. \tag{5.10}$$

We know that the position of the C.P. in a plane remains unchanged, if the plane is rotated about its line of intersection with the effective surfaces. Hence, eqn. (5.10) holds even if the ellipse is rotated about LM, thus (5.10) gives the position of the C.P. of every elliptic plane completely immersed in a liquid.

As a special case, the ellipse may be vertical. Thus, *if the ellipse* $\frac{x^2}{a^2} + \frac{y^2}{b^2} = 1$ *is immersed vertically, with its major axis vertical and centre at a depth* h, *the C.P. is at a vertical distance* $\frac{a^2}{4h}$ *from the C.G.*

5.13 If a Plane Area turns in its Own Plane about a Fixed Point, the C.P. describes a Curve in that Plane

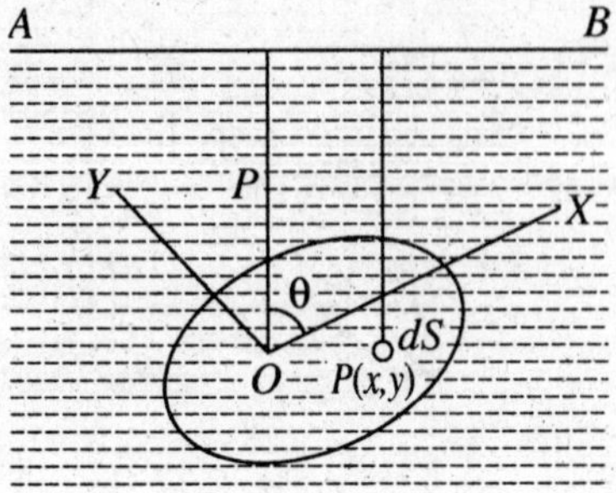

Fig. 5.13

Let O be the fixed point. Taking the plane as vertical, h be the depth of O. Let the principal axes of the area of O be taken as the axes of x and y. If OX makes an angle θ with the vertical, then equation of AB is

$$x\cos\theta + y\sin\theta = h. \qquad (5.11)$$

Now we consider an elementary area ds enclosing the point P, then depth of this element from the free surface

$$\begin{aligned} &= \text{distance of } (x, y) \text{ from } (5.11) \\ &= h - x\cos\theta - y\sin\theta. \end{aligned}$$

Thrust on the element $= \rho g(h - x\cos\theta - y\sin\theta)\, ds$.
Let $(\bar{x}, \bar{y})$ be the coordinates of C.P. of the area, then

$$\begin{aligned} \bar{x} &= \frac{\int x(h - x\cos\theta - y\sin\theta)\, ds}{\int (h - x\cos\theta - y\sin\theta)\, ds} \\ &= \frac{h\int x ds - \cos\theta \int x^2 ds - \sin\theta \int xy ds}{h\int ds - \cos\theta \int x ds - \sin\theta \int y ds}. \end{aligned}$$

Now, if (x_0, y_0) be the coordinates of the C.G. of area S, then

$$\begin{aligned} \frac{\int x ds}{S} &= x_0, \quad \frac{\int y ds}{S} = y_0 \\ \text{or,} \quad \int x ds &= S x_0, \quad \int y ds = S y_0 \\ \int x^2 ds &= \text{M.I. about axis} = S a^2 \\ \int y^2 ds &= \text{M.I. about } y\text{-axis} = S b^2 \\ \text{and} \quad \iint xy ds &= \text{product of inertia about principal axis} = 0. \end{aligned}$$

Hence,

$$x = \frac{hSx_0 - \cos\theta S a^2}{hS - \cos\theta S x_0 - \sin\theta S y_0}. \qquad (5.12)$$

Similarly,

$$\bar{y} = \frac{hSy_0 - \sin\theta Sb^2}{hS - \cos\theta Sx_0 - \sin\theta Sy_0}. \quad (5.13)$$

From (5.12), we have

$$hS\bar{x} - \cos\theta Sx_0\bar{x} - \sin\theta Sy_0\bar{x} = hSx_0 - \cos\theta Sa^2$$

or, $h(x_0 - \bar{x}) + \cos\theta(x_0\bar{x} - a^2) + \sin\theta y_0\bar{x} = 0$

or, $\cos\theta(x_0\bar{x} - a^2) + \sin\theta y_0\bar{x} + h(x_0 - \bar{x}) = 0.$

Similarly from (5.13), we get $\cos\theta x_0\bar{y} + \sin\theta(y_0\bar{y} - b^2) + h(y_0 - \bar{y}) = 0.$
Solving these for $\sin\theta$ and $\cos\theta$, we get

$$\frac{\cos\theta}{h[y_0\bar{x}(y_0 - \bar{y}) - (x_0 - \bar{x})(y_0\bar{y} - b^2)]} = \frac{\sin\theta}{h[(x_0 - \bar{x})x_0\bar{y} - (y_0 - \bar{y})(x_0\bar{x} - a^2)]} = \frac{1}{(x_0\bar{x} - a^2)(y_0\bar{y} - b^2) - x_0y_0\bar{x}\bar{y}}.$$

Since, $\sin^2\theta + \cos^2\theta = 1.$
Therefore

$$h^2\left[(x_0 - \bar{x})x_0\bar{y} - (y_0 - \bar{y})(x_0\bar{x} - a^2)\right]^2 + h^2\left[y_0\bar{x}(y_0 - \bar{y}) - (x_0 - \bar{x})(y_0\bar{y} - b^2)\right]^2 = \left[(x_0\bar{x} - a^2)(y_0\bar{y} - b^2) - x_0y_0\bar{x}\bar{y}\right]^2. \quad (5.14)$$

Locus of $(\bar{x}, \bar{y})$ is clearly an equation of second degree, which is a conic.

Corollary 5.13.1 *If the fixed point O be the C.G. of the area, then* $x_0 = y_0 = 0.$

Thus, eqn. (5.14) becomes

$$(a^2b^2)^2 = h^2[(-\bar{x}b^2)^2 + (-\bar{y}a^2)^2], \quad \text{or,} \quad \frac{a^4b^4}{h^2} = \bar{x}^2b^4 + \bar{y}^2a^4.$$

Hence, locus of $(\bar{x}, \bar{y})$ is $\dfrac{x^2}{a^4} + \dfrac{y^2}{b^4} = \dfrac{1}{p^2}$ which is an ellipse.

5.14 Formula for the Depth of the Centre of Pressure of a Plane Area

[L.N.M.U. 95H, R.U. 97H]

Let a plane area S be immersed vertically in a liquid. Take OY, the line of intersection of the plane of the area with the free surface, as the y-axis and a line perpendicular to OY, lying in the plane of the area, as x-axis.

Let (x, y) be the coordinates of any point P in the given area and let $(\bar{x}, \bar{y})$ be the coordinates of the centre of pressure. Surrounding P, consider an element of area ds. If p be the fluid pressure at P, then the fluid thrust over this element of area $= p\,ds$ and the resultant thrust on the whole area S is $\int p\,ds$ which acts at the C.P. $(\bar{x}, \bar{y})$. Moment of the resultant thrust about y-axis is $\bar{x}\int p\,ds$.

Also the moment of $p\,ds$ about y-axis is $xp\,ds$.

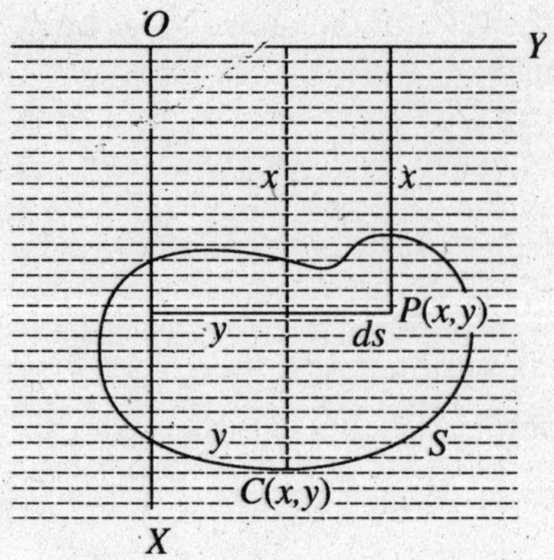

Fig. 5.14

Now, the moment of the resultant on the whole area about y-axis must be equal to the algebraic sum of the moments of the thrusts on all elements of type ds of the whole area about y-axis.

Therefore, taking moment about y-axis, we get

$$\bar{x}\int p\,ds = \int xp\,ds,$$

the integration extends over the whole area

$$\bar{x} = \frac{\int xp\,ds}{\int p\,ds}.$$

Similarly, by taking moment about x-axis, we get

$$\bar{y} = \frac{\int yp\,ds}{\int p\,ds}.$$

If ds = dx dy, then

$$\bar{x} = \frac{\iint xp\,dx\,dy}{\iint p\,dx\,dy} \quad \text{and} \quad \bar{y} = \frac{\iint yp\,dx\,dy}{\iint p\,dx\,dy},$$

where the limit covers the whole area.

5.15 Centre of Pressure of an Area immersed in a Number of Liquids

Suppose there are n liquids of densities $\rho_1, \rho_2, \ldots, \rho_n$ in ascending order of magnitude, which do not mix when placed one over the other. Let an area A be immered vertically, such that the parts $A_1, A_2, \ldots, A_n$ of the whole area A are in contact with different liquids, respectively.

Now, we know that the thrust on the whole area remains the same as if we consider that the whole area. $A = A_1 + A_2 + \cdots + A_n$ is in contact with the liquid of density ρ_1, the area $A_2 + A_3 + \cdots + A_n$ is in contact with the liquid of density $(\rho_2 - \rho_1)$, the area $A_3 + A_4 + \cdots + A_n$ is in contact with the liquid of density $(\rho_3 - \rho_2)$ and so on and finally the area A_n is in contact with the liquid of density $(\rho_n - \rho_{n-1})$.

Let $P_1, P_2, \ldots, P_n$ be the thrusts on the areas $A_1 + A_2 + \cdots + A_n$, $A_2 + A_3 + \cdots + A_n$, $A_3 + A_4 + \cdots + A_n$, A_n in contact with the liquid of densities $\rho_1, \rho_2 - \rho_1,$

$\rho_3 - \rho_2, \ldots, \rho_n - \rho_{n-1}$, respectively, and let the C.P.'s of the above areas be $z_1, z_2, \ldots, z_n$ below the surface, respectively.

Again, the resultant thrust of the above thrusts is $P_1 + P_2 + \cdots + P_n$ acting at a depth $\bar{z}$ (*say*), below the surface.

Now, taking moments about a line on the surface, we get $(P_1 + P_2 + \cdots + P_n)$

$$\bar{Z} = P_1Z_1 + P_2Z_2 + \cdots + P_nZ_n$$
$$\bar{Z} = \frac{P_1Z_1 + P_2Z_2 + \cdots + P_nZ_n}{P_1 + P_2 + \cdots + P_n}.$$

Illustrative Examples

► **Ex. 1.** The lengths of two parallel sides of a trapezium are a and b, and the distance between them is h. If the trapezium be immersed in water with its plane vertical and the side a in the surface, prove that the centre of pressure will be at a depth $\frac{a+3b}{a+2b} \cdot \frac{h}{2}$ below to the surface. **[Bch. 81, P.U. 70A, R.U. 81H, 83H, 93H, Mith. 75H]**

Solution: Let ABCD be the trapezium so that side AB $= a$ is in the free surface and parallel side CD $= b$ is at a depth h below the free surface. Join BD.

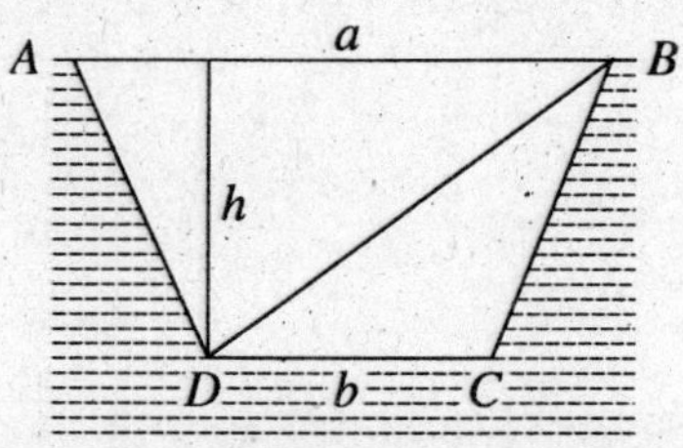

Fig. 5.15

If T_1 and T_2 be the fluid thrusts on $\triangle ABD$ and $\triangle BCD$, respectively, then

$$T_1 = \rho g \frac{(0+0+h)}{3} \triangle ABD = \frac{\rho g h}{3} \triangle ABD = \frac{\rho g h}{3} \frac{ah}{2}$$
$$\text{and} \quad T_2 = \rho g \frac{(0+h+h)}{3} \triangle BCD = \rho g \frac{2h}{3} \triangle BCD = \frac{\rho g 2h}{3} \frac{bh}{2}.$$

Also, if Z_1 and Z_2 be the depths of the C.P.'s of $\triangle ABD$ and $\triangle BCD$ below the free surface, then $Z_1 = \frac{h}{2}$ and $Z_2 = \frac{3}{4}h$.

If $\bar{Z}$ be the depth of C.P. of the trapezium, then

$$\bar{Z} = \frac{T_1Z_1 + T_2Z_2}{T_1 + T_2} = \frac{\frac{\rho g h}{3}\frac{ah}{2}\frac{h}{2} + \frac{\rho g 2h}{3}\frac{bh}{2}\frac{3}{4}h}{\frac{\rho g h}{3}\frac{ah}{2} + \frac{\rho g 2h}{3}\frac{bh}{2}}$$
$$= \frac{\rho g h^3\left(\frac{a}{12} + \frac{b}{4}\right)}{\rho g h^2\left(\frac{a}{6} + \frac{b}{3}\right)} = \frac{h\frac{a+3b}{12}}{\frac{a+2b}{6}} = \frac{(a+3b)h}{2(a+2b)} \quad \textbf{Proved.}$$

▶ **Ex. 2.** Find the centre of pressure of a regular hexagon of side a with one side in the surface: **[I.A.S. 78, R.U. 97H]**

Solution: Let the regular hexagon ABCDEF be immersed vertically in water with the side AB in the surface. Centre O of the hexagon is also the centre of gravity of the hexagon.

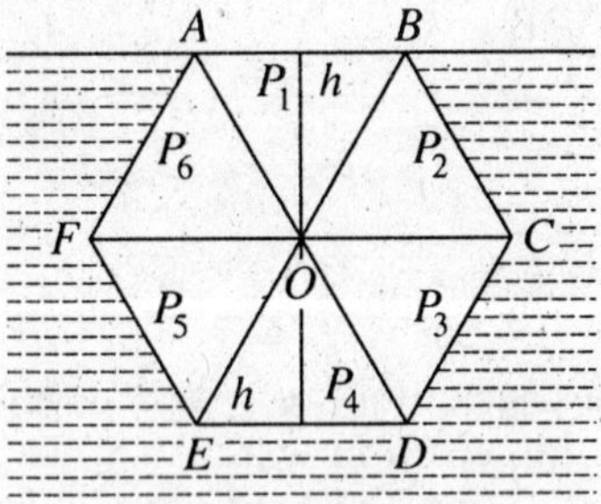

Fig. 5.16

Join O with all the vertices A, B, C, D, E, F. Now the hexagon is divided into six equilateral triangles of equal area $\sqrt{\frac{3}{4}}a^2 = S$ (*say*), where AB = a is the length of each side of hexagon. Let the depth of G below the free surface be h.

Let P_1, P_2, P_3, P_4, P_5 and P_6 be the fluid thrusts and Z_1, Z_2, Z_3, Z_4, Z_5 and Z_6 be the depths of the C.P.'s of triangles AOB, BOC, COD, DOE, EOF and AOF, respectively.

Then,

$$P_1 = \rho g S\left(\frac{0+0+h}{3}\right) = \frac{\rho g h S}{3}$$

$$P_2 = \rho g\frac{2h}{3}S = 2P_1 = \rho g S\left(\frac{h+h+U}{3}\right) = \frac{\rho g 2hS}{3} = 2P_1$$

$$P_3 = \rho g\left(\frac{h+h+2h}{3}\right)S = \frac{4}{3}\rho g h S = 4P_1$$

$$P_4 = \rho g\left(\frac{h+2h+2h}{3}\right)S = \rho g\frac{5h}{3}S = 5P_1$$

$$P_5 = P_3 \text{ (by symmetry)} = 4P_1$$

$$P_6 = P_2 \text{ (by symmetry)} = 2P_1.$$

Also,

$$Z_1 = \text{depth of C.P of } \triangle OAB = \frac{1}{2}h$$

$$Z_2 = \text{depth of C.P of } \triangle OBC = \frac{3}{4}h$$

Z_3 = depth of C.P of $\triangle OCD$ whose three vertices O, C and D are at depths h, h and 2h from the free surface

$$= \frac{h^2 + h^2 + (2h)^2 + h \cdot h + h \cdot 2h + h \cdot 2h}{2(h+h+2h)} = \frac{11h^2}{8h} = \frac{11h}{8}$$

$$Z_4 = \text{depth of C.P. of } \triangle OED \text{ whose three vertices O, E and D are at depths h, 2h and 2h from the free surface}$$

$$= \frac{h^2 + (2h)^2 + (2h)^2 + h.2h + 2h.2h + 2h.h}{2(h + 2h + 2h)}$$

$$= \frac{17h^2}{10h} = \frac{17h}{10}$$

$$Z_5 = \text{depth of C.P. of } \triangle OEF = Z_3 \text{ (by symmetry)}$$

$$Z_6 = \text{depth of C.P. of } \triangle OAF = Z_2 \text{ (by symmetry).}$$

Hence, the depths of C.P. of the hexagon ABCDEF below the free surface

$$\bar{z} = \frac{P_1Z_1 + P_2Z_2 + P_3Z_3 + P_4Z_4 + P_5Z_5 + P_6Z_6}{P_1 + P_2 + P_3 + P_4 + P_5 + P_6}$$

$$= \frac{\frac{h}{2}P_1 + 2P_1\frac{3}{4}h + 4P_1\frac{11h}{8} + 5P_1\frac{17h}{10} + 4P_1\frac{11h}{8}h + 2P_1\frac{3}{4}h}{P_1 + 2P_1 + 4P_1 + 5P_1 + 4P_1 + 2P_1}$$

$$= \frac{\frac{P_1h}{2}[1 + 3 + 11 + 11 + 3 + 17]}{18P_1} = \frac{23h}{18}.$$

Also, h = depth of the C.G. below the free surface $= a \sin 60 = \frac{\sqrt{3}}{2}a$.

$$\bar{Z} = \frac{23}{18}\frac{\sqrt{3}}{2}a = \frac{23\sqrt{3}}{36}a. \quad \textbf{Ans.}$$

► **Ex. 3.** A lamina in the shape of a quadrilateral ABCD has its side CD in the surface of a liquid and the sides AD, BC vertical and equal to α, β, respectively. Show that the depth of C.P. is

$$\frac{1}{2}\frac{(\alpha^2 + \beta^2)(\alpha + \beta)}{\alpha^2 + \alpha\beta + \beta^2}.$$

[Bih. 67A, R.U. 74H, 75S, 80, M.U. 75A, T.M.B.U. 95H]

Solution: Let the side AB be produced to meet the free surface in E. Let EC = x.

From the similar triangles ADE and BCE, we have

$$\frac{AD}{BC} = \frac{ED}{EC}, \frac{\alpha}{\beta} = \frac{ED}{x} \quad \text{or,} \quad ED = \frac{\alpha x}{\beta}.$$

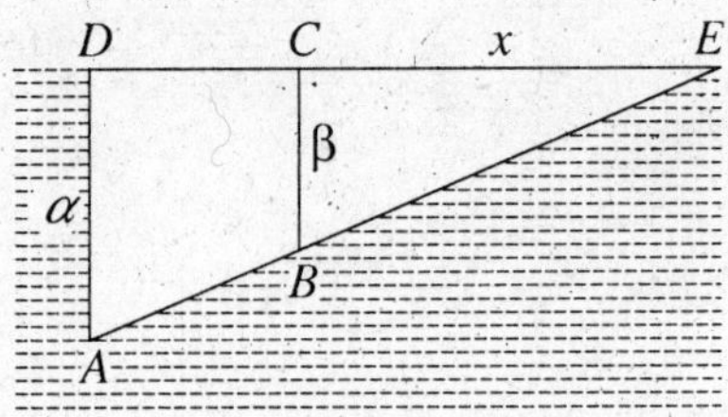

Fig. 5.17

Now the thrust on $\triangle ADE$

$$= \rho g \frac{(\alpha + 0 + 0)}{3} \frac{1}{2} DE \cdot AD$$

$$= \frac{\rho g \alpha}{3} \frac{1}{2} \frac{\alpha x}{\beta} \alpha = \frac{\rho g \alpha^3 x}{6\beta} = T_1 \quad (say).$$

Again, the thrust on $\triangle BCE$

$$= \rho g \frac{(\beta + 0 + 0)}{3} \frac{1}{2} CE \cdot BC = \frac{\rho g \beta}{\sigma} x \cdot \beta$$

$$= \frac{\rho g x \beta^2}{6} = T_2 \ (say).$$

If Z_1 and Z_2 be the depths of C.P.'s of $\triangle ADE$ and $\triangle BCE$, then

$$Z_1 = \frac{\alpha}{2} \quad \text{and} \quad Z_2 = \frac{\beta}{2}.$$

If $\bar{z}$ be the required depth of the C.P. of the quadrilateral ABCD

$$\bar{z} = \frac{T_1 Z_1 - T_2 Z_2}{T_1 - T_2} = \frac{\frac{\rho g \alpha^3 x}{6\beta} \cdot \frac{\alpha}{2} - \frac{\rho g x \beta^2}{6} \cdot \frac{\beta}{2}}{\frac{\rho g \alpha^3 x}{6\beta} - \frac{\rho g x \beta^2}{6}} = \frac{\frac{\rho g x}{12}\left(\frac{\alpha^4}{\beta} - \beta^3\right)}{\frac{\rho g x}{6}\left(\frac{\alpha^3}{\beta} - \beta^2\right)}$$

$$= \frac{\alpha^4 - \beta^4}{(\alpha^3 - \beta^3)2} = \frac{(\alpha^2 - \beta^2)(\alpha^2 + \beta^2)}{2(\alpha - \beta)(\alpha^2 + \alpha\beta + \beta^2)}$$

$$= \frac{(\alpha - \beta)(\alpha + \beta)(\alpha + \beta)}{2(\alpha - \beta)(\alpha^2 + \alpha\beta + \beta^2)} = \frac{(\alpha + \beta)(\alpha^2 + \beta^2)}{2(\alpha^2 + \alpha\beta + \beta^2)}.$$

▸ **Ex. 4.** Find the centre of pressure of a square lamina immersed in a fluid with one vertex in the surface and the diagonal vertical.

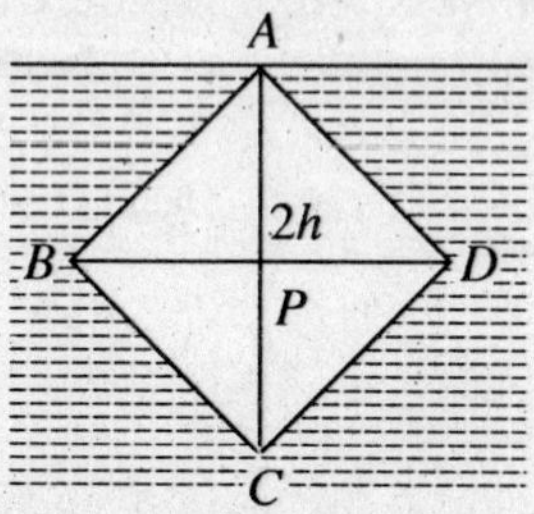

Fig. 5.18

Solution: Let ABCD be a square with the vertex A in the free surface and the diagonal AC vertical.

The lamina is symmetrical about AC, hence C.P. will lie on it and is at the same depth as the C.P. of $\triangle ABC$.

Let $AC = 2h$, then depth of $B = h$.

If $\bar{x}$ = be the depth of C.P. of $\triangle ABC$, then

$$\bar{x} = \frac{0^2 + h^2 + 4h^2 + 0 \cdot h + 0 \cdot 2h + h \cdot 2h}{2(0 + h + 2h)}$$

$$= \frac{5h^2 + 2h^2}{6h^2} = \frac{7h}{6}.$$

If P is the C.P. of rectangle, then

$$AP = \bar{x} = \frac{7}{6}h.$$

Also, $PC = AC - AP = 2h - \frac{7}{6}h = \frac{5}{6}h$

$$\therefore \frac{AP}{PC} = \frac{7}{5}.$$

Thus, the C.P. divides the vertical diagonal in the ratio 7 : 5.

▶ **Ex. 5.** A square lamina of side a, has a portion of it in the form of the inscribed circle removed from it and the remaining figure is immersed vertically in the water with one side of the square in the surface. Prove that the depth of C.P. is

$$\frac{a}{24}\frac{64 - 15\pi}{4 - \pi}.$$

Solution: Let ABCD be a square lamina. From ABCD, an inscribed circle with centre O has been removed. The side AB is in the surface. Let ρ be the density of the water.

Let T_1 and T_2 be the thrusts on square ABCD and circle, respectively.

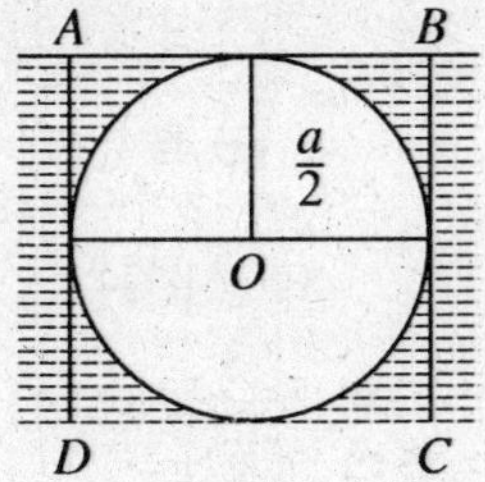

Fig. 5.19

Hence,

$$T_1 = \rho g \cdot \frac{a}{2} a^2 = \frac{\rho g a^3}{2}$$

$$\text{and } T_2 = \rho g \cdot \frac{a}{2}\pi\left(\frac{a}{2}\right)^2 = \frac{\rho g \pi a^3}{8}.$$

If Z_1 and Z_2 be the depths of the square ABCD and circle from free surface, respectively, then

$$Z_1 = \frac{2}{3}a \quad \text{and} \quad Z_2 = \frac{a}{2} + \frac{\frac{a^2}{4}}{4 \cdot \frac{a}{2}} = \frac{a}{2} + \frac{a}{8} = \frac{5a}{8}.$$

If $\bar{z}$ be the depth of the C.P. of the remaining figure, then

$$\bar{z} = \frac{T_1 Z_1 - T_2 Z_2}{T_1 - T_2} = \frac{\frac{\rho g a^3}{2} \cdot \frac{2}{3}a - \frac{\rho g \pi a^3}{8} \cdot \frac{5a}{8}}{\frac{\rho g a^3}{2} - \frac{\rho g \pi a^3}{8}} = \frac{\rho g a^4 \left(\frac{1}{3} - \frac{5\pi}{64}\right)}{\frac{\rho g a^3}{2}\left(1 - \frac{\pi}{4}\right)}$$

$$= \frac{8a(64 - 15\pi)}{3 \cdot 64 \cdot (4 - \pi)} = \frac{a(64 - 15\pi)}{24(4 - \pi)} \quad \textbf{Proved.}$$

▶ **Ex. 6.** A square lamina is just immersed vertically in water and is then lowered through a depth b. If a be the length of the edge of the square, prove that the distance of the centre of pressure from the centre of square is $\frac{a^2}{6a+12b}$.

[P.U. 74A, Bih. 71A, L.N.M.U. 94H, M.U. 03H, R.U. 98H]

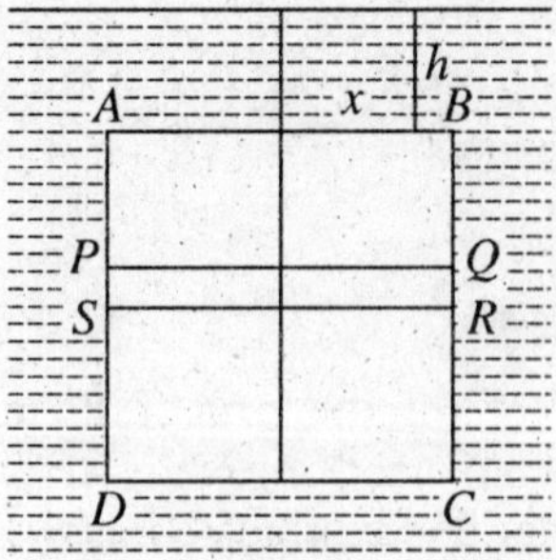

Fig. 5.20

Solution: Let ABCD be the square in its lowered position. We consider an elementary strip PQRS of breadth δx at a depth x from the free surface ds = area of the strip $= a\,dx$.

Pressure p at a point at a depth $x = \rho g x$.

If $\bar{x}$ be the depth of C.P. below the free surface, then

$$\bar{x} = \frac{\int xp\,ds}{\int p\,ds} = \frac{\int_h^{a+h} x\rho g x a\,dx}{\int_h^{a+h} \rho g x a\,dx} = \frac{\rho g a \int_h^{a+h} x^2 dx}{\rho g a \int_h^{a+h} x\,dx}$$

$$= \frac{\left[\frac{x^3}{3}\right]_h^{a+h}}{\left[\frac{x^2}{2}\right]_h^{a+h}} = \frac{2[(a+h)^3 - h^3]}{3[(a+h)^2 - h^2]} = \frac{2[a^3 + 3a^2h + 3ah^2]}{3(a^2 + 2ah)}.$$

Hence, the depth of C.P. below the centre of the square

$$= \frac{2}{3}\frac{a^3 + 3a^2h + 3ah^2}{a^2 + 2ah} - \left(h + \frac{a}{2}\right)$$

$$= \frac{4a^3 + 12a^2h + 12ah^2 - 6ha^2 - 12ah^2 - 3a^3 - 6a^2h}{6(a^2 + 2ah)}$$

$$= \frac{a^3}{6(a^2 + 2ah)} = \frac{a^2}{6a + 12h}$$ **Proved.**

▶ **Ex. 7.** Prove that the depth of the centre of pressure of a parallelogram; two of whose sides are horizontal and at depths h, k below the surface of a liquid whose density varies as the depth below the surface is

$$\frac{3}{4}\frac{h^3 + h^2k + hk^2 + k^3}{h^2 + hk + k^2}.$$ [P.U. 62H, T.M.B.U. 68H, 97H]

Solution: Let ABCD be the parallelogram whose sides AB and CD are parallel. Let $AB = a$.

We consider an elementary strip PQRS of thickness δx at a depth x below the free surface.

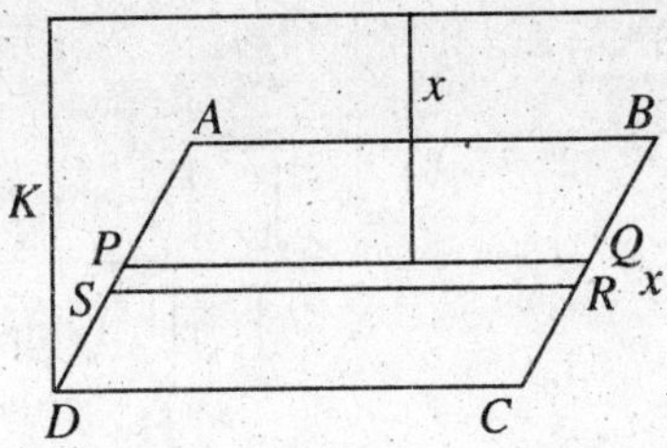

Fig. 5.21

Then,

$$ds = \text{area of the strip} = a\,dx$$

$$p = \text{pressure at a point at a depth } x = \rho g x.$$

Given $\rho \propto x$. Hence, $\rho = \lambda x$, where λ is a constant, $p = \lambda g x^2$.

If $\bar{x}$ be the depth of C.P. of the parallelogram below the free surface, then

$$\bar{x} = \frac{\int xp\,ds}{\int p\,ds} = \frac{\int_h^k x\lambda g x^2 dx}{\int_h^k k\lambda g x^2 dx} = \frac{\int_h^k x^3 dx}{\int_h^k x^2 dx} = \frac{\left[\frac{x^4}{4}\right]_h^k}{\left[\frac{x^3}{3}\right]_h^k}$$

$$= \frac{k^4 - h^4}{4} \frac{3}{(k^3 - h^3)} = \frac{3(k^2 - h^2)(k^2 + h^2)}{4(k - h)(k^2 + kh + h^2)}$$

$$= \frac{3(k + h)(k^2 + h^2)}{4(k^2 + kh + h^2)} = \frac{3}{4} \frac{k^3 + kh^2 + hk^2 + h^3}{k^2 + kh + h^2}$$ **Proved.**

▶ **Ex. 8.** A square lamina is vertically immersed in a homogeneous fluid with one corner in the surface, if it be turned in its own plane about this corner and remains always totally immersed, show that the locus of C.P. is a straight line.

[P.U. 62H, 66H, Bhag. 63H, C.U. 88H, 91H, 2000H, T.M.B.U. 91H, R.U. 2000H]

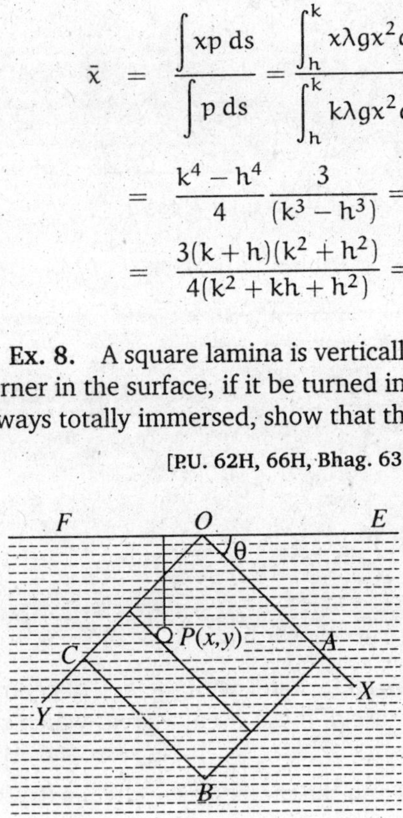

Fig. 5.22

Solution: Let OABC be the square lamina with the angular point O in the surface. Let OA and OC be taken as axes of reference.

Let EF be the free surface and let $\angle EOA = \theta$.

Then equation of EF is $y = -x \tan\theta$

$$\text{or,} \quad y = -\frac{x \sin\theta}{\cos\theta}$$

$$\text{or,} \quad x \sin\theta + y \cos\theta = 0. \qquad (1)$$

We consider an element $dx\,dy$ at $P(x, y)$. Depth of P below the free surface

$$= \text{perpendicular distance of P from (1)}$$
$$= x \sin\theta + y \cos\theta.$$

Let $OA = 2a$. If $(\bar{x}, \bar{y})$ be the coordinates of C.P. of the square, then

$$\bar{x} = \frac{\int_0^{2a}\int_0^{2a} x(x\sin\theta + y\cos\theta)\,dx\,dy}{\int_0^{2a}\int_0^{2a} (x\sin\theta + y\cos\theta)\,dx\,dy}$$

$$= \frac{\sin\theta\int_0^{2a}\int_0^{2a} x^2 dx\,dy + \cos\theta\int_0^{2a}\int_0^{2a} xy\,dx\,dy}{\sin\theta\int_0^{2a}\int_0^{2a} x\,dx\,dy + \cos\theta\int_0^{2a}\int_0^{2a} y\,dx\,dy}$$

$$= \frac{\sin\theta\left[\frac{x^3}{3}\right]_0^{2a}[y]_0^{2a} + \cos\theta\left[\frac{x^2}{2}\right]_0^{2a}\left[\frac{y^2}{2}\right]_0^{2a}}{\sin\theta\left[\frac{x^2}{2}\right]_0^{2a}[y]_0^{2a} + \cos\theta[x]_0^{2a}\left[\frac{y^2}{2}\right]_0^{2a}}$$

$$= \frac{\sin\theta\frac{8a^3}{3}2a + \cos\theta\frac{4a^2}{2}\cdot\frac{4a^2}{2}}{\sin\theta\frac{4a^2}{2}2a + \cos\theta 2a\frac{4a^2}{2}} = \frac{16a^4\left(\frac{\sin\theta}{3} + \frac{\cos\theta}{4}\right)}{\frac{8a^3}{2}(\sin\theta + \cos\theta)}$$

$$\bar{x} = \frac{a(4\sin\theta + 3\cos\theta)}{3(\sin\theta + \cos\theta)}. \tag{2}$$

Similarly,

$$\bar{y} = \frac{a(4\cos\theta + 3\sin\theta)}{3(\sin\theta + \cos\theta)}. \tag{3}$$

Adding (2) and (3), we get

$$\bar{x} + \bar{y} = \frac{7a(\sin\theta + \cos\theta)}{3(\sin\theta + \cos\theta)} = \frac{7a}{3} \quad \text{or,} \quad 3(\bar{x} + \bar{y}) = 7a.$$

Hence, locus of $(\bar{x}, \bar{y})$ is $3(x + y) = 7a$, which is a straight line.

▶ **Ex. 9.** A system of coaxial circles is immersed in water with the line of centres at a given depth. Prove that the centre of pressure of those circular areas which are completely immersed, lie on a parabola. **[P.U. 68H, M.U. 65H]**

Solution: With the line of centres as the x-axis and the radial axis as the y-axis, the equation of any circle of the coaxial system can be taken as

$$x^2 + y^2 - 2gx + c = 0, \tag{1}$$

where c is a constant and g is a variable for each circle.

Centre C of the circle (1) is $(g, 0)$ and its radius is given by

$$r^2 = g^2 - c. \qquad (2)$$

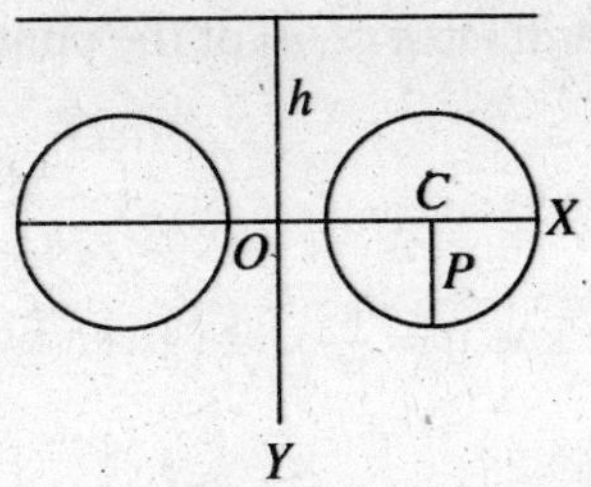

Fig. 5.23

Now, let h be the depth of the line of centres below the free surface, therefore, C.P. P is at a depth C.P. $= \frac{r^2}{4h}$.

If (x, y) be the coordinates of C.P., then

$$x = g,\ y = \frac{r^2}{4h} = \frac{g^2 - c}{4h} = \frac{x^2 - c}{4h}$$

$$\text{or,}\quad 4hy = x^2 - c, \quad \text{or,}\quad x^2 = 4hy + c, \text{ which is a parabola.}$$

▶ **Ex. 10.** An ellipse, semi-axis (a, b) is immersed in fluid in a vertical plane so that the axes make equal angles with the vertical. Show that its centre of pressure lies in a line through the centre making with the vertical an angle $\tan^{-1}\frac{a^2-b^2}{a^2+b^2}$ and at a depth $h + \frac{a^2+b^2}{8b}$, h being the depth of the centre of the ellipse.

[P.U. 65H, M.U. 73H, C.U. 93H]

Solution: Let the equation of the ellipse be $\frac{x^2}{a^2} + \frac{y^2}{b^2} = 1$ with major axis ACA' and minor axis BCB' equally inclined to the vertical MC.

$$\therefore \quad \angle ACM = \angle BCM = \frac{\pi}{4}.$$

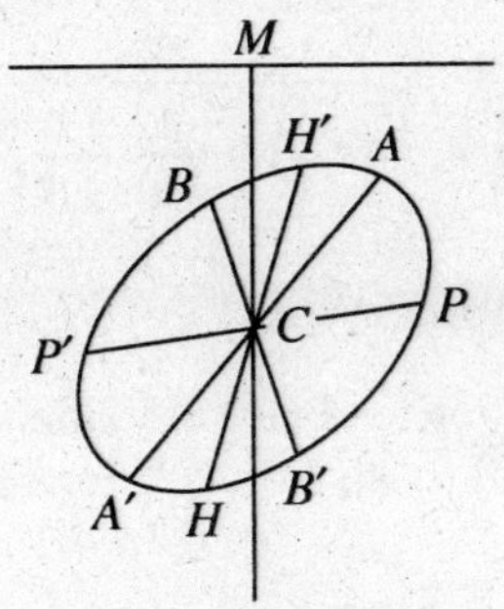

Fig. 5.24

Hence, equation of CM is

$$y = x \qquad (1)$$

and the equation of the horizontal line PCP' is $y = -x$.

The C.P. will lie on a line HCH' passing through the centre C and conjugate to PCP'. The equation of this line is

$$y = \frac{b^2}{a^2}x. \qquad (2)$$

If HCH' is inclined at an angle θ to the vertical, then

$$\tan\theta = \frac{1 - \frac{b^2}{a^2}}{1 + \frac{b^2}{a^2}} = \frac{a^2 - b^2}{a^2 + b^2} \quad \text{[From (1) and (2)]} \qquad (3)$$

The coordinates (x, y) of the points H, H′ of intersection of the ellipse

$$\frac{x^2}{a^2} + \frac{y^2}{b^2} = 1$$

and the line $y = \frac{b^2}{a^2}x$ are given by

$$\frac{x^2}{a^2} + \frac{1}{b^2}\frac{b^4}{a^4}x^2 = 1, \quad \text{or,} \quad \frac{x^2}{a^2}\left(1 + \frac{b^2}{a^2}\right) = 1,$$

$$\text{or,} \quad x^2 = \frac{a^4}{a^2 + b^2}.$$

Hence,

$$y^2 = \frac{a^4}{a^4}x^2 = \frac{b^4}{a^4}\frac{a^4}{a^2 + b^2} = \frac{b^4}{a^2 + b^2}.$$

Thus, coordinates of H are

$$\left(\frac{a^4}{a^2 + b^2}, \frac{a^4}{a^2 + b^2}\right)$$

$$\therefore \qquad CH^2 = x^2 + y^2 = \frac{a^4 + b^4}{(a^2 + b^2)} = a^2 \quad (say).$$

If F be the C.P. and h be the depth of the C.G., then

$$CO = \frac{a^2}{4h}\cos\theta.$$

∴ Depth of the C.P. below C.G.

$$= \frac{a^2}{4h}\cos^2\theta = \frac{a^2}{4h^2\sec^2\theta} = \frac{a^2}{4h^2(1 + \tan^2\theta)} = \frac{a^2}{4h\left[1 + \frac{a^2 - b^2)^2}{(a^2 + b^2)^2}\right]}$$

$$= \frac{a^2(a^2 + b^2)^2}{4h[(a^2 + b^2)^2 + (a^2 - b^2)^2]} \quad \text{[From (3)]}$$

$$= \frac{a^2(a^2 + b^2)^2}{8h(a^4 + b^4)} = \frac{(a^4 + b^4)(a^2 + b^2)^2}{8h(a^2 + b^2)(a^4 + b^4)} = \frac{a^2 + b^2}{8h}.$$

Hence, depth of the C.P. below the free surface $= h + \frac{a^2+b^2}{8h}$.

► **Ex. 11.** A semicircular lamina is immersed in a liquid with the diameter in the surface. Find the centre of pressure.

[I.A.S. 88, Bhag. 72A, P.U. 73A, M.U. 67H, 71H, Mith. 88H, M.U. 98H, 02H]

Solution: Let O be the centre and a the radius of the semicircular area, with diameter AB in the free surface.

If OC be the vertical radius, then it is the line of symmetry and C.P. will lie on it. Referred to OX and OY as coordinate axes, equation of circle is $x^2 + y^2 = a^2$.

We divide the area into a large number of elementary strips parallel to the free surface. Let PQRS be one of such elementary strip of thickness dx at a depth x below the free surface.

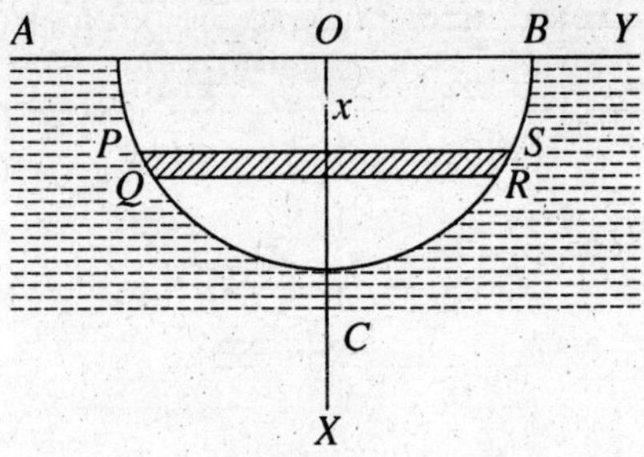

Fig. 5.25

Then,

$$ds = \text{area of the elementary strip} = 2y\,dx$$

$$\text{and } p = \text{pressure at any point on the elementary strip} = \rho g x.$$

If $\bar{x}$ be the depth of C.P. below the free surface, then

$$\bar{x} = \frac{\int xp\,ds}{\int p\,ds} = \frac{\int_0^a x\rho g x 2y\,dx}{\int_0^a \rho g x 2y\,dx} = \frac{2\rho g \int_0^a x^2 y\,dx}{2\rho g \int_0^a xy\,dx} = \frac{\int_0^a x^2\sqrt{a^2 - x^2}\,dx}{\int_0^a x\sqrt{a^2 - x^2}\,dx}.$$

Let $x = a\sin\theta$. $\therefore$ $dx = a\cos\theta\,d\theta$.

When $x = a$, $\theta = \frac{\pi}{2}$ when $x = 0$, $\theta = 0$.

$$\bar{x} = \frac{\int_0^{\pi/2} a^2\sin^2\theta a\cos\theta a\cos\theta\,d\theta}{\int_0^{\pi/2} a\sin\theta a\cos\theta a\cos\theta\,d\theta} = \frac{a\int_a^{\pi/2}\sin^2\theta\cos^2\theta\,d\theta}{\int_0^{\pi/2}\sin\theta\cos^2\theta\,d\theta}$$

$$= \frac{a\cdot\frac{1}{4}\cdot\frac{1}{2}\cdot\frac{\pi}{2}}{-\int_0^{\pi/2}\cos^2\theta d(\cos\theta)\,d\theta} = \frac{\frac{\pi a}{16}}{-\left[\frac{\cos^3\theta}{3}\right]_0^{\pi/2}} = \frac{\pi a}{16} \times 3 = \frac{3\pi a}{16} \quad \textbf{Ans.}$$

▶ **Ex. 12.** A semicircular area of radius a is immersed vertically with its diameter horizontal at a depth b. If the circumference be below the centre, prove that the depth of centre of pressure is

$$\frac{3\pi(a^2 + 4b^2) + 32ab}{4(3b\pi + 4a)}.$$ **[I.A.S. 92]**

Solution: Let O be the centre and a be the radius of the semicircular area, with diameter AB in the surface.

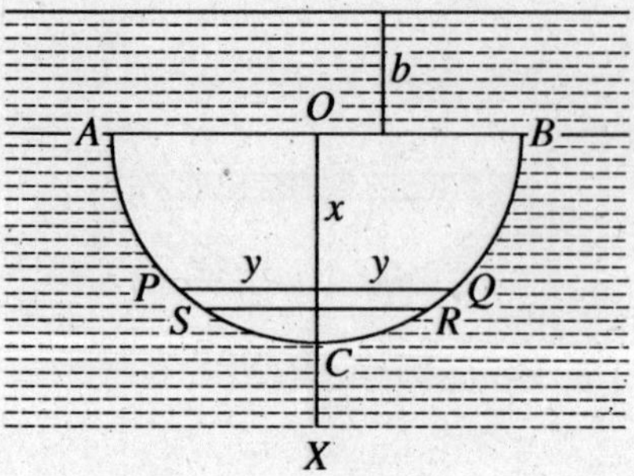

Fig. 5.26

If OC be the vertical radius, then it is the line of symmetry and C.P. will lie on OC. Referred to OX and OY as coordinate axes, equation of circle is

$$x^2 + y^2 = a^2. \quad (1)$$

We divide the area into a large number of elementary strip parallel to the free surface. Let PQRS be one of such elementary strips of thickness dx at a depth $(b + x)$ from the free surface.

Then

$$ds = \text{area of the strip} = 2y\,dx = 2\sqrt{a^2 - x^2}\,dx \quad \text{[From (1)]}$$

$$\text{and} \quad p = \text{pressure at any point on the strip} = \rho g(b + x).$$

If $\bar{x}$ be the depth of C.P. below the free surface, then

$$\bar{x} = \frac{\int xp\,ds}{\int xds} = \frac{\int_0^a x\rho g(b + x)2\sqrt{a^2 - x^2}\,dx}{\int_0^a \rho g(b + x)2\sqrt{a^2 - x^2}\,dx}$$

$$= \frac{2\rho g \int_0^a x(b + x)\sqrt{a^2 - x^2}\,dx}{2\rho g \int_0^a (b + x)\sqrt{a^2 - x^2}\,dx}.$$

Putting $x = a\sin\theta$, $dx = a\cos\theta d\theta$.

When $x = 0$, $\theta = \frac{\pi}{2}$, when $x = 0$, $\theta = 0$.

$$\bar{x} = \frac{\int_0^{\pi/2} a\sin\theta(b + a\sin\theta)\sqrt{a^2 - a^2\sin^2\theta}a\cos\theta ds}{\int_0^{\pi/2} (b + a\sin\theta)\sqrt{a^2 - a^2\sin^2\theta}a\cos\theta\,d\theta}$$

$$= \frac{a^2 \int_0^{\pi/2} \sin\theta(b + a\sin\theta)\cos^2\theta\,d\theta}{a^2 \int_0^{\pi/2} (b + a\sin\theta)\cos^2\theta\,d\theta}$$

$$= \frac{a \int_0^{\pi/2} (b\sin\theta\cos^2\theta + a\sin^2\theta\cos^2\theta)\,d\theta}{\int_0^{\pi/2} (b\cos^2\theta + a\sin\theta\cos^2\theta)\,d\theta}$$

$$= \frac{a\left[-b\int_0^{\pi/2}\cos^2\theta d(\cos\theta)\text{ C.P. } + a\frac{1}{4}\cdot\frac{1}{4}\cdot\frac{\pi}{2}\right]}{b\frac{2-1}{2}\frac{\pi}{2} - a\int_0^{\pi/2}\cos^2\theta d(\cos\theta)}$$

$$= \frac{\frac{-ab}{3}\left[\cos^3\theta\right]_0^{\pi/2} + \frac{\pi a^2}{16}}{\frac{\pi b}{4} - \frac{a}{3}\left[\cos^3\theta\right]_0^{\pi/2}} = \frac{\frac{ab}{3} + \frac{\pi a^2}{16}}{\frac{\pi b}{4} + \frac{a}{3}} = \frac{3\pi a^2 + 16ab}{4(3\pi b + 4a)}.$$

Hence, depth of C.P. from the free surface

$$= \frac{3a^2\pi + 16ab}{4(3\pi b + 4a)} + b = \frac{3\pi(a^2 + 4b^2) + 32ab}{4(4a + 3b\pi)}$$ **Proved.**

▸ **Ex. 13.** Find the coordinates of the centre of pressure of the area enclosed between the curve $\sqrt{x} + \sqrt{y} = \sqrt{a}$ and the axes taking the axes to be rectangular and one of them in the surface.

[R.U. 65H, 68H, 70H, M.U. 77H, R.U. 94H, P.U. 95H, M.U. 97H, T.M.B.U. 97H]

Solution: The shape of the curve is as shown in the figure. The curve cuts the axes at $A(a, 0)$ and $B(0, a)$, respectively. We consider an elementary strip PQRS of thickness dx at a distance x from the origin.

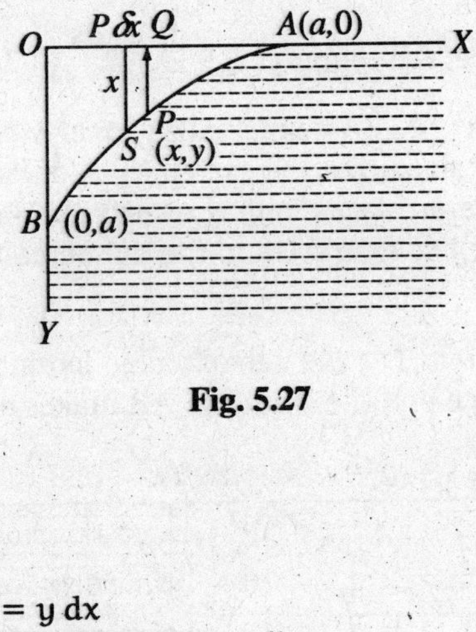

Fig. 5.27

Then,

ds = area of the strip $= y\,dx$

and p = pressure at the C.G. of the strip $= \frac{y}{2}\rho g$.

Also the C.P. of the strip is $(x, \frac{2}{3}y)$.

Hence, if $(\bar{x}, \bar{y})$ be the required coordinates of the C.P., then

$$\bar{x} = \frac{\int xp\,ds}{\int p\,ds} = \frac{\int_0^a x\frac{y}{2}\rho g y\,dx}{\int_0^a \frac{y}{2}\rho g y\,dx} = \frac{\int_0^a xy^2 dx}{\int_0^a ay^2 dx}.$$

Putting $x = a\sin^4\theta$, $= a\cos^4\theta$, $dx = 4a\sin^3\theta\cos\theta\,d\theta$.

When $x = a, \theta = \frac{\pi}{2}$, when $x = 0, \theta = 0$.

$$\bar{x} = \frac{\int_0^{\pi/2} a \sin^4 \theta a^2 \cos^8 \theta 4a \sin^3 \theta \cos\theta \, d\theta}{\int_0^{\pi/4} a^2 \cos^8 \theta 4a \sin^3 \theta \cos\theta \, d\theta} = \frac{a\int_0^{\pi/2} \sin^7 \theta \cos^{11} \theta \, d\theta}{\int_0^{\pi/2} \sin^3 \theta \cos^9 \theta \, d\theta}$$

$$= a\frac{\frac{6}{16}\cdot\frac{4}{14}\cdot\frac{2}{12}\cdot\frac{1}{10}}{\frac{2}{12}\cdot\frac{1}{10}} = \frac{3a}{28}$$

$$\text{and} \quad \bar{y} = \frac{\int yp\, ds}{\int p\, ds} = \frac{\int_0^a \frac{2}{3}y\frac{y}{2}\rho g y\, dx}{\int_0^a \frac{y}{2}\rho g y\, dx} = \frac{2}{3}\frac{\int_0^a y^3 dx}{\int_0^a y^2 dx}.$$

$$= \frac{2}{3}\frac{\int_0^{\pi/2} a^3 \cos^{12} \theta 4a \sin^3 \theta \cos\theta \, d\theta}{\int_0^{\pi/2} a^2 \cos^8 \theta 4a \sin^3 \theta \cos\theta \, d\theta} = \frac{2}{3}\frac{a\int_0^{\pi/2} \cos^{13} \theta \sin^3 \theta \, d\theta}{\int_0^{\pi/2} \cos^9 \theta \sin^3 \theta \, d\theta}$$

$$= \frac{2a}{3}\frac{\frac{2}{16}\cdot\frac{1}{14}}{\frac{2}{12}\cdot\frac{1}{10}} = \frac{5a}{14}.$$

Hence, the required C.P. is $\left(\frac{3a}{28}, \frac{5a}{14}\right)$.

► **Ex. 14.** A semicircular area is completely immersed in water with its plane vertical, so that the extremity A of its bounding diameter is in the surface, and the diameter makes with the surface an angle α. Prove that if E be the C.P. and θ the angle between AE and the diameter, $\tan\theta = \frac{3\pi + 16\tan\alpha}{16 + 15\pi\tan\alpha}$.

[I.A.S. 74, 93, R.U. 62H, 67H, Bhag. 66H]

Solution: Let the semicircular lamina ACB be immersed in water with O in the surface and the diameter AB makes an angle α with the surface of the water.

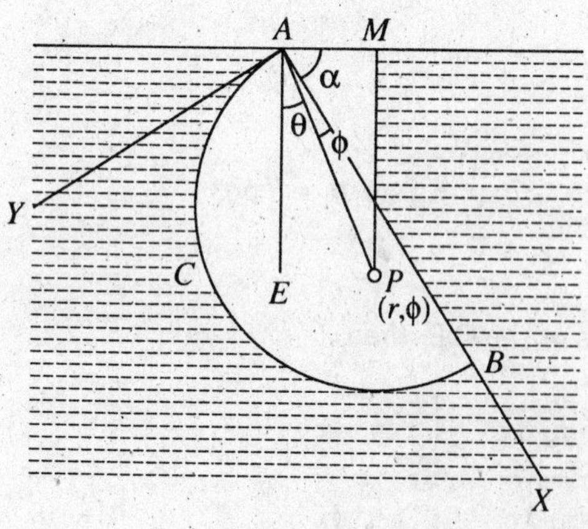

Fig. 5.28

Taking A as origin and the diameter AB as the initial line we consider a small elementary area $r\,d\phi\,dr$, surrounding the point $P(r, \phi)$.

Depth of P below the free surface

$$= PM = AP\sin(\phi + \alpha)$$
$$= r\sin(\phi + \alpha).$$

Pressure p at the point

$$P = \rho g PM = \rho g r \sin(\phi + \alpha).$$

Equation of the semicircle is

$$r = 2a\cos\phi.$$

If $(\bar{x}, \bar{y})$ be the coordinates of C.P. E of the semicircular lamina, then

$$\bar{x} = \frac{\int xp\,ds}{\int p\,ds} = \frac{\int_{\theta=0}^{\pi/2}\int_{r=0}^{2a\cos\phi} r\cos\phi\rho g r\sin(\phi+\alpha)r d\phi dr}{\int_{\theta=0}^{\pi/2}\int_{r=0}^{2a\cos\phi} \rho g r\sin(\phi+\alpha)r d\phi dr}$$

$$= \frac{\rho g\int_0^{\pi/2}\int_0^{2a\cos\phi} r^3\sin(\phi+\alpha)\cos\phi\, d\phi dr}{\rho g\int_0^{\pi/2}\int_0^{2a\cos\phi} r^2\sin(\phi+\alpha)\, d\phi dr}$$

$$= \frac{\int_0^{\pi/2}\left(\frac{2a\cos\phi}{4}\right)^4\sin(\phi+\alpha)\cos\phi\, d\phi}{\int_0^{\pi/2}\frac{(2a\cos\phi)^3}{3}\sin(\phi+\alpha)\, d\phi}$$

$$= \frac{\frac{2a}{4}\int_0^{\pi/2}\sin(\phi+\alpha)\cos^5\phi\, d\phi}{\frac{1}{3}\int_0^{\pi/2}\cos^3\phi\sin(\phi+\alpha)\, d\phi}.$$

Also,

$$\bar{y} = \frac{\int yp\,ds}{\int p\,ds} = \frac{\int_0^{\pi/2}\int_0^{2a\cos\phi} r\sin\phi\rho g r\sin(\phi+\alpha)r d\phi dr}{\int_0^{\pi/2}\int_0^{2a\cos\phi} \rho g r\sin(\phi+\alpha)r d\phi dr}$$

$$= \frac{\rho g\int_0^{\pi/2}\int_0^{2a\cos\phi} r^3\sin(\phi+\alpha)\sin\phi\, d\phi dr}{\rho g\int_0^{\pi/2}\int_0^{2a\cos\phi} r^2\sin(\phi+\alpha)\, d\phi dr}$$

$$= \frac{\int_0^{\pi/2}\frac{(2a\cos\phi)^4}{4}\sin(\phi+\alpha)\sin\phi\, d\phi}{\int_0^{\pi/2}\frac{(2a\cos\phi)^3}{3}\sin(\phi+\alpha)\, d\phi}$$

$$= \frac{\frac{2a}{4}\int_0^{\pi/2}\sin(\phi+\alpha)\cos^4\phi\sin\phi\, d\phi}{\frac{1}{3}\int_0^{\pi/2}\sin(\phi+\alpha)\cos^3\phi\, d\phi}.$$

By question,

$$\tan\theta = \frac{\bar{y}}{\bar{x}} = \frac{\frac{2a}{4}\int_0^{\pi/2} \sin(\phi+\alpha)\cos^4\phi\sin\phi\, d\phi}{\frac{2a}{4}\int_0^{\pi/2} \cos^5\phi\sin(\phi+\alpha)\, d\phi}$$

$$= \frac{\int_0^{\pi/2} \cos^4\phi\sin\phi(\sin\phi\cos\alpha+\cos\phi\sin\alpha)\, d\phi}{\int_0^{\pi/2} \cos^5\phi(\sin\phi\cos\alpha+\cos\phi\sin\alpha)\, d\phi}$$

$$= \frac{\int_0^{\pi/2} d\phi(\cos^4\phi\sin^2\phi\cos\alpha+\cos^5\phi\sin\phi\sin\alpha)}{\int_0^{\pi/2} (\cos^5\phi\sin\phi\cos\alpha+\cos^6\phi\sin\alpha)\, d\phi}$$

$$= \frac{\cos\alpha\int_0^{\pi/2} \cos^4\phi(1-\cos^2\phi)\, d\phi + \sin\alpha\int_0^{\pi/2} \cos^5\phi\sin\phi\, d\phi}{\cos\alpha\int_0^{\pi/2} \cos^5\phi\sin\phi\, d\phi + \sin\alpha\int_0^{\pi/2} \cos^6\phi\, d\phi}$$

$$= \frac{\cos\alpha\left[\frac{3}{4}\cdot\frac{1}{2}\cdot\frac{\pi}{2} - \frac{5}{6}\cdot\frac{3}{4}\cdot\frac{1}{2}\cdot\frac{\pi}{2}\right] - \sin\alpha\int_0^{\pi/2} \cos^5\phi d(\cos\phi)}{-\cos\alpha\int_0^{\pi/2} \cos^5\phi d(\cos\phi) + \sin\alpha\frac{6-1}{6}\cdot\frac{3}{4}\cdot\frac{1}{2}\cdot\frac{\pi}{2}}$$

$$= \frac{\cos\alpha\left(\frac{3\pi}{16}-\frac{5\pi}{32}\right) - \frac{\sin\alpha}{6}\left[\cos^6\phi\right]_0^{\pi/2}}{-\frac{\cos\alpha}{6}\left[\cos^6\phi\right]_0^{\pi/2} + \sin\alpha\frac{5}{6}\cdot\frac{3}{4}\cdot\frac{1}{2}\cdot\frac{\pi}{2}}$$

$$= \frac{\cos\alpha\frac{\pi}{32}+\frac{\sin\alpha}{6}}{\frac{\cos\alpha}{6}+\frac{5\pi}{32}\sin\alpha} = \frac{3\pi\cos\alpha+16\sin\alpha}{16\cos\alpha+15\pi\sin\alpha}$$

$$\tan\theta = \frac{3\pi+16\tan\alpha}{16+15\pi\tan\alpha}$$ **Proved.**

Corollary 5.15.1 *When* $\alpha = \frac{\pi}{4}$, *then* $\tan\theta = \frac{3\pi+16}{16+15\pi}$. [I.A.S. 86]

▶ **Ex. 15.** A semi-ellipse bounded by its minor axis is just immersed in a liquid the density of which varies as the depth, if the minor axis be in the surface, find the eccentricity in order that the focus may be the centre of pressure.

[L.N.M.U. 93H, I.A.S. 95]

Solution: Let the equation of the ellipse be

$$\frac{x^2}{a^2}+\frac{y^2}{b^2} = 1.$$

We consider an elementary strip PQRS at a depth x below the free surface of thickness dx.

Area of the strip PQRS $= ds = 2y\,dx$, where (x, y) is the coordinate of Q.

Also p = pressure at any point of the strip $= \rho g x$.

Given $\rho \alpha x, \rho = kx$. Hence, $p = kgx^2$.

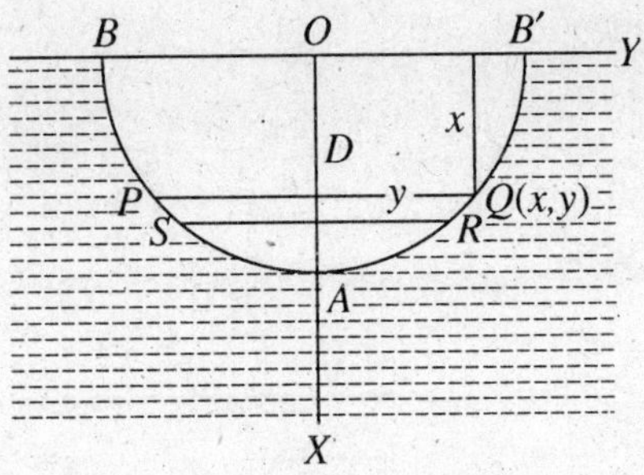

Fig. 5.29

If x be the depth of C.P. below the free surface, then

$$\bar{x} = \frac{\int xp\,ds}{\int p\,ds} = \frac{\int_0^a xkgx^2 2y\,dx}{\int_0^a kgx^2 2y\,dx} = \frac{\int_0^a x^3 y\,dx}{\int_0^a x^2 y\,dx}.$$

Putting $x = a\cos\theta,\ y = b\sin\theta,\ dx = -a\sin\theta\,d\theta$.

When $x = a, \theta = 0$, when $x = 0, \theta = \frac{\pi}{2}$

$$\bar{x} = \frac{-\int_{\frac{\pi}{2}}^{0} a^3\cos^3\theta a\sin\theta b\sin\theta\,d\theta}{-\int_{\frac{\pi}{2}}^{0} a^2\cos^2\theta b\sin\theta a\sin\theta\,d\theta} = \frac{a\int_0^{\pi/2}\cos^3\theta\sin^2\theta\,d\theta}{\int_0^{\pi/2}\cos^2\theta\sin^2\theta\,d\theta}$$

$$= \frac{a\frac{2}{5}\frac{1}{3}}{\frac{1}{4}\cdot\frac{1}{2}\cdot\frac{\pi}{2}} = \frac{2a}{15}\frac{16}{\pi} = \frac{32a}{15\pi}.$$

But the C.P. coincides with focus $(ae, 0)$.

$$\therefore\ ae = \frac{32a}{15\pi},\ e = \frac{32}{15\pi} \quad \textbf{Ans.}$$

▶ **Ex. 16.** When the depth of the liquid is increased by a, the depth of the C.P. is found to increase by y, and when instead the depth of the liquid is increased by b, that of the C.P. is found, to be increased by z. Show that the depth of the centre of gravity in the original state of the liquid is $\frac{ab(b-a+y-z)}{az-by}$.

[R.U. 69H, Bhag. 68H, C.U. 88H, 95H, 04H]

Solution: We know that the distance between the C.P. and C.G. of an area varies inversely as the depth of the C.G. Hence, if λ, μ be the depths of the C.G. and the C.P. originally, we have

$$(\mu + y) - (\lambda + a) \propto \frac{1}{\lambda + a}, \quad \text{or,} \quad (\mu + y) - (\lambda + a) = \frac{k}{\lambda + a},$$

where k is a constant.

Similarly, $(\mu + z) - (\lambda + b) = \frac{k}{\lambda+b}$.

Also, $\mu - \lambda = \frac{k}{\lambda}$.

Hence,

$$\frac{(\mu + y - \lambda - a)(\lambda + a)}{1} = [(\mu + z) - (\lambda + b)](\lambda + b) = (\mu - \lambda)\lambda$$

$$\frac{\mu + y - \lambda - a}{\frac{1}{\lambda+a}} = \frac{\mu + z - \lambda - b}{\frac{1}{\lambda+b}} = \frac{\mu - \lambda}{\frac{1}{\lambda}}$$

or, $$\frac{\mu + y - \lambda - a - (\mu - \lambda)}{\frac{1}{\lambda+a} - \frac{1}{\lambda}} = \frac{\mu + z - \lambda - b - (\mu - \lambda)}{\frac{1}{\lambda+b} - \frac{1}{\lambda}}$$

or, $$\frac{y - a}{\frac{\lambda-\lambda-a}{\lambda(\lambda+a)}} = \frac{z - b}{\frac{\lambda-\lambda-b}{\lambda(\lambda+b)}}$$

or, $$\frac{(y - a)(\lambda + a)}{a} = \frac{(z - b)(\lambda + b)}{b}$$

or, $(y - a)(\lambda + a)b = (\lambda + b)(z - b)a$

or, $(y\lambda + ya - a\lambda - a^2)b = (z\lambda + bz - b\lambda - b^2)a$

or, $y\lambda b + yab - \lambda ab - a^2 b = z\lambda a + abz - ab\lambda - ab^2$

or, $\lambda(by - az) = ab(z - b - y + a)$

$$\lambda = \frac{ab(b - a + y - z)}{az - by}$$ **Proved.**

▶ **Ex. 17.** A circular area is immersed in a liquid with its plane vertical and its centre at a depth a. If the area is lowered slowly so that the centre is at a depth $f(t)$ at time t; show that the velocity of the centre by pressure at that instant is

$$f'(t)\left[1 - \frac{a^2}{4\{f(t)\}^2}\right].$$

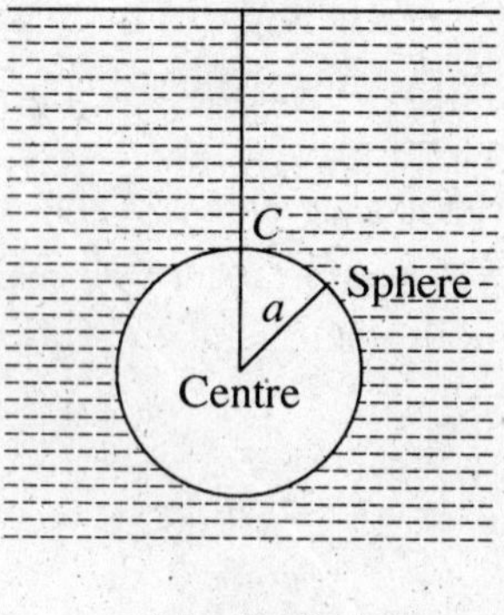

Fig. 5.30

Solution: The depth of the C.P. of the circular area of radius a with its centre at a depth c, below the free surface of the liquid is $c + \frac{a^2}{4c}$.

It is given that the depth of centre of circle at time t is $f(t)$.

If x be the distance of the C.P. of the circle from the free surface; at time t, then

$$x = f(t) + \frac{a^2}{4f(t)}.$$

Differentiating both the sides w.r.t., we have

$$\frac{dx}{dt} = f'(t) + \frac{a^2}{4}\left[\frac{-f'(t)}{\{f(t)\}^2}\right],$$

$$\text{or,}\quad V = f'(t)\left[1 - \frac{a^2}{4\{f(t)\}^2}\right].\quad \textbf{Proved.}$$

▶ **Ex. 18.** A triangle has its base in the surface of water and its vertex downwards. If the atmospheric pressure be equivalent to a height h of water, prove that the centre of pressure will be higher by a distance $\frac{hd}{2(h+d)}$, than it is when the atmospheric pressure is neglected, where d is the distance of centre of gravity of the triangle below the surface of the water. [P.U. 73H, 79H]

Solution: Let ABC be the triangle immersed in water whose side BC is in the surface.

Let when the atmospheric pressure is neglected the C.G. and C.P. be at G and P, respectively.

When the atmospheric pressure, equivalent to a height h of water, be taken into consideration, its effect may be supposed to be equivalent to the further lowering of the triangle through a distance h. Let P′ be the position of the C.P. after lowering the triangle.

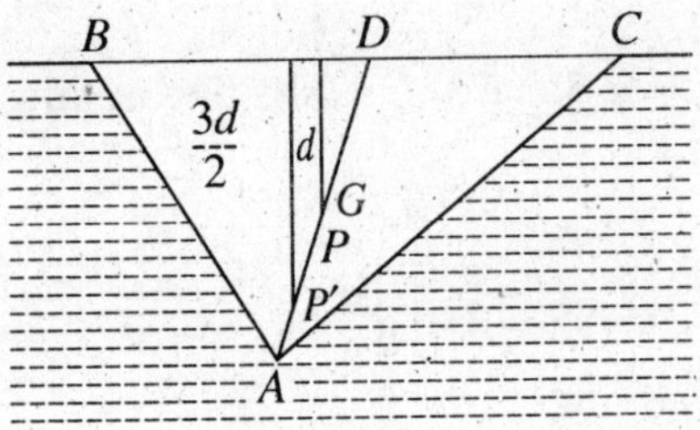

Fig. 5.31

Let h′ be the depth of the vertex below the free surface,

$$\therefore \text{Depth of C.G.} = \frac{h'}{3} = d \text{ (given)}$$
$$\text{or, } h' = 3d$$

Here, h = Depth of C.G. in original position $= \frac{h'}{3} = d$

h′ = Depth of C.P. in original position $= \frac{h'}{2} = \frac{3d}{2}$

H = Distance through which the area is assumed to move downwards = h.

If $\bar{x}$ be the depth of C.P. in new position, then by applying the formula

$$\bar{x} = \frac{H^2 + 2hH + hh'}{H + h}$$

we get,

$$\bar{x} = \frac{h^2 + 2dh + d\frac{3d}{2}}{h + d} = \frac{2h^2 + 4dh + 3d^2}{2(h + d)}.$$

$\therefore$ Distance through which C.P. is raised

$$= (h' + H) - \bar{x} = \left(\frac{3d}{2} + h\right) - \frac{2h^2 + 4dh + 3d^2}{2(h + d)}$$

$$= \frac{(3d + 2h)(h + d) - (2d^2 + 4dh + 3d^2)}{2(d + h)}$$

$$= \frac{3dh + 3d^2 + 2h^2 + 2dh - 2d^2 - 4dh - 3d^2}{2(d + h)}$$

$$= \frac{dh}{2(d + h)} \quad \textbf{Proved.}$$

▶ **Ex. 19.** ABC is a triangular lamina vertically immersed in a fluid with A in the surface and BC horizontal. M and N are the mid-points of AB and AC. Prove that the depth of centre of pressure of the area MNCB is $\frac{45}{56}$ d, where d is the depth of B.

[P.U. 81H]

Solution: Since MN is the line joining the mid-points of the sides AB and AC.

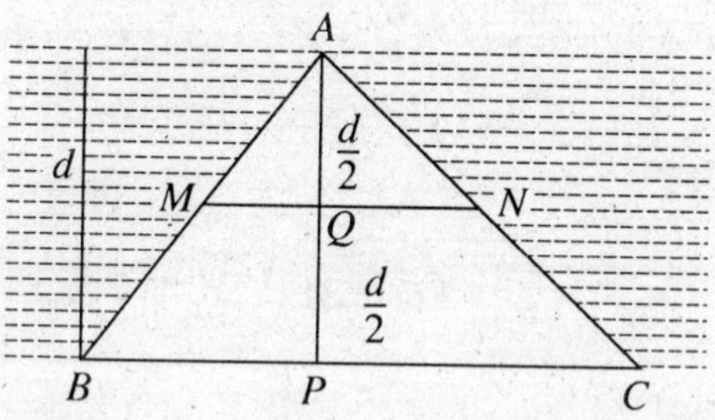

Fig. 5.32

Hence, MN is parallel to BC and MN = $\frac{1}{2}$BC.

From A we draw AP perpendicular to BC. Hence,

$$AQ = \frac{1}{2}AP = \frac{d}{2}.$$

Let T_1 and T_2 be the fluid thrusts on $\triangle ABC$ and $\triangle AMN$, respectively.

Let ρ be the density of the fluid, then

$$T_1 = \rho g \frac{2d}{3} \frac{1}{2} BCd = \rho g \frac{d^2 a}{3}$$

$$\text{and } T_2 = \rho g \frac{d}{3} \frac{1}{2} MN \cdot \frac{d}{2} = \rho g \frac{d}{3} \frac{1}{2} \frac{a}{2} \frac{d}{2} = \rho g \frac{ad^2}{24}.$$

Let Z_1 and Z_2 be the depths of C.P. of $\triangle ABC$ and $\triangle AMN$.

Then $Z_1 = \frac{3}{4}$ d and $Z_2 = \frac{3}{4}\frac{d}{2} = \frac{3d}{8}$.

If $\bar{z}$ be the required depth of the C.P. of the area below the free surface, then

$$\bar{z} = \frac{T_1 Z_1 - T_2 Z_2}{T_1 - T_2} = \frac{\rho g \frac{d^2 a}{3} \frac{3}{4} d - \frac{\rho g a d^2}{24} \cdot \frac{3d}{8}}{\rho g \frac{d^2 a}{3} - \rho g \frac{a d^2}{24}} = \frac{\rho g a d^3 \left(\frac{1}{4} - \frac{1}{64}\right)}{\rho g a d^2 \left(\frac{1}{3} - \frac{1}{24}\right)}$$

$$= d\frac{15}{64} \times \frac{24}{7} = d\frac{15 \times 3}{8 \times 7} = \frac{45}{56} d. \quad \textbf{Proved.}$$

▶ **Ex. 20.** A triangular lamina is immersed in a heavy homogeneous liquid with its plane vertical. Prove that the distance between the centre of pressure and the centre of mass of the triangular lamina is $\frac{(b-c)^2+(c-a)^2+(a-b)^2}{12(a+b+c)}$, a, b, c being the depths of the angular points of lamina below the surface of the liquid.

Solution: We know that the C.P. and the C.G. of a plane lamina are along a line of greatest slope, which is vertical here. Hence, the required distance is the difference between the depths of the C.P. and the C.G. and is

$$= \frac{a^2+b^2+c^2+bc+ca+ab}{2(a+b+c)} - \frac{a+b+c}{3}$$

$$= \frac{3a^2+3b^2+3c^2+3bc+3ca+3ab-2(a^2+b^2+c^2+2ab+2bc+2ca)}{6(a+b+c)}$$

$$= \frac{a^2+b^2+c^2-ab-bc-ca}{6(a+b+c)}$$

$$= \frac{2a^2+2b^2+2c^2-2ab-2bc-2ca}{12(a+b+c)}$$

$$= \frac{a^2-2ab+b^2+b^2-2ac+c^2+c^2-2ca+a^2}{12(a+b+c)}$$

$$= \frac{(a-4)^2+(b-c)^2+(c-a)^2}{12(a+b+c)}.$$

▶ **Ex. 21.** An ellipse is completely immersed with minor axis horizontal and at a depth h, find the position of the centre of pressure.

or

Find the position of centre of pressure of an ellipse which is immersed vertically with its major axis vertical at depth h. **[M.U. 95H, T.M.B.U. 99H, 92H]**

Solution: Let equation of ellipse be $\frac{x^2}{a^2}+\frac{y^2}{b^2}=1$, where $AA' = 2a$ and $BB' = 2b$.

We consider an elementary strip of thickness δx at a depth x below the centre C of the ellipse.

Then,

$$\delta s = \text{area of the strip}$$
$$= 2y\delta x = 2\frac{b}{a}\sqrt{a^2-x^2}\,\delta x$$

and p = pressure at any point at a depth x below C
$$= \rho g(h+x).$$

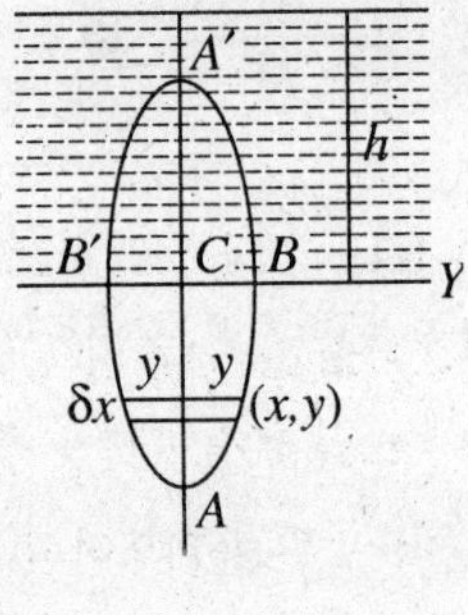

Fig. 5.33

If $\bar{x}$ be the depth of C.P. of the ellipse below the centre of the ellipse, on the major axis (due to symmetry), then

$$\bar{x} = \frac{\int xp\,ds}{\int p\,ds} = \frac{\int_{-a}^{a} x\rho g(h+x)\frac{2b}{a}\sqrt{a^2-x^2}\,dx}{\int_{-a}^{a} \rho g(h+x)\frac{2b}{a}\sqrt{a^2-x^2}\,dx}$$

$$= \frac{h\int_{-a}^{a} x\sqrt{a^2-x^2}\,dx + \int_{-a}^{a} x^2\sqrt{a^2-x^2}\,dx}{h\int_{-a}^{a} \sqrt{a^2-x^2}\,dx + \int_{-a}^{a} x\sqrt{a^2-x^2}\,dx}$$

$$= \frac{0 + 2\int_{0}^{a} x^2\sqrt{a^2-x^2}\,dx}{2h\int_{0}^{a} \sqrt{a^2-x^2}\,dx + 0}.$$

$\left[\because \int_{-a}^{a} f(x)\,dx = 0 \text{ if } f(x) \text{ is an odd function and } \int_{-a}^{a} f(x)\,dx = 2\int_{0}^{a} f(x)\,dx \text{ if } f(x)\right.$
$\left. \text{is an even function.}\right]$

Let $x = a\sin\theta$. $\quad\therefore\ dx = a\cos\theta\,d\theta$.

When $x = 0$, $\theta = 0$ and when $x = a$, $\theta = \frac{\pi}{2}$.

$$= \frac{2\int_{0}^{\pi/2} a^2\sin^2\theta\, a\cos\theta\, a\cos\theta\,d\theta}{2h\int_{0}^{\pi/2} a\cos\theta\, a\cos\theta\,d\theta} = \frac{a^2\int_{0}^{\pi/2} \sin^2\theta\cos^2\theta\,d\theta}{h\int_{0}^{\pi/2} \cos^2\theta\,d\theta}$$

$$= \frac{\frac{a^2}{4}\int_{0}^{\pi/2} \sin^2 2\theta\,d\theta}{h\int_{0}^{\pi/2} \frac{1+\cos 2\theta}{2}\,d\theta} = \frac{a^2}{2h}\frac{\int_{0}^{\pi/2} \frac{1-\cos 4\theta}{2}\,d\theta}{\left[\theta + \frac{\sin 2\theta}{2}\right]_0^{\pi/2}} = \frac{a^2}{4h}\frac{\left[\frac{\pi}{2}-0\right]}{\left[\frac{\pi}{2}+0\right]} = \frac{a^2}{4h}. \quad \textbf{Ans.}$$

▶ **Ex. 22.** A quadrant of a circle is immersed in a liquid with a bounding radius in the surface, find the position of its centre of pressure.

[M.U. 2000H, C.U. 2001H, T.M.B.U. 91H, 99H]

or

A quadrant of a circle just immersed vertically in a heavy homogeneous liquid, with one edge in the surface. Find the centre of pressure. **[L.N.M.U. 95H]**

Solution: Let OAB be the quadrant of the circle with radius OB in the surface. We take the radii OA and OB as x and y axes, respectively.

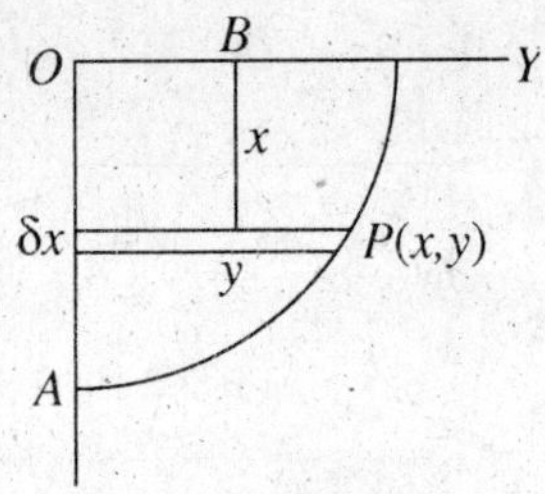

Fig. 5.34

Then, referred to these axes, the equation of the circle is

$$x^2 + y^2 = a^2, \tag{1}$$

where a is the radius of the circle.

We consider an elementary strip of thickness δx at a depth x below the free surface.

$$\begin{aligned} \text{Then } \delta s &= \text{the area of the strip} = y\delta x = \sqrt{a^2 - x^2}\delta x. \\ p &= \text{pressure at any point at a depth } x = \rho g x. \end{aligned}$$

The coordinates of C.P. of this elementary strip are $\left(x, \frac{y}{2}\right)$.

Hence, if $(\bar{x}, \bar{y})$ be the coordinates of the required C.P. of the quadrant, then

$$\begin{aligned} \bar{x} &= \frac{\int xp\, ds}{\int p\, ds} = \frac{\int_0^a x\rho g x\sqrt{a^2 - x^2}\, dx}{\int_0^a \rho g x\sqrt{a^2 - x^2}\, dx} \\ &= \frac{\int_0^a x^2\sqrt{a^2 - x^2}\, dx}{\int_0^a x\sqrt{a^2 - x^2}\, dx}. \end{aligned}$$

Let $x = a\sin\theta$. $\quad\therefore\ dx = a\cos\theta d\theta$.

When $x = 0, \theta = 0$, when $x = a, \theta = \frac{\pi}{2}$.

$$\begin{aligned} \therefore\ \bar{x} &= \frac{\int_0^{\pi/2} a^2\sin^2\theta\sqrt{a^2 - a^2\sin^2\theta}\, a\cos\theta\, d\theta}{\int_0^{\pi/2} a\sin\theta\sqrt{a^2 - a^2\sin^2\theta}\, a\cos\theta\, d\theta} = \frac{a^4\int_0^{\pi/2}\sin^2\theta\cos^2\theta\, d\theta}{a^3\int_0^{\pi/2}\sin\theta\cos^2\theta\, d\theta} \\ &= \frac{\frac{a}{4}\int_0^{\pi/2}\sin^2 2\theta\, d\theta}{-\int_0^{\pi/2}\cos^2\theta d(\cos\theta)} = \frac{\frac{a}{4}\int_0^{\pi/2}\frac{1-\cos 4\theta}{2}\, d\theta}{-\left[\frac{\cos^3\theta}{3}\right]_0^{\pi/2}} = \frac{\frac{a}{8}\left[\theta - \frac{\sin 4\theta}{4}\right]_0^{\pi/2}}{-\left[0 - \frac{1}{3}\right]} \\ &= \frac{a}{8}\times\frac{\pi}{2}\times 3 = \frac{3\pi a}{16} \end{aligned}$$

and
$$\bar{y} = \frac{\int yp\, ds}{\int p\, ds} = \frac{\int_a^0 \frac{y}{2}\rho g x\sqrt{a^2-x^2}\, dx}{\int_0^a \rho g x\sqrt{a^2-x^2}\, dx} = \frac{\frac{1}{2}\int_0^a x(a^2-x^2)\, dx}{\int_0^a x\sqrt{a^2-x^2}\, dx}$$

$$= \frac{\frac{1}{2}\int_0^{\pi/2} a\sin\theta \cdot a^2\cos^2\theta\, a\cos\theta\, d\theta}{\int_0^{\pi/2} a\sin\theta\, a\cos\theta\, a\cos\theta\, d\theta} = \frac{\frac{1}{2}a\int_0^{\pi/2}\cos^3\theta\sin\theta\, d\theta}{\int_0^{\pi/2}\cos^2\theta\sin\theta\, d\theta}$$

$$= \frac{-\frac{a}{2}\int_0^{\pi/2}\cos^3\theta\, d(\cos\theta)}{-\int_0^{\pi/2}\cos^2\theta\, d(\cos\theta)} = \frac{\frac{a}{2}\left[\frac{\cos^4\theta}{4}\right]_0^{\pi/2}}{\left[\frac{\cos^3\theta}{3}\right]_0^{\pi/2}} = \frac{3a}{8}.$$

Hence, the coordinates of C.P. of the quadrant are $\left(\frac{3a\pi}{16}, \frac{3a}{8}\right)$. **Ans.**

Exercises 5(IV)

1. Show that the depth below the surface of a liquid of the centre of pressure of a rectangle, two of whose sides are horizontal and at depths a and b is
$$\frac{2(a^2+ab+b^2)}{3(a+b)}.$$
[R.U. 96H]

2. A regular hexagon is immersed in water with one side in the surface, find the C.P. of the upper half.
[*Ans.* $\frac{7\sqrt{3}a}{20}$, where a is the length of each side of the hexagon]

3. A rhombus ABCD is completely immersed in a liquid with the vertex A in the surface and the diagonal AC through the vertex vertical. Prove that the C.P. divides the diagonal AC in the ratio 7 : 5.
[Bih. 58A, P.U. 58A, 65S, 76A, Bh.U. 69A, M.U. 63A, R.U. 67A, Gorakh. 59, Agra 52, 56, Nag. 54, Raj. 55, Alld. 54, Vikram 66, Sagar 62, Karnataka 62, Mith. 88H]

4. P is the C.P. of the rectangle ABCD, the side AB being on the surface. Prove that the line through A and P divides the area into two portions the pressures on which are equal.
[M.U. 70A, P.U. 41A, Delhi 54]

5. A triangle is wholly immersed in a liquid with its base in the surface. Show that a horizontal straight line drawn through the centre of pressure of the triangle divides it into two parts, the pressures on which are equal.
[C.U. 37, Delhi 56, Banaras 59, Agra 83]

6. A quadrilateral is immersed in water with two angular points in the surface and the other two at

depths a and b. If x and y are the depths below the surface of its centre of gravity and pressure, respectively, show that

$$6xy + ab = 3x(a + b).$$

[Gorakhpur 74, 78, Lucknow 74, 82, Magadh 74, Allahabad 62]

7. A rectangular lamina ABCD is vertically immersed in water with AB in the surface. Let O be the point of intersection of its diagonals. If the triangle AOB be removed, find the C.P. of the remainder. **[B.U. 54A, P.U. 64S, M.U. 65S]**

[*Ans.* The depth of the C.P. is $\frac{3}{4}BC$]

8. A square is immersed, with its diagonal vertical and its lowest point as deep again as its highest point. Find the depth of its C.P.

[M.U. 73A, P.U. 73S, R.U. 63A, B.U. 54A, T.M.B.U. 94H, 98H, 04H]

[*Ans.* $\frac{19}{12}$ times the depth of the highest point]

9. A plane triangular area is immersed in liquid of uniform density with its plane vertical, one side horizontal and the opposite corner downwards. Its vertical altitude is h and the horizontal side is at a depth h below the effective surface. Show that its C.P. is at a depth $\frac{11}{8}h$ below the surface.

10. Show that the depth of C.P. of a rhombus totally immersed, with one diagonal vertical and its centre at a depth h, is $h + \frac{a^2}{24h}$, where a is the length of the vertical diagonal.

[B.U. 70A, Bh.U. 65S, L.N.M.U. 93H]

11. Show that the C.P. of a rhombus immersed in two liquids which do not mix, with a vertex in the upper surface and a diagonal in the common surface, divides the other diagonal in the ratio 5 : 3, if the density of the lower liquid be three times that of the upper.

[Allahabad 58, Gorakhpur 63]

12. A circular area is immersed vertically in water, prove that if the depth of its centre is doubled, the distance between its centre and the C.P. will be halved.

[R.U. 69A, 72H, M.U. 76H]

13. A system of coaxial circles is immersed in water with the line of centres at a given depth. Prove that the centres of pressure of those circular areas which are completely immersed lie on a parabola.

[B.U. 57H, P.U. 42H, 68H, M.U. 65H]

14. A segment of a parabola, cut off by a double ordinate at a distance b from the vertex, is immersed with this ordinate in the surface of a liquid, find the depth of the C.P. of the lamina. Also find the resultant thrust.

[Bh.U. 72H, Agra 61, 81, Raj. 55, 74, Lucknow 76, Rohilkhand 78, 81]

[*Ans.* $\frac{4}{7}f$, $\frac{16}{15}\rho g b^{5/2}\sqrt{a}$]

15. One asymptote of a hyperbola lies in the surface of a fluid, find the depth of the C.P. of the area included between the immersed asymptote, the curve and two given horizontal lines in the plane of the hyperbola.

[*Ans.* The A.M. of the depths of the given horizontal lines]

[P.U. 66H, M.U. 64H, 79H]

16. If a quadrilateral area ABCD be entirely immersed in water, and $\alpha, \beta, \gamma, \delta$ be the depths of its four corners, and h that of its centre of gravity, show that the depth of the C.P. is

$$\frac{1}{2}(\alpha+\beta+\gamma+\delta+) - \frac{1}{6h}(\alpha\beta+\beta\gamma+\gamma\alpha + \alpha\delta+\beta\delta+\gamma\delta).$$

[P.U. 52H, B.U. 61H, 66H, 72H, R.U. 64H, B.P.S.C 99]

17. The C.P. of a parallelogram completely immersed is in one of its diagonals. Prove that the other diagonal is horizontal.

[P.U. 57H, 63H, 67H, M.U. 66H, 51H, 79H, B.U. 80H, M.U. 91H, 2004, L.N.M.U. 96H, T.M.B.U. 92H, 2000H, 03H]

18. Prove that the depth of the C.P. of a parallelogram, two of whose sides are horizontal and at depths h, k below the surface of a liquid whose density varies as the depth below the surface is

$$\frac{3}{4}\frac{h^3+h^2k+hk^2+k^3}{h^2+hk+k^2}.$$

[P.U. 48H, 50H, 62H, Bh.U. 68H, Agra 55]

19. A parallelogram whose plane is vertical and centre at a depth h below the surface, is totally immersed in a homogeneous fluid. Show that, if a, b be the lengths of the projections of the sides on a vertical line, depth of its C.P. will be

$$h+\frac{a^2+b^2}{12h}.$$

[I.A.S. 75]

20. A plane of quadrilateral ABCD is entirely immersed in water with side AB in the surface. If the depths of C and D are γ and δ, respectively and that of the C.P. is h, prove that the depth of the C.P. is

$$\frac{1}{2}(\gamma+\delta)-\frac{1}{6}\frac{\gamma\delta}{h}.$$

[C.U. 89H, 96H]

21. An isosceles triangle of vertical angle 2α is placed vertically with its vertex in the surface of a liquid and its base inclined at an angle θ to the horizon. Show that the depth of its C.P. is

$$l\cdot\frac{3\cos^2\theta\cos^2\alpha+\sin^2\theta\sin^2\alpha}{4\cos\theta\cos\alpha}$$

l being one of the equal sides.

22. A parallelogram has its corners at depths h_1, h_2, h_3, h_4 below the surface of a liquid and its centre at a depth h, show that the depth of its C.P. is

$$\frac{h_1^2+h_2^2+h_3^2+h_4^2+8h^2}{12h}.$$

[Vikram 65, Delhi 65, Lucknow 62, M.U. 74H, R.U. 79H, M.U. 04H, T.M.B.U. 94H]

23. A triangular lamina ABC, right angled at C; is just immersed in a fluid with the vertex C in the surface and the side CA inclined at 30° with the surface. If the C.P. is vertically below C, prove that tan B is a root of the equation $x^2+\sqrt{3}x=1$.

24. A rectangular lamina of which the sides are 2a and 2b is completely immersed in water which is rotating about a vertical axis with uniform angular velocity $\sqrt{\frac{ng}{b}}$, the plane of the lamina is vertical and one of its sides 2b touches the

surface of the water at its lowest point, the point of contact being the middle point of the side, and the lamina rotates with the same angular velocity as the water. Show that the depth of the C.P. of the lamina is

$$\frac{8a+6h+nb}{6a+6h+nb}a,$$

where h denotes the height of the water barometer.

25. Prove that, if a circular area of radius a is immersed with its centre at a depth h in a liquid in which the density varies as the depth, the plane of the area being vertical, the C.P. is at a depth $\dfrac{2a^2h}{a^2+4h^2}$ below the centre of the circle.

26. An area is bounded by two concentric semicircles with the common bounding diameter in the surface, show that the depth of its C.P. is

$$\frac{3\pi}{16}\frac{(a+b)(a^2+b^2)}{a^2+ab^2+b^2},$$

where a and b are the radii of semicircles.

[Raj. 56, Delhi Hons. 53, M.U. 75H, Agra 58, 63, 66]

27. From a semicircle whose diameter is in the surface of a liquid, a circle is cut out, whose diameter is the vertical radius of the semicircle. Show that the depth of the C.P. of the remainder is $\dfrac{9\pi a}{8(16-3\pi)}$.

28. A circular disc of radius a is completely immersed with its plane vertical in homogeneous fluid. If h is the depth of the centre below the free surface of the liquid, prove that the distance between the centre of pressures of the two semicircles into which the disc is divided by its horizontal diameter is $6\pi a\dfrac{4h^2-a^2}{9\pi^2h^2-16a^2}$.

[B.U. 68H, 76H]

29. Show that, if a lamina totally immersed in a liquid is a quadrant of a circle of radius a of which the centre is in the surface, the locus of the C.P.; in the lamina lies on a straight line of length $\dfrac{3a\sqrt{2}(\pi-2)}{16}$.

[P.U. 62H, 82H, B.U. 62H, 67H, Bh.U. 62H, C.U. 92H, 03H]

30. A circular sector of centre O and radius a, symmetrically about the radius ON, is completely immersed with N in the surface and ON vertical. Determine the depth of its centre of pressure.

[*Ans.* $\frac{a(15\alpha-16\sin\alpha+3\sin\alpha\cos\alpha)}{4(3\alpha-2\sin\alpha)}$]

31. Find the C.P. of a semicircular area, the diameter being horizontal and at depth h the radius being a.

[B.U. 73H, M.U. 82H]

[*Ans.* $\frac{16ah+3\pi a^2}{4(3\pi h+4a)}$]

32. The axis of a cylindrical vessel containing liquid is inclined to the vertical at an angle. Prove that the distance of the C.P. of the base from the centre of the base is $\frac{r^2}{4h}\tan\alpha$, where r is the radius of the base and πr^2h is the volume of the liquid inside.

33. An ellipse is just immersed in water with its major axis vertical. If the C.P. is at the focus, find the eccentricity of the ellipse.

[B.U. 64H, 72A, M.U. 68H, 69H, P.U. 77H, Agra 80, Bhopal 80, Lucknow 80, Raj. 75, 77, R.U. 81, I.A.S. 99, T.M.B.U. 02H]

[*Ans.* $\frac{1}{4}$]

34. A uniform elliptic lamina whose axes are $2a$ and $2b$, is half immersed in water, the axis $2b$ in the surface. Find the depth of the C.P.

[Bh.U. 70H, Indore 70, Raj. 63, Nag. 51, Alld. 53, 76, Bhopal 82]

[*Ans.* $\frac{3\pi a}{16}$]

35. An elliptic disc is just immersed in a homogeneous liquid with its plane vertical; show that the locus of the disc of its C.P. is another ellipse similar, and similarly situated, to the given ellipse.

[M.U. 91H]

36. An area bounded by the curve $ay^2 = x^2$, an abscissa of length h and the ordinate at its extremity is placed in water with the ordinate in the surface. Prove that the depth of the C.P. is $\frac{4h}{9}$.

[B.U. 53H, R.U. 66H, 71H]

37. Find the coordinates of the C.P. of the area bounded by $ay^2 = x^2$, the x-axis and the ordinate $x = a$, is immersed in water with the axis in the surface. **[Lucknow 77]**

[*Ans.* $\left(\frac{4a}{5}, \frac{16a}{33}\right)$]

38. An inclined plane surface totally immersed in a liquid is lowered, always remaining parallel to the former position, with its centroid moving along a vertical straight line. Show that the locus of the C.P. is a hyperbola. **[B.U. 69H]**

39. A parallelogram is immersed vertically in a liquid with a corner in the surface. If a and b be the depths of the adjacent corners, prove that the depth of the C.P. is

$$\frac{2a^2 + 3ab + 2b^2}{3(a+b)}.$$

[M.U. 70H, Bh.U. 71A, B.B.A.B.U. 92H]

40. Show that the depth of the centre of pressure of the area included between the arc and the asymptote of the curve $(r-a)\cos\theta = b$ is $\frac{a}{4}\frac{3\pi a + 16b}{3\pi b + 4a}$, the asymptote being in the surface and the plane of the curve vertical.

[Vikram 64, Agra 57, 52, 67, Raj. 65]

41. A plane lamina consists of a circular disc (radius a and centre O) from which a circular portion (radius $\frac{a}{2}$ and centre P) has been cut. The lamina is completely immersed in a homogeneous fluid with its plane vertical and P vertically below O. If OP is equal to $\frac{a}{2}$ and the C.P. of the lamina is at O, prove that O is at a depth $\frac{11a}{8}$ below the surface of the fluid.

[P.U. 80H]

42. A quadrilateral ABCD, with sides AB, CD parallel (and AD, BC equal) is immersed in a homogeneous liquid with AB in the free surface. Show that the centre of pressure will be the intersection of the diagonals, if $AB = \sqrt{3}\,CD$.

[Delhi Hons. 60]

43. The angular points of a triangle immersed in a liquid whose density varies as the depth, are at distances α, β, γ, respectively below the surface. Show that the centre of pressure is at a depth

$$\frac{3}{5}\frac{(\alpha+\beta+\gamma)(\alpha^2+\beta^2+\gamma^2)+\alpha\beta\gamma}{\alpha^2+\beta^2+\gamma^2+\alpha\beta+\beta\gamma+\gamma\alpha}.$$

[M.U. 80H]

44. If the centre of pressure of a triangle ABC, completely immersed, coincide with the circumcentre;

show that the depth of the angular points are as

$$1 - 2\cot B \cot C :$$
$$1 - 2\cot C \cot A :$$
$$1 - 2\cot A \cot B.$$

45. Find the C.P. of a quadrant of a circle immersed with its plane vertical in a heavy homogeneous liquid with its bounding radius on the free surface, the radius of the circle being a. **[C.U. 87H, 90H]**

[*Ans.* $\left(\frac{3}{16}\pi a, \frac{3}{8}a\right)$]

46. A rectangle is immersed in fluids of density $\rho, 2\rho, \ldots, n\rho$, the top of the rectangle being in the surface of the first liquid and the area immersed in each being the same. Show that the depth of the C.P. of the rectangle is $\frac{3n+1}{2n+1}\frac{h}{2}$, where h is the depth of the lower side.

[C.U. 84H]

47. A parallelogram is immersed vertically on a liquid with a corner in the surface. If x and y be the depths of the adjacent corners, find the depth of its C.P.

[Cal 82H, 85H, M.U. 73H]

[*Ans.* $\frac{2x^2+3xy+2y^2}{3(x+y)}$]

CHAPTER 6

Resultant Thrust on Curved Surface

6.1 Introduction

When a curved surface is in contact with a fluid, the thrusts on different elements of the surface into which it is divided are in different directions at different points and hence, unlike the thrust on plane surface, the resultant thrust cannot reduce to a single force, but is equivalent to a single force together with a single couple. By resolving the thrust on each element of the surface into vertical and horizontal components, we can find forces to which the resultant thrust on the surface is equivalent.

6.2 The Resultant Thrust on a Curved Surface

[C.U. 99H, R.U. 91H, 93H]

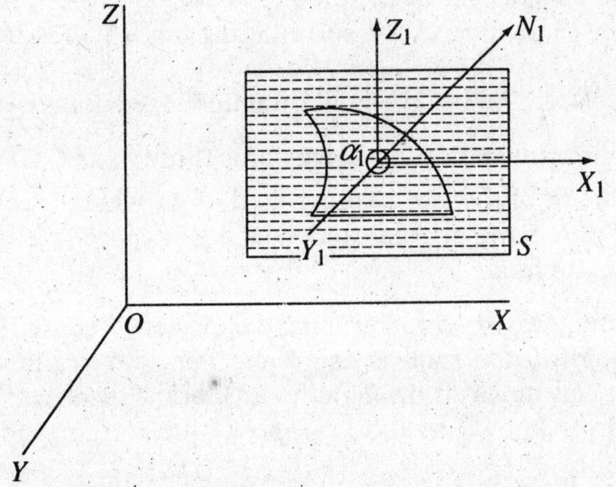

Fig 6.1

Let us divide the curved surface S into a number of elements $\alpha_1, \alpha_2, \alpha_3, \ldots$. Let N_1 be the normal thrust on element α_1.

Since the normal pressures at various points of the curved surface form a system of force in a three-dimensional space, each of these can be resolved into three-components in the direction of three-mutually perpendicular lines, say OX, OY, OZ, OX and OY being horizontal and OZ being vertical.

So, we resolve N_1 in the directions of OX, OY and OZ. Let components of N_1 be X_1, Y_1 and Z_1. Similarly, we can resolve the normal thrusts at other points of the surface into components parallel to the assigned directions OX, OY and OZ.

The vertical components being all parallel can be compounded into a single resultant thrust Z. This thrust is known as *Resultant Vertical Thrust.* **[R.U. 94H]**

Similarly, the components parallel to OX and OY can be compounded into single resultant thrusts X and Y. The resultants X and Y are called the *Resultant Horizontal Thrust* on the surfaces parallel to the given directions.

If these two horizontal resultant thrusts X and Y meet in a point, they can be combined into a single resultant thrust known as *Resultant Horizontal Thrust*. If the line of action of the resultant vertical thrust Z also passes through the point of intersection of resultant horizontal thrusts, then they can be compounded by the Law of Parallelepiped of Forces into a single resultant $R = \sqrt{X^2 + Y^2 + Z^2}$ called the *Resultant Thrust* on the curved surface.

If R be inclined at an angle α to the vertical, then

$$\tan \alpha = \frac{\sqrt{X^2 + Y^2}}{Z}.$$

In general, the forces X, Y and Z are not concurrent and hence a single resultant thrust is not possible.

6.3 Resultant Thrust on a Solid Wholly or Partially Immersed in Heavy Fluid at Rest

Let us consider the equilibrium of a solid body wholly or partially immersed in a fluid at rest. We imagine the body to be removed and its space is occupied by the same fluid of the same kind as the surrounding fluid. This extra fluid is called the *displaced fluid.*

Now the displaced fluid is in equilibrium under two forces:

- weight acting vertically downwards through its C.G. and
- the thrust of the surrounding fluid on it, which is equal to the thrust of the fluid on the original solid which is supposed to have been replace by this extra fluid.

For equilibrium these two forces must balance each other. Hence, the resultant thrust of the fluid on the solid is equal and opposite to the weight of the fluid displaced and it acts upwards through the C.G. of the displaced fluid.

This theorem is known as the *Principle of Archimedes* and may be stated as follows.

The resultant thrust on a solid body, wholly or partially immersed in a fluid at rest, is equal to the weight of the fluid displaced by the body and acts vertically upwards through the centre of gravity of the fluid displaced.

[R.U. 92H, 95H, C.U. 91H, 98H, 200H, 03H, R.U. 92H, 95H]

Note 1. *This theorem is true whatever be the nature of the fluid in which the body is immersed. The fluid may be of varying density. In such a case we take the displaced fluid of the same density as the surrounding one.*

Force of Buoyancy

The resultant fluid thrust on a body wholly or partially immersed in a fluid is called the *Force of Buoyancy.* Thus, the force of buoyancy is equal to the weight of the liquid displaced.

Centre of Buoyancy

The centre of gravity of the displaced fluid is called the *Centre of Buoyancy.* This is the point through which the force of buoyancy acts.

6.4 To Find the Resultant Vertical Thrust on any Surface exposed to the Pressure of Heavy Fluid at Rest under Gravity

[R.U. 93H]

We will consider three cases:

Case I. *When the fluid is above the surface.*

Let the curved surface PQRS be immersed in a homogeneous liquid at rest under gravity.

We draw vertical lines through every point of the bounding curve of the surface so as to meet the free surface of the liquid in a curve ABCD. We get a cylinder PQRS, ABCD. The liquid in this cylinder is called the *Superincumbent Liquid.*

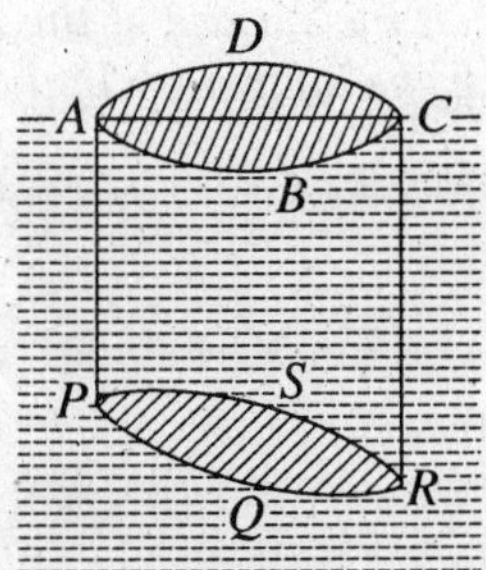

Fig. 6.2

Let us consider the vertical equilibrium of the cylinder of superincumbent liquid. It is in equilibrium under the following vertical forces:

- The weight W of the liquid cylinder (i.e., the superincumbent liquid) acting vertically downwards through its centre of gravity.
- The vertical component R of the reaction between liquid and the curved surface PQRS, acting vertically upwards.

But the reaction of the surface on the fluid is equal and opposite to the thrust of the fluid on the surface.

Therefore, for equilibrium, the vertical component of the reaction which is equal and opposite to the thrust of the liquid, is equal to the weight of the superincumbent liquid acting through its centre of gravity.

Hence, *the Resultant Vertical Thrust on any surface immersed in a liquid is equal to the weight of the superincumbent liquid and acts through the centre of gravity of this super incumbent liquid.* [C.U. 89H, 91H, M.U. 90H, R.U. 80H]

Case II. *When the liquid is below the surface.*

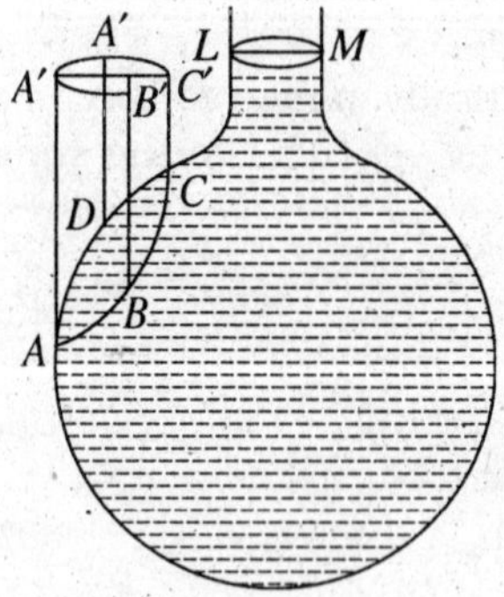

Fig. 6.3

Here we suppose that the fluid is contained in a vessel and we are required to find the vertical thrust on the portion ABCD in the given figure. In this case, the fluid presses the surface upwards.

We draw vertical lines through every point of the boundary of the surface ABCD to meet the effective surface LM in A′B′C′D′. Since the pressure at a point depends upon its depth below the free surface. It is clear that the resultant vertical thrust on ABCD is equal to the weight of the liquid which would fill the cylinder ABCD· A′B′C′D′.

Hence, in this case, *the resultant vertical thrust on a surface being pressed by the liquid from below, is equal to the weight of the liquid that could lie upon it up to the level of the surface of the liquid and acts vertically upwards through the centre of gravity of the imaginary liquid.*

Case III. *When the liquid is partly above and partly below the surface.*

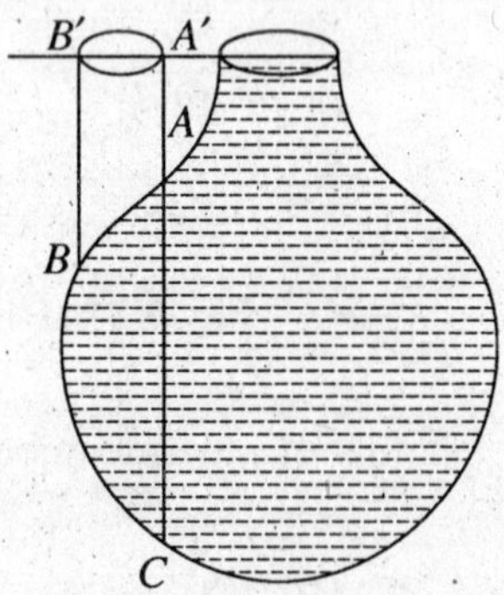

Fig. 6.4

Let ABC be the curved surface, the portion AB pressed upwards and the portion BC pressed downwards: by the liquid above and below these portions, respectively.

On the portion AB the vertical thrust is upwards and is equal to the weight of the liquid contained in the cylinder AB· B′A′.

On the portion BC the vertical thrust is downwards and is equal to the weight of the liquid contained in the cylinder CB· B′A′. Therefore, the resultant vertical thrust on the portion ABC is the difference of these two weight, i.e.,

= difference of the above two weights,
= weight of the liquid ABC,
= weight of the liquid contained in ABC, acting downwards.

Hence, *the Resultant Vertical Thrust on a surface pressed partly upwards and partly downwards by a liquid is equal to the weight of the liquid contained in that surface and acts vertically downwards through the C.G. of the liquid.*

6.5 To Find the Resultant Horizontal Thrust on a Current Surface in a Given Direction

Let ABCD be the curved surface in contact with the liquid, for which the resultant thrust in horizontal direction is to be determined.

Through every point of the boundary of the curved surface ABCD we draw horizontal lines so as to meet the vertical plane perpendicular to their direction in the closed curve $A'B'C'D$. Then closed curve $A'B'C'D'$ is the projection of the closed surface ABCD on the vertical plane.

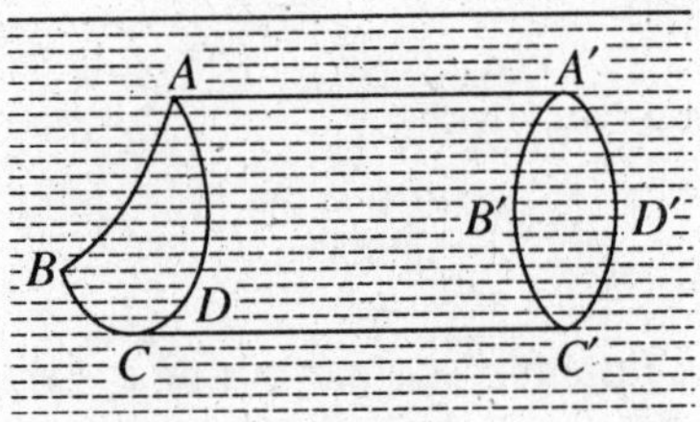

Fig. 6.5

Let us consider the equilibrium of the fluid enclosed between the curved surface ABCD, its projection $A'B'C'D'$ and the cylindrical surface generated by the horizontal lines. The only horizontal forces acting on this portion of the liquid are:

- The horizontal thrust on the curved surface ABCD,
- The resultant thrust on the plane area $A'B'C'D'$ acting at its centre of pressure in the direction of its normal (i.e., in the given horizontal direction).

For equilibrium these two thrusts must balance one another for which they must be equal in magnitude and opposite in direction.

Hence, *the Resultant Horizontal Thrust on a curved surface in contact with a liquid in a given direction is equal in magnitude and line of action to the whole pressure on the projection of the surface upon a vertical plane perpendicular to the given direction.*

6.6 To Find the Resultant Pressure on any Surface of a Fluid at Rest under the Action of any Given Forces.

Let MN be any curved surface and $P(x, y, z)$ be any point on it. Let p be the pressure at P so that thrust on an elementary area ds at (x, y, z) is $p\,ds$.

These thrusts on elementary areas are normal to the surface at their respective points but these normals are not all parallel as the surface is curved, hence these thrusts may not necessarily have a single resultant. There may be several couples and several forces when resolved along suitable directions. Let l, m, n be the direction cosine of the normal at (x, y, z), then the components X_1, Y_1, Z_1 of the pressure on this element parallel to axes are: $X_1 = p\,ds \cdot l = pl\,ds$, $Y_1 = pm\,ds$, $Z_1 = pn\,ds$.

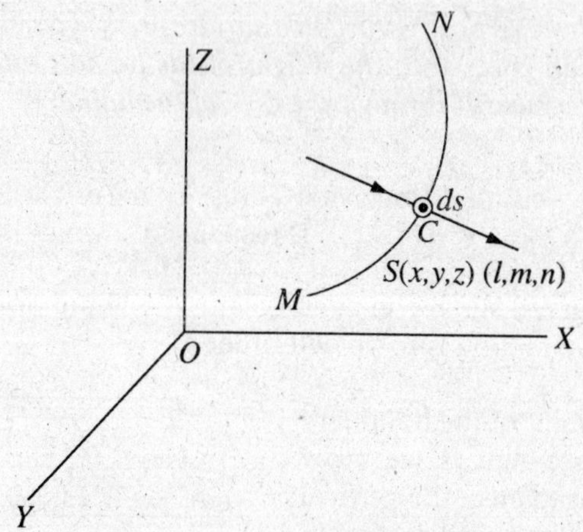

Fig. 6.6

All the forces constituting the thrust at separate points may be combined to form three forces X, Y, Z and the three couples L, M, N along three rectangular axes, then

$$\begin{aligned} X &= \Sigma X_1 = \Sigma pl\,ds = \int pl\,ds \\ Y &= \int pm\,ds \\ Z &= \int pn\,ds \\ L &= \Sigma(y_1 Z_1 - z_1 Y_1) = \Sigma(ypn\,ds - zpm\,ds) = \int p(yn - zm)\,ds \\ M &= \int p(zl - xn)\,ds \\ N &= \int p(xm - yl)\,ds \end{aligned}$$

the integration being taken such as to include the whole surface under consideration.

The resultant thrust will be a single force if $X^2 + Y^2 + Z^2$ is not equal to zero but $LX + MY + NZ = 0$.

It will reduce to a wrench whose pitch $= \frac{LX+MY+NZ}{X^2+Y^2+Z^2}$ and the equation to central axis is

$$\frac{L - yZ + zY}{X} = \frac{M - zX + xZ}{Y} = \frac{N - xY + yX}{Z}.$$

6.7 Resultant Thrust on the Curved Surface of a Solid bounded by a Plane Base

[R.U. 94H]

Let PQR be a curved surface bounded by a plane area PSR. Then the resultant thrust on PQR together with the thrust on the base PSR will be equivalent to the resultant thrust on the whole body which, according to Archimedes principle, balances the weight W of the displaced liquid.

Therefore, if the vertical and horizontal components of the thrust on the plane base be X and Y, respectively, the thrusts on the curved surface PQR will be $W - X$ and $-Y$ in the horizontal and vertical directions.

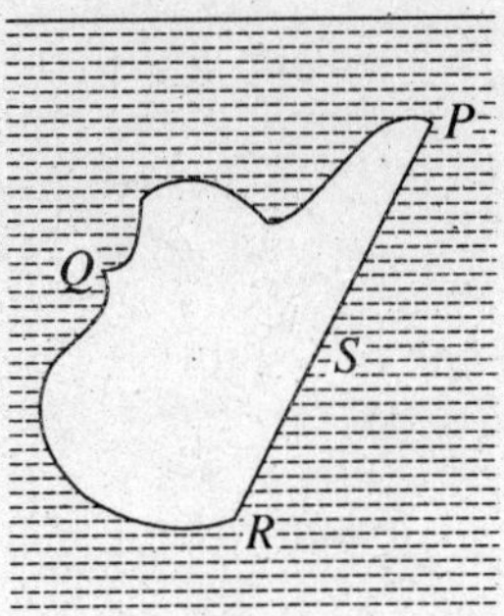

Fig. 6.7

$\therefore$ the resultant is $\sqrt{(W-X)^2 + Y^2}$ inclined to the vertical at an angle $\tan^{-1} \frac{Y}{W-X}$.

Illustrative Examples

▶ **Ex. 1.** A hollow cone is placed with its vertex upwards on a horizontal table and a liquid is poured in through a small hole in the vertex, if the cone begins to rise when the weight of the liquid poured in it is equal to its own weight, prove that its weight is to the weight of the liquid to fill the cone as $9 - 3\sqrt{3} : 4$.

[P.U. 54A, 61S, 81H, Bh.U. 63A, R.U. 69H, Delhi 62, Raj. 62, Allahabad 62, R.U. 96H]

Solution: Let OAB be the cone of height h and semivertical angle α. Through a small hole at O liquid is poured in. The cone begins to rise when the vertical thrust on the surface of the cone in contact with the liquid is equal to the weight of the liquid poured. Let x be the height of the liquid poured, i.e., $OQ = x$.

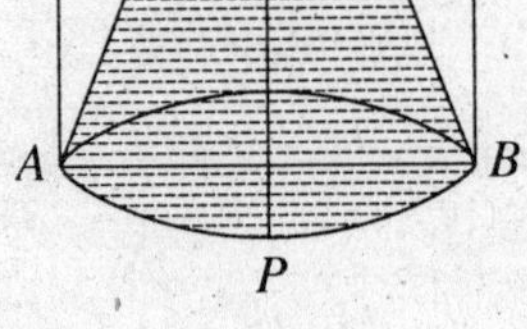

Fig. 6.8

The weight of the liquid poured in

$$= \text{the weight of the liquid which would fill the cone OAB} - \text{weight of the liquid which would fill the cone OCD}$$

$$= \frac{\pi}{3} \tan^2 \alpha (h^3 - x^3)\rho g, \text{ where } \rho \text{ is the density of the liquid.}$$

And the thrust on the curved surface ABCD

= weight of the superincumbent liquid

= weight of the liquid contained in right circular cylinder ABEF −the weight of the liquid poured in

$= \pi h^2 \tan^2 \alpha (h - x)\rho g - \frac{\pi}{3} \tan^2 \alpha (h^3 - x^3)\rho g.$

For the cone to begin rising weight of the liquid poured in = thrust on the curved surface ABCD

$$\frac{\pi}{3} \tan^2 \alpha (h^3 - x^3)\rho g = \rho h^2 \tan^2 \alpha (h - x)\rho g - \frac{\pi}{3} \tan^2 \alpha (h^3 - x^3)\rho g$$

or, $\frac{(h^3 - x^3)}{3} = h^2(h - x) - \frac{(h^3 - x^3)}{3}$

or, $\frac{2}{3}(h^3 - x^3) = h^2(h - x)$

or, $2(h^3 - x^3) - 3h^2(h - x) = 0$

or, $2(h - x)(h^2 + hx + x^2) - 3h^2(h - x) = 0$

or, $(h - x)[2h^2 + 2hx + 2x^2 - 3h^2] = 0$

or, $(h - x)(2x^2 + 2hx - h^2) = 0$

$\therefore \quad 2x^2 + 2hx - h^2 = 0$ since $h \neq x$

$\therefore \quad x = \frac{-2h \pm \sqrt{4h^2 + 8h^2}}{2 \cdot 2} = \frac{-h \pm \sqrt{3}h}{2} = \frac{(\sqrt{3} - 1)h}{2}$

[$\because$ x cannot be negative].

Hence,

$$\frac{x}{h} = \frac{\sqrt{3} - 1}{2}. \quad (1)$$

Hence,

$$\frac{\text{weight of the liquid poured in}}{\text{weight of the liquid to fill the cone}}$$

$$= \frac{\frac{\pi}{3} \tan^2 \alpha (h^3 - x^3)\rho g}{\frac{\pi}{3} \tan^2 \alpha h^3 \rho g} = \frac{h^3 - x^3}{h^3} = 1 - \left(\frac{x}{h}\right)^3$$

$$= 1 - \left(\frac{\sqrt{3} - 1}{2}\right)^3 \quad \text{From eqn. (1)}$$

$$= 1 - \frac{1}{8}\left[3\sqrt{3} - 3.3 + 3\sqrt{3} - 1\right] = 1 - \frac{1}{8}\left(6\sqrt{3} - 10\right)$$

$$= \frac{8 - 6\sqrt{3} + 10}{8} = \frac{18 - 6\sqrt{3}}{8} = \frac{9 - 3\sqrt{3}}{4}.$$ **Proved.**

▶ **Ex. 2.** A right circular cone, closed by a plane base is held with its axis horizontal and is full of water, find the resultant vertical thrust on the lower and upper halves of the curved surface.

[P.U. 51S, 72A, M.U. 66S, 69A, R.U. 69A, 72A, R.U. 91H, Raj. 64, Gorakhpur 60]

Solution: Let CEO′ be the cone of height h and radius r. The horizontal plane CAB, through the vertex C and the axis CO of the cone, divides the cone into two halves.

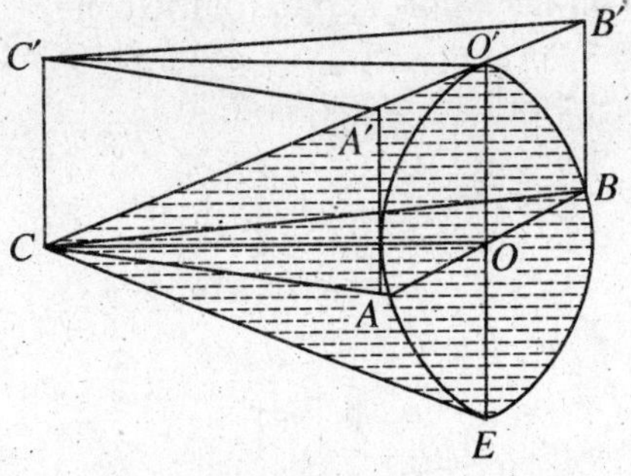

Fig. 6.9

$\therefore$ Vertical thrust on the upper half CAO′B of the curved surface

= weight of the super incumbent liquid which would fill the space between the circumscribing prism CABB′A′C′ and the upper half CO′AB of the cone.

$$\begin{aligned}
&= \text{weight of the prism on } \triangle CAB - \text{weight of upper half of the cone}\\
&= \triangle CAB\, OO'\rho g - \frac{1}{2}\cdot\frac{1}{3}\pi r^2 h\rho g\\
&= \frac{1}{2}AB\cdot CO\, OO'\rho g - \frac{1}{6}\pi r^2 h\rho g\\
&= = \frac{1}{2}2rhr\rho g - \frac{1}{6}\pi r^2 h\rho g\\
&= r^2\rho gh - \frac{\pi r^2 h}{6}\rho g = r^2\rho gh\left(1-\frac{\pi}{6}\right) \quad \text{acting vertically upwards.}
\end{aligned}$$

Similarly, vertical thrust on the lower half CAEB of the curved surface

$$\begin{aligned}
&= \text{weight of the superincumbent liquid } CC'A'B'E\\
&= \text{weight of the prism on } \triangle CAB + \text{weight of lower half of the cone}\\
&= r^2\rho gh + \frac{1}{6}\pi r^2\rho gh = \rho ghr^2\left(1+\frac{\pi}{6}\right).
\end{aligned}$$

▶ **Ex. 3.** A solid right circular cone of vertical angle 2α is immersed in water with one generator in the surface of the liquid. Prove that the resultant pressure on the curved surface of the cone is to the weight of the fluid displaced by the cone as $\sqrt{1+3\sin^2\alpha} : 1$.

Further, show that it is inclined to the axis of the cone of an angle $\cot^{-1}(2\tan\alpha)$ while its inclination to the vertical is given by

$$\tan^{-1}\frac{3\tan\alpha}{1-2\tan^2\alpha}.$$

[I.A.S. 83, P.U. 68H, Bh.U. 68H, B.U. 55A, 62H, C.U. 88H, 04H]

Solution: Let h be the height of the cone, then the radius of the base of the cone will be $h\tan\alpha$.

Now the resultant thrust on the whole surface of the cone = W = the weight of the liquid displaced

$$= \frac{1}{3}\pi h^2 \tan^2 \alpha \rho g h = \frac{1}{3}\pi h^3 \tan^2 \alpha \rho g.$$

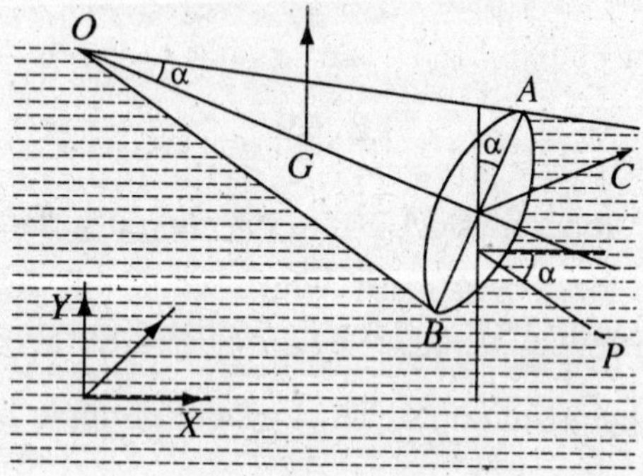

Fig. 6.10

Which acts vertically upwards through the C.G. of the liquid displaced. The thrust P on the plane base of the cone

$$\begin{aligned} &= \pi h^2 \tan^2 \alpha CA \cos \alpha \rho g \\ &= \pi h^3 \tan^2 \alpha h \tan \alpha \cos \alpha \rho g \\ &= \pi h^3 \tan^2 \alpha \rho g \cos \alpha \end{aligned}$$

which acts at right angle to the plane base of the cone.

If X and Y be the horizontal and vertical components of the thrusts on the curved surface of the cone, then

$$\begin{aligned} X &= P \cos \alpha = \pi h^3 \tan^3 \alpha \cos^2 \alpha \rho g \\ Y &= W - P \sin \alpha \\ &= \frac{\pi h^3}{3} \tan^2 \alpha \rho g - \pi h^3 \tan^3 \alpha \cos \alpha \rho g \sin \alpha \\ &= \frac{\pi}{3} h^3 \tan^2 \alpha \rho g \left(1 - \frac{3 \sin \alpha}{\cos \alpha} \cos \alpha \sin \alpha\right) \\ &= \frac{\pi h^3}{3} \tan^2 \alpha \rho g (1 - 3 \sin^2 \alpha). \end{aligned}$$

Therefore, the resultant thrust on the curved surface

$$
\begin{aligned}
&= \sqrt{X^2 + Y^2} \\
&= \sqrt{(\pi h^3 \tan^2 \alpha \rho g)^2 \left\{ \tan^2 \alpha \cos^4 \alpha + \frac{1}{9}(1 - 3\sin^2 \alpha)^2 \right\}} \\
&= \frac{\pi h^3 \tan^2 \alpha \rho}{3} g \sqrt{9 \sin^2 \alpha \cos^2 \alpha + 1 - 6 \sin^2 \alpha + 9 \sin^4 \alpha} \\
&= \frac{\pi h^3}{3} \tan^2 \alpha \rho g [9 \sin^2 \alpha (1 - \sin^2 \alpha) + 1 - 6 \sin^2 \alpha + 9 \sin^4 \alpha]^{1/2} \\
&= \frac{\pi h^3}{3} \tan^2 \alpha \rho g [3 \sin^2 \alpha + 1]^{1/2}.
\end{aligned}
$$

Hence, required ratio

$$
\begin{aligned}
&= \frac{\text{resultant pressure on the curved surface}}{\text{weight of the liquid displaced by the cone}} \\
&= \frac{\frac{\pi h^3}{3} \tan^2 \alpha \rho g \sqrt{1 + 3 \sin^2 \alpha}}{\frac{\pi}{3} h^3 \tan^2 \rho g} \\
&= \frac{\sqrt{1 + 3 \sin^2 \alpha}}{1}.
\end{aligned}
$$

Again, if θ be the inclination with the horizontal, of the resultant thrust on the curved surface, then

$$
\begin{aligned}
\tan \theta &= \frac{Y}{X} = \frac{\pi h^3 \tan^2 \alpha \rho g (1 - 3 \sin^2 \alpha)}{3 \pi h^3 \tan^3 \alpha \cos \alpha \rho g} \\
&= \frac{1 - 3 \sin^2 \alpha}{3 \tan \alpha \cos^2 \alpha} = \frac{(1 - 3 \sin^2 \alpha) \tan \alpha}{3 \tan^2 \alpha \cos^2 \alpha} \\
&= \frac{(1 - 3 \sin^2 \alpha) \tan \alpha}{3 \sin^2 \alpha} = \frac{1}{3} \tan \alpha [\operatorname{cosec}^2 \alpha - 3] \\
&= \frac{1}{3} \tan \alpha [1 + \cot^2 \alpha - 3] = \frac{1}{3} \tan \alpha [\cot^2 \alpha - 2] \\
&= \frac{1}{3} (\cot \alpha - 2 \tan \alpha). \qquad (1)
\end{aligned}
$$

Hence, the inclination of the resultant, to the axis of the cone is $\theta + \alpha$, which is given by

$$
\begin{aligned}
\tan(\theta + \alpha) &= \frac{\tan \theta + \tan \alpha}{1 - \tan \theta \tan \alpha} = \frac{\frac{1}{3}(\cot \alpha - 2 \tan \alpha) + \tan \alpha}{1 - \frac{\cot \alpha - 2 \tan \alpha}{3} \cdot \tan \alpha} \\
&= \frac{\cot \alpha - 2 \tan \alpha + 3 \tan \alpha}{3 - (1 - 2 \tan^2 \alpha)} = \frac{\cot \alpha + \tan \alpha}{2(1 + \tan^2 \alpha)} \\
\tan(\theta + \alpha) &= \frac{\frac{1}{\tan \alpha} + \tan \alpha}{2(1 + \tan^2 \alpha)} = \frac{1}{2 \tan \alpha}.
\end{aligned}
$$

$$\therefore \quad \cot(\theta + \alpha) = 2\tan\alpha$$
$$\theta + \alpha = \cot^{-1}(2\tan\alpha).$$

If the resultant thrust makes an angle with the vertical, then

$$\tan\beta = \frac{X}{Y} = \frac{3}{\cos\alpha - 2\tan\alpha} \quad \text{[From eqn. (1)]}$$
$$= \frac{3\tan\alpha}{1 - 2\tan^2\alpha}.$$

▶ **Ex. 4.** A hemisphere of radius a is entirely immersed in a liquid of density ρ so that its diametral plane makes an angle θ with the horizontal and its centre is at a depth h. Find the resultant thrust on the curved surface and determine its direction also. **[I.A.S. 89, Agra 62, Lucknow 60, Delhi 72]**

Solution:

Case I. *Let the base of the hemisphere be upwards.*

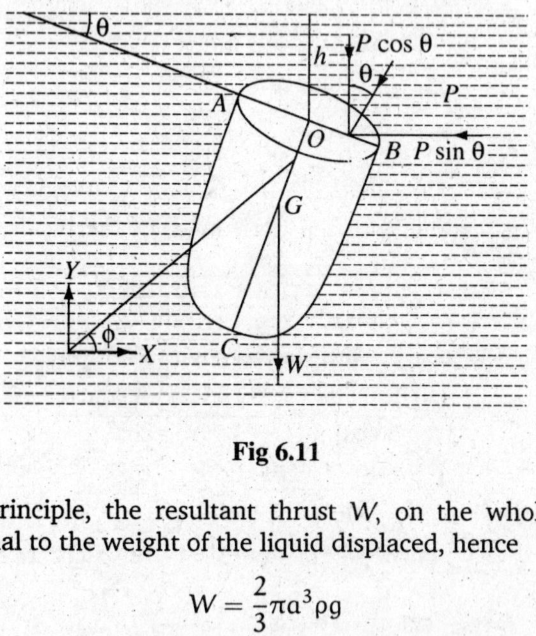

Fig 6.11

By Archimedes principle, the resultant thrust W, on the whole surface of the hemisphere is equal to the weight of the liquid displaced, hence

$$W = \frac{2}{3}\pi a^3 \rho g$$

which acts vertically upwards through the C.G. G of the hemisphere.

Let P be the thrust on the plane base of the hemisphere $= \rho g h \pi a^2$ acting at C.P. of the plane base.

Let X and Y be the horizontal and vertical components of the thrust on the curved surface of the hemisphere, then

$$X = P\sin\theta = \pi a^2 \rho g h \sin\theta$$

$$\text{and} \quad Y = W + P\cos\theta = \frac{2}{3}\pi a^3 \rho g + \rho g h \pi a^2 \cos\theta$$
$$= \rho g \frac{\pi a^2}{3}(2a + 3h\cos\theta).$$

Hence, resultant thrust on the curved surface

$$= R = \sqrt{X^2 + Y^2}$$

$$= \rho g \pi a^2 \sqrt{h^2 \sin^2\theta + \frac{1}{9}(2a + 3h\cos\theta)^2}$$

$$= \rho g \pi a^2 \sqrt{h^2 \sin^2\theta + \frac{1}{9}(4a^2 + 12ah\cos\theta + 9h^2\cos^2\theta)}$$

$$= \rho g \pi a^2 \left[h^2 + \frac{4}{9}a^2 + \frac{4}{3}ah\cos\theta\right]^{1/2}$$

If ϕ be the angle which the direction of the resultant thrust makes with the horizontal, then

$$\tan\phi = \frac{Y}{X} = \frac{\pi a^2 \rho g(2a + 3h\cos\theta)}{3\pi a^2 \rho g h\sin\theta} = \frac{2a + 3h\cos\theta}{h3\sin\theta}$$

$$\therefore \quad \phi = \tan^{-1}\frac{2a + 3h\cos\theta}{3h\sin\theta}.$$

Case II. *Let the base of the hemisphere be downwards.*

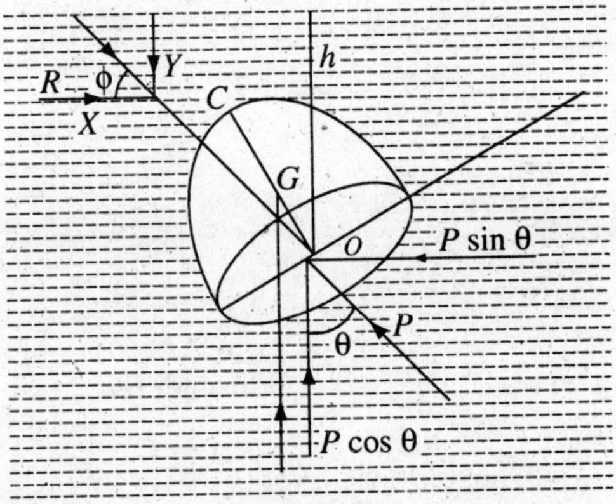

Fig. 6.12

In this case, we have

$$X = P\sin\theta = \pi a^2 h\rho g\sin\theta$$

$$\text{and} \quad Y = P\cos\theta - W = \pi a^2 g\rho h\cos\theta - \frac{2}{3}\pi a^3 \rho g$$

$$= \frac{\pi a^2}{3}\rho g(3h\cos\theta - 2a).$$

Hence, resultant thrust on the curved surface

$$= R = \sqrt{X^2 + Y^2}$$

$$= \pi a^2 \rho g \left[h^2 + \frac{4}{9}a^2 - \frac{4}{3}ah\cos\theta\right]^{1/2} \quad \text{[as in Case I]}$$

and inclination of R is given by

$$\tan\phi = \frac{Y}{X} = \frac{3h\cos\theta - 2a}{3h\sin\theta} \quad \text{[as in Case I]}.$$

$$\therefore \quad \phi = \tan^{-1}\frac{3h\cos\theta - 2a}{3h\sin\theta}.$$

▶ **Ex. 5.** A portion of a sphere cut off by two planes through its centre inclined at an angle $\frac{\pi}{4}$ is just immersed in a liquid with one face in the surface. Find the resultant thrust on the curved surface and show that it makes an angle θ with the horizontal so that $\tan\theta = \frac{\pi}{2} - 1$.

[I.A.S. 74, P.U. 65H, 67H, B.U. 64H, R.U. 65H, 68H, P.U. 95H]

Solution: Let OAB be a section of the portion of the sphere of radius r by a plane perpendicular to the given planes and dividing the portion in two equal parts.

Area of the plane face

$$OB = \frac{1}{2}\pi r^2.$$

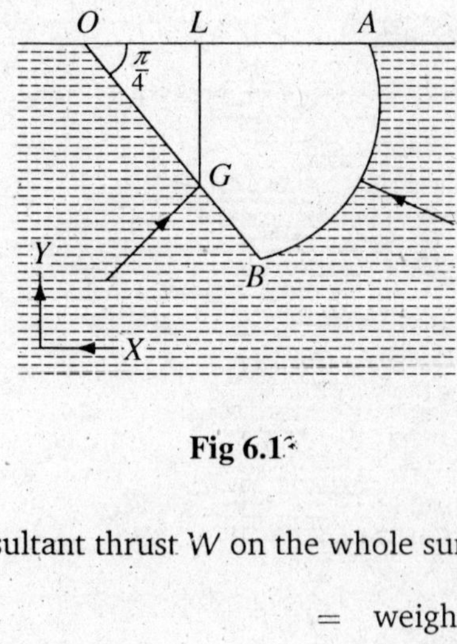

Fig 6.1?

$\therefore$ The thrust on the plane face OB

$$= \quad P = \frac{1}{2}\pi r^2 \rho g LG = \frac{1}{2}\pi r^2 \rho g OG \sin\frac{\pi}{4}$$

$$= \quad \frac{1}{2}\pi r^2 \rho g OG \frac{1}{\sqrt{2}} = \frac{1}{2}\pi r^2 \rho g \frac{4r}{3\pi}\frac{1}{\sqrt{2}}$$

$$= \quad \frac{\sqrt{2}}{3} r^3 \rho g$$

which act normally to the plane face OB, through the C.G. of the face OB.

Resultant thrust W on the whole surface of the portion

$$= \quad \text{weight of the liquid displaced}$$

$$= \quad \frac{2}{3}\cdot\frac{1}{4}\pi r^3 \rho g = \frac{1}{6}\pi r^3 \rho g$$

which act vertically upwards through the C.G. of the liquid displaced.

Let X and Y be the horizontal and vertical components of the thrusts on the curved surface of the portion, then

$$X = P\cos\frac{\pi}{4} = \frac{\sqrt{2}}{3} r^3 \rho g \frac{1}{\sqrt{2}} = \frac{1}{3} r^3 \rho g$$

$$\text{and} \quad Y = W - P\sin\frac{\pi}{4} = \frac{1}{6}\pi r^3 \rho g - \frac{\sqrt{2}}{3} r^3 \rho g \frac{1}{\sqrt{2}}$$

$$= \quad \frac{1}{6}\pi r^3 \rho g - \frac{r^3 \rho g}{3} = \frac{r^3}{6}\rho g(\pi - 2).$$

Hence, resultant thrust on the curved surface

$$= \quad R = \sqrt{X^2+Y^2} = \sqrt{\left(\frac{r^3}{3}\rho g\right)^2 + \left(\frac{r^3}{6}\rho g\right)^2 \cdot (\pi-2)^2}$$

$$= \quad \frac{r^3\rho g}{3} \cdot \sqrt{1+\frac{1}{4}(\pi-2)^2} = \frac{r^3}{6}\rho g\sqrt{4+(\pi-2)^2}$$

$$= \quad \frac{r^3}{6}\rho g[\pi^2-4\pi+8]^{1/2}$$

and inclination of R is given by

$$\tan\theta = \frac{Y}{X} = \frac{\frac{r^3}{6}(\pi-2)\rho g}{\frac{r^3}{3}\rho g} = \frac{\pi-2}{2} = \frac{\pi}{2} - 1.$$

▶ **Ex. 6.** A solid right circular cone is divided into two parts by a plane through its axis and one of these portions is just immersed, vertex downwards, in water. Find the resultant thrust on its curved surface and show that it is inclined at an angle $\tan^{-1}\left(\frac{\pi}{2}\tan\alpha\right)$ to the horizontal, where α is the semi-vertical angle of the cone.

[Bhag. 62S, Lucknow 75, Indore 71, Agra 69, B.P.S.C. 82]

Solution: Let the vertical plane OAB, through the axis divides the cone of height h into two parts. Let one of these two parts OACB be immersed in a liquid of density ρ.

If X and Y be the horizontal thrusts on the curved surface of the portion OACB in two mutually perpendicular directions DC and DB, respectively.

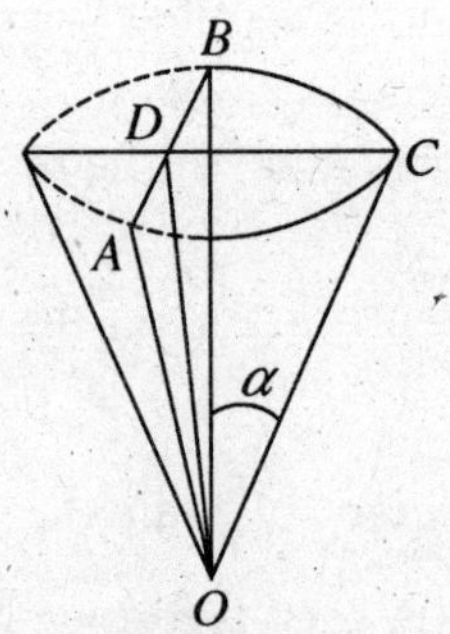

Fig 6.14

X = horizontal thrust on the projection of the curved surface OACB on a vertical plane perpendicular to CD.

= thrust on the $\triangle OAB$

$$= \quad \rho g\frac{h}{3}\left(\frac{1}{2}2h\tan\alpha h\right) = \rho g\tan\alpha\frac{h^3}{3}$$

and Y = horizontal thrust on the projection of the curved surface OACB on a vertical plane perpendicular to DB.

= 0; since half cone is symmetrical about the plane ODC which is perpendicular to DB.

Also,

$$\begin{aligned} Z &= \text{vertical thrust on the curved surface OACB} \\ &= \text{weight of the liquid filling half cone} \\ &= \frac{1}{2}\cdot\left(\frac{1}{3}\pi h^2 \tan^2 \alpha h\right)\rho g = \frac{1}{6}\pi h^3 \tan^2 \alpha \rho g. \end{aligned}$$

Now, we know that X acts normal to the plane OAB at its centre of pressure lying on OD and so the line of action of X is in the plane ODC.

Again, Z acts vertically upwards through the C.G. of the half cone OABC, which due to symmetry lies in the plane ODC.

Therefore, both X and Z lie in one plane and so they must intersect somewhere.

Hence,

$$\begin{aligned} \text{Resultant thrust} &= \sqrt{X^2 + Y^2 + Z^2} \\ &= \sqrt{\left(\frac{1}{3}h^3 \tan \alpha \rho g\right)^2 + \left(\frac{1}{6}\pi h^3 \tan^2 \alpha \rho g\right)^2} \quad [\because Y = 0] \\ &= \frac{1}{3}h^3 \tan \alpha \rho g \sqrt{1 + \tan^2 \alpha \frac{\pi}{4}} \\ &= \frac{h^3}{6}\tan^2 \alpha \rho g \sqrt{\pi + 4\cot^2 \alpha}. \end{aligned}$$

If θ be the resultant of the resultant thrust to the horizontal, then we have

$$\tan\theta = \frac{Z}{X} = \frac{\frac{\pi h^3}{6}\tan^2 \alpha \rho g}{\frac{h^3}{3}\tan \alpha \rho g} = \frac{\pi}{2}\tan\alpha.$$

Hence, $\theta = \tan^{-1}\left(\frac{\pi}{2}\tan\alpha\right)$.

► **Ex. 7.** A solid is formed by the revolution of a semicircle of radius a about its bounding diameter through an angle α and the solid is immersed with one plane face in the surface of a liquid, prove that magnitude of the resultant thrust on the curved surface of the liquid is $\frac{2}{3}\rho g a^3[\alpha - \sin\alpha\cos\alpha)^2 + \sin^4\alpha]^{1/2}$.

[P.U. 51H, I.A.S. 84]

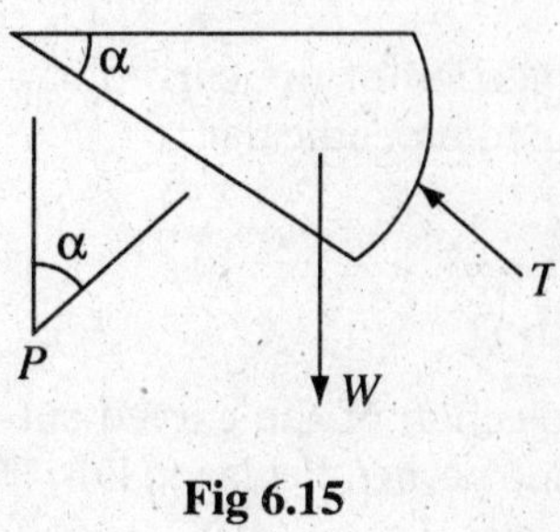

Fig 6.15

Solution: Let the adjoining figure represents the cross-section perpendicular to the plane faces. The resultant of the thrust P on the immersed plane face and the thrust T on the curved surface balances W, so if X and Y be the horizontal and vertical components of W, then $X = P\sin\alpha$, $Y = W - P\cos\alpha$.

Now,

$$W = \rho g \frac{4}{3}\pi a^3 \frac{\alpha}{2\pi} = \frac{2\rho g a^3 \alpha}{3}.$$

The C.G. of the semi-circular immersed face is at a distance $\frac{4a}{3\pi}$ from the centre, so its depth below the free surface is $\frac{4a}{3\pi}\sin\alpha$.

$$P = \rho g \frac{4a}{3\pi} \sin\alpha \frac{1}{2}\pi a^2 = \frac{2\rho g a^3 \sin\alpha}{3}.$$

Hence,

$$\begin{aligned} T^2 &= P^2 \sin^2\alpha + (W - P\cos\alpha)^2 \\ &= \frac{4\rho^2 g^2 a^6 \sin^2\alpha}{9}\sin^2\alpha + \left(\frac{2\rho g a^3 \sin\alpha}{3} - \frac{2\rho g a^3 \sin\alpha}{3}\cos\alpha\right)^2 \\ &= \frac{4^2 \rho g^2 a^6 \sin^4\alpha}{9} + \frac{4\rho^2 g^2 a^6}{9}(\alpha - \sin\alpha\cos\alpha)^2 \\ &= \frac{4\rho^2 g^2 a^6}{9}[\sin^4\alpha + (\alpha - \sin\alpha\cos\alpha)^2] \\ \therefore\quad T &= \frac{2}{3}\rho g a^3 [\sin^4\alpha + (\alpha - \sin\alpha\cos\alpha)^2]^{1/2}. \end{aligned}$$

► **Ex. 8.** A weightless sphere is divided by a vertical plane into two halves which are hinged together at their lowest point and is just filled with water, show that tension of a string which ties together the highest points of the two halves is three-eights of the weight of the water that the sphere would contain.

[I.A.S. 53, 58, B.U. 62H, R.U. 62H, Bh.U. 66H, M.U. 70H]

Solution: Let a be the radius of the sphere ABCD in the vertical plane dividing the sphere into two parts which are hinged at the lowest point B.

If X and Y be the horizontal thrusts on the curved surface of the hemisphere AEB in two mutually perpendicular directions OE and OD, respectively, then

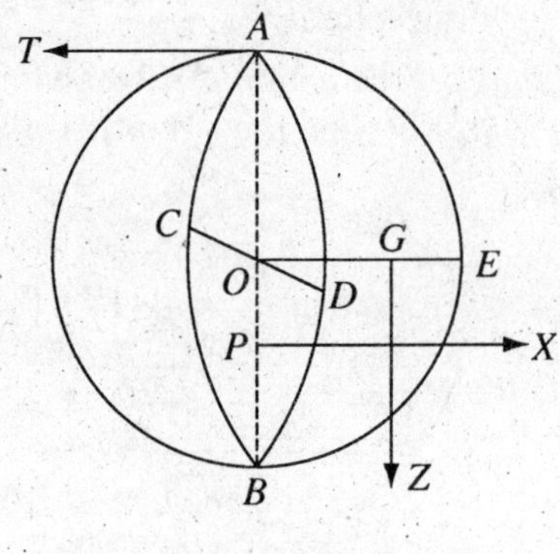

Fig 6.16

$$\begin{aligned} X &= \text{horizontal thrust in direction of OE} \\ &= \text{thrust on the circle ACBD} = \rho g a \pi a^2 = \rho g \pi a^3, \end{aligned}$$

which act through P, the centre of pressure of the circle ABCD, such that

$$AP = a + \frac{a^2}{4a} = \frac{5}{4}a \quad \text{and} \quad BP = AB - AP = 2a - \frac{5}{4}a = \frac{3a}{4}$$

and Y = horizontal thrust in the direction of OD
= O, by symmetry of the surface about the plane AOBE,
Z = vertical thrust = weight of the liquid contained
= $\frac{2}{3}\pi a^3 \rho g$,

acting through G, the C.G. of the liquid contained, such that $OG = \frac{3}{8}a$.

Let T be the tension of the string at A, then taking moments about B, in order to avoid the reaction of the hinge at B, we have

$$T \cdot BA - X \cdot BP - Z \cdot OG = 0$$

$$\text{or,} \quad T \cdot 2a - X \cdot \frac{3}{4}a - Z \cdot \frac{3}{8}a = 0$$

$$\text{or,} \quad T \cdot 2a = X \cdot \frac{3}{4}a + Z \cdot \frac{3}{8}a$$

$$= \pi a^3 \rho g \frac{3}{4}a + \frac{2}{3}\pi a^3 \rho g \frac{3}{8}a = \frac{3}{4}\pi a^4 \rho g\left(1 + \frac{1}{3}\right)$$

$$T \cdot 2a = \frac{3}{4}\pi a^4 \rho g \frac{4}{3}.$$

$$\therefore T = \frac{3}{8}\left(\frac{4}{3}\pi a^3 \rho g\right) = \frac{3}{8} \text{ (weight of water that the sphere would contain).}$$

▶ **Ex. 9.** A solid octant of a sphere is immersed in water with a plane base in the surface, prove that the resultant pressure on the curved surface is $\sqrt{1 + \frac{8}{\pi^2}}$ times the weight of the water displaced by the octant.

[B.U. 63H, M.U. 63H, 69H, 76H, 81H, Bh.U. 65H, R.U. 70H, I.A.S. 57, Raj. 53, Vikram 65, P.U. 82H]

Also show that the depth of the centre of pressure for the curved surface is $\frac{\pi}{\sqrt{\pi^2+8}} \times$ radius of the sphere. **[Bh.U. 65H, C.U. 85H, 87H]**

Solution: Let the octant OABC be immersed in the liquid, with the plane face OAB in the surface, then the two other plane faces OAC and OBC will be vertical.

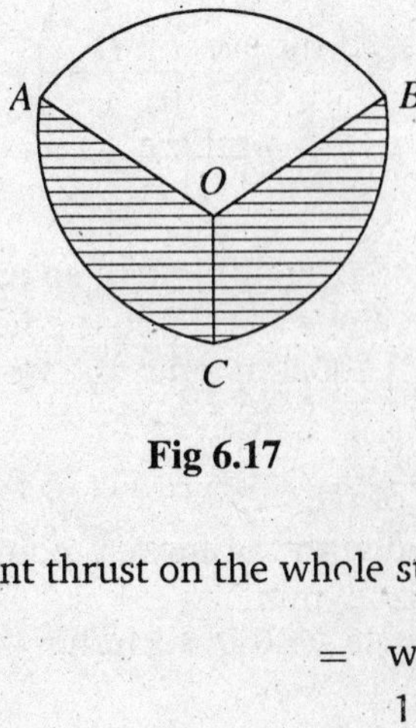

Fig 6.17

Let the radius of the original sphere be a.

Now the thrust on the plane face OAC

$$= \frac{1}{4}\pi a^2 \frac{4a}{3\pi}\rho g = \frac{1}{3}a^3 \rho g$$

= thrust on the plane face OBC.

Resultant thrust on the whole surface of the octant

= weight of the water displaced

$$= \frac{1}{4}\left(\frac{2}{3}\pi a^3 \rho g\right) = \frac{1}{6}\pi a^3 \rho g,$$

acting vertically upwards through the centre of gravity of the water displaced.

If X and Y be the horizontal thrusts on the curved surface of the octant, in two mutually perpendicular directions, and if Z be the resultant vertical thrust on the curved surface, then

$$X = \frac{1}{3}\rho g a^3, \quad Y = \frac{1}{3}\rho g a^3, \quad Z = \frac{1}{6}\rho g \pi a^3.$$

Since the thrusts on the plane faces OAC and OBC are horizontal and normal to the respective faces, and there will be no vertical thrust on these plane faces. The vertical thrust on the whole octant is due to the thrust on the curved surface.

Hence, resultant thrust R on the curve surface

$$= \sqrt{X^2 + Y^2 + Z^2} = \frac{\rho g a^3}{3}\sqrt{2 + \frac{\pi^2}{4}}$$

$$= \frac{\rho g a^3}{6}\sqrt{8 + \pi^2} = \frac{\rho g a^3}{6}\pi\sqrt{1 + \frac{8}{\pi^2}} = \sqrt{1 + \frac{8}{\pi^4}}$$

times the weight of the water displaced by the octant.

If R be inclined at an angle θ to the horizontal, then

$$\tan\theta = \frac{\sqrt{X^2 + Y^2}}{Z} = \frac{\frac{\rho g a^3}{3}\sqrt{2}}{\frac{\rho g \pi a^3}{6}} = \frac{2\sqrt{2}}{\pi}.$$

Now,

$$\operatorname{cosec}\theta = \sqrt{1 + \cot^2\theta} = \sqrt{1 + \frac{8}{\pi^2}} = \sqrt{\frac{\pi^2 + 8}{\pi}}.$$

Hence,

$$\sin\theta = \frac{\pi}{\sqrt{\pi^2 + 8}}.$$

Hence, the depth of the centre of pressure for the curved surface

$$= a\sin\theta = \frac{a\pi}{\sqrt{\pi^4 + 8}} = \frac{\pi}{\sqrt{\pi^4 + 8}} \text{ radius of the sphere.}$$

▶ **Ex. 10.** A solid hemisphere is immersed in a liquid with highest point of its plane base in the surface and the base inclined at an angle $\tan^{-1} 2$ to the horizon, show that the resultant thrust on the curved surface is equal to twice the weight of the displaced liquid.

[I.A.S. 73, Jodhpur 78, Raj. 74, Vikram 80, 63, Rohilkhand 78, Delhi 61]

Solution: Let ABC be the solid hemisphere immersed with the highest point A in the surface and AB inclined at an angle α to the horizon, then $\tan\alpha = 2$.

Hence,

$$\sin\alpha = \frac{2}{\sqrt{5}}, \quad \cos\alpha = \frac{1}{\sqrt{5}}.$$

Let a be the radius of the sphere and ρ be the density of the liquid.
If P be the normal thrust on the plane base of the hemisphere, then

$$P = \rho g OM \pi a^2 = \rho g a \sin \alpha \pi a^2 = \rho g a \frac{2}{\sqrt{5}} \pi a^2 = \frac{2\pi}{\sqrt{5}} \rho g a^3.$$

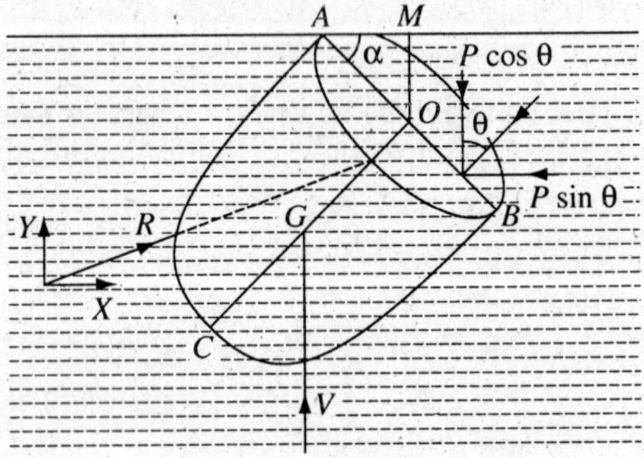

Fig 6.18

If V be the resultant vertical thrust on the hemisphere, then

$$\begin{aligned} V &= \text{the weight of the liquid displaced} \\ &= \frac{2}{3}\pi a^3 \rho g, \text{ acting vertically upwards.} \end{aligned}$$

If X and Y be the horizontal and vertical components of the resultant thrust, say R, on the curved surface of the hemisphere, then

$$X = P \sin \alpha \text{ and } Y - P \cos \alpha = V \quad \text{or,} \quad Y = P \cos \alpha + V.$$

Hence, if R be the resultant thrust on the curved surface, then

$$\begin{aligned} R &= \sqrt{X^2 + Y^2} = \sqrt{P^2 \sin^2 \alpha + (V + P \cos \alpha)^2} \\ &= \sqrt{P^2 \sin^2 \alpha + P^2 \cos^2 \alpha + 2VP \cos \alpha + V^2} \\ &= \sqrt{P^2 + 2VP \cos \alpha + V^2} \\ &= \sqrt{\frac{4\pi^2}{5}(\rho g a^3)^2 + 2\frac{2\pi}{\sqrt{5}} \rho g a^3 \frac{2}{3} \pi a^3 \rho g \frac{1}{\sqrt{5}} + \frac{4}{5}\pi^2 (\rho g a^3)^2}. \\ &= \pi \rho g a^3 \sqrt{\frac{4}{5} + \frac{8}{15} + \frac{4}{9}} = \pi a^3 \rho g \sqrt{\frac{36 + 24 + 20}{45}} \\ &= \pi a^2 \rho g \sqrt{\frac{80}{45}} = \pi a^3 \rho g \sqrt{\frac{16}{9}} = \frac{4}{3} \pi a^3 \rho g \\ &= 2\left(\frac{2}{3}\pi a^3 \rho g\right) \\ &= 2 \text{ (weight of the displaced liquid).} \end{aligned}$$

▶ **Ex. 11.** A vessel full of homogeneous liquid is in the form of the octant of the ellipsoid $\frac{x^2}{a^2} + \frac{y^2}{b^2} + \frac{z^2}{c^2} = 1$, where x, y, z are all ≥ 0 and the axis of Z is drawn vertically upwards. If the atmospheric pressure be neglected, prove that the resultant fluid pressure on the curved surface is

$$\frac{\rho g}{3}\left[b^2c^4 + a^2c^4 + \frac{1}{4}\pi^2a^2b^2c^2\right]^{1/2},$$

where ρ is the density of the liquid. **[R.U. 63H, Vikram 64, Agra 51]**

Solution: The equation of the ellipsoid is

$$\frac{x^2}{a^2} + \frac{y^2}{b^2} + \frac{z^2}{c^2} = 1. \quad (1)$$

Now the vertical component Z of the thrust is given by

$$\begin{aligned} Z &= \text{weight of the superincumbent liquid} \\ &= \rho g\left(\frac{1}{8}\text{ volume of the ellipsoid}\right) \\ &= \frac{1}{8}\rho g\frac{4\pi}{3}abc\rho = \frac{\pi}{6}abc\rho g, \end{aligned}$$

acting vertically downwards through the centre of gravity of the octant of the ellipsoid.

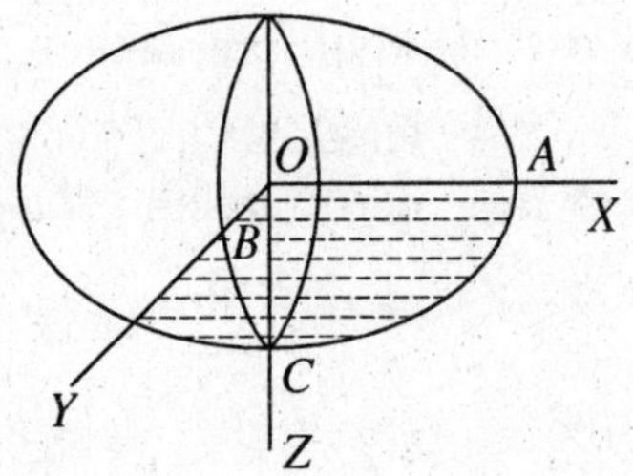

Fig 6.19

Again,

$$\begin{aligned} X &= \text{horizontal thrust along x-axis} = \text{thrust on the face OBC} \\ &= \rho g\left(\frac{1}{4}\pi bc\right)\frac{4c}{3\pi} = \frac{\rho g}{3}bc^2, \text{ acting through the C.P. of face OBC.} \end{aligned}$$

$$\begin{aligned} \text{and } Y &= \text{horizontal thrust along y-axis} = \text{thrust on face OAC} \\ &= \rho g\left(\frac{1}{4}\pi ac\right)\frac{4c}{3\pi} = \frac{\rho g}{3}ac^2, \text{ acting through the C.P. of the face OAC.} \end{aligned}$$

Since, X, Y, Z intersect at a common point, these can be reduced to a single force of magnitude

$$\begin{aligned} &= \sqrt{X^2 + Y^2 + Z^2} \\ &= \sqrt{\left(\frac{\rho g}{3}bc^2\right)^2 + \left(\frac{\rho g}{3}ac^2\right)^2 + \left(\frac{\rho g\pi abc}{6}\right)^2} \\ &= \frac{\rho g}{3}\sqrt{b^2c^4 + a^2c^4 + \frac{\pi^2a^2b^2c^2}{4}} \\ &= \frac{\rho g}{3}\left[a^2c^4 + b^2c^4 + \frac{1}{4}\pi^2a^2b^2c^2\right]^{1/2} \end{aligned}$$

► **Ex. 12.** A hemisphere of radius a is immersed in a liquid of density ρ. The plane base of the hemisphere is vertical and its centre at a depth $a\sqrt{5}$ below the surface. Show that the resultant thrust on the curved surface is $\frac{7}{3}\pi\rho g a^3$ and its direction makes with the horizontal an angle θ, where $\tan\theta = \frac{2}{3\sqrt{5}}$.

[M.U. 80H, Lucknow 80, R.U. 72H]

Solution: Let O be the centre of the hemisphere.

Given $OM = a\sqrt{5}$.

Let OA and OB be two mutually perpendicular horizontal directions.

X = horizontal thrust along OA

= whole pressure on the projection of the curved surface on a vertical plane perpendicular to OA

= whole pressure on the circle BDCE

$= \rho g \pi a^2 a\sqrt{5} = \pi a^3 \sqrt{5}\rho g$

Y = horizontal thrust along $OB = 0$ (by symmetry)

Z = vertical thrust = weight of the liquid contained $= \frac{2}{3}\pi a^3 \rho g$.

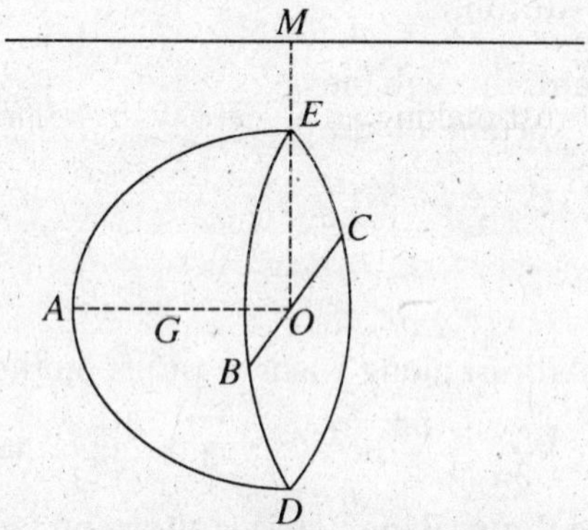

Fig 6.20

Let R be the resultant thrust making an angle θ with the horizontal, then

$$R = \sqrt{X^2+Y^2+Z^2} = \sqrt{\left(\pi a^3\rho g\sqrt{5}\right)^2 + \left(\frac{2}{3}\pi a^3\rho g\right)^2}$$

$$= \pi a^3\rho g\sqrt{5+\frac{4}{9}} = \pi a^3 \rho g\sqrt{\frac{49}{9}} = \frac{7}{3}\pi a^3\rho g$$

and $$\tan\theta = \frac{Z}{\sqrt{X^2+Y^2}} = \frac{\frac{2}{3}\pi a^3\rho g}{\pi a^3\sqrt{5}\rho g} = \frac{2}{3\sqrt{5}}.$$

► **Ex. 13.** A closed cylindrical vessel with hemispherical ends is filled with water and placed with its axis horizontal. Find the resultant thrust on each of the ends and determine its line of action. **[R.U. 98H]**

Solution: Let us consider the hemispherical end.

$$\begin{aligned} X &= \text{horizontal component along } OA \\ &= \text{whole pressure on the circle } BCDE \\ &= \pi r^2 r\rho g = \pi r^3 \rho g, \end{aligned}$$

where r is the radius of the base of the cylinder and ρ is the density of the liquid

$$\begin{aligned} Y &= \text{horizontal thrust along } OB = 0, \text{ by symmetry} \\ Z &= \text{vertical thrust} = \text{weight of the liquid contained} \\ &= \frac{2}{3}\pi r^3 \rho g. \end{aligned}$$

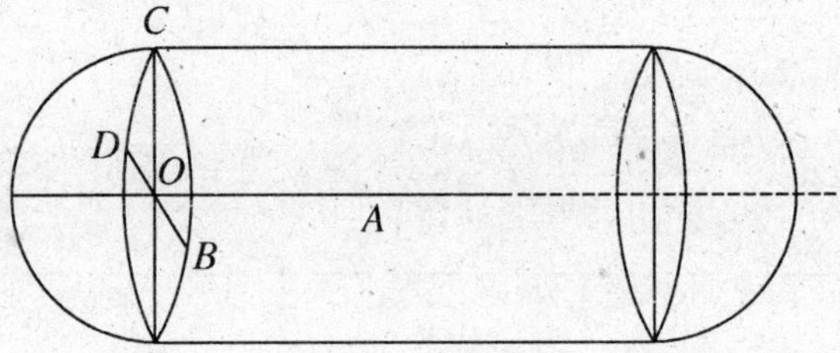

Fig 6.21

Let R be the resultant thrust making an angle θ with the horizontal, then

$$\begin{aligned} R &= \sqrt{X^2 + Y^2 + Z^2} = \sqrt{\pi^2 r^6 \rho^2 g^2 + \frac{4}{9}\pi^2 r^6 \rho^2 g^2} \\ &= \pi r^3 \rho g \sqrt{1 + \frac{4}{9}} = \pi r^3 \rho g \frac{\sqrt{13}}{3}. \end{aligned}$$

If the resultant thrust be inclined at an angle θ to the horizontal, then

$$\tan\theta = \frac{Z}{X} = \frac{\frac{2}{3}\pi r^3 \rho g}{\pi r^3 \rho g} = \frac{2}{3}.$$

Hence, $\theta = \tan^{-1}\frac{2}{3}$.

▶ **Ex. 14:** A solid hemisphere of radius a is placed with its centre at a distance h below the surface of water and has its plane face vertical. Find the horizontal thrust on the curved surface. Find also the resultant thrust on it.

[P.U. 65S, R.U. 68H, 74S, 75S, 95H, R.U. 97H, 94H, Delhi 58, 69, B.U. 73A]

Solution: Let ABC be the solid hemisphere of radius a.

$$\begin{aligned} X &= \text{horizontal thrust parallel to } OB \\ &= \text{whole pressure on the circle of diameter } AC \\ &= \rho g(\pi a^2)h = \rho g \pi a^2 h \\ Y &= \text{horizontal thrust along } OD = 0, \text{ by symmetry.} \end{aligned}$$

Hence, the resultant horizontal thrust $= \sqrt{X^2 + Y^2} = \rho g \pi a^2 h$.

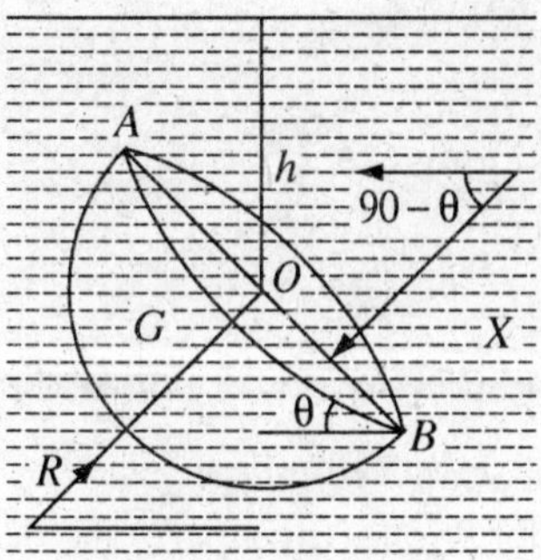

Fig 6.22

Again, Z = vertical thrust = weight of the liquid displaced = $\rho g \frac{2}{3}\pi a^3$.

Hence, if R is the resultant thrust making an angle θ with the horizontal, then

$$\begin{aligned} R &= \sqrt{X^2 + Y^2 + Z^2} \\ &= \sqrt{\rho^2 g^2 \pi^2 a^4 h^2 + \rho^2 g^2 \frac{4}{9}\pi^2 a^6} \\ &= \rho g \pi a^2 \sqrt{h^2 + \frac{4}{9}a^2} \\ &= \frac{\rho g \pi a^2}{3}\sqrt{9h^2 + 4a^2}. \end{aligned}$$

Also,

$$\tan\theta = \frac{Z}{X} = \frac{\frac{2}{3}\pi a^3 \rho g}{\pi a^2 h \rho g} = \frac{2}{3}\frac{a}{h}, \quad \text{i.e.,} \quad \theta = \tan^{-1}\left(\frac{2a}{3h}\right).$$

Exercises 6(V)

1. A conical wine glass is fulled with water and placed in an inverted position upon a table, show that the resultant vertical thrust of the water on the glass is two-thirds that on the table.

   [B.U. 55S, 69S, Agra 75, 60, Bhopal 83, Lucknow 83, Sagar 62]

2. A conical cup whose weight is $\frac{5}{8}$th of the weight of water, which would just fill it, is placed vertex upwards on a smooth table and water is gradually poured in through a hole made in the top. Show that the cup will on the point of rising from the table when the water reaches half the height of the cup.

   [Agra 56, Allahabad 60, Sagar 49, 63]

3. A hollow sphere is just filled with water, prove that the resultant vertical pressures on the upper and lower halves on the internal surface are $\frac{1}{4}W$ and $\frac{5}{4}W$, respectively, where W is the weight of water.

   [R.U. 64H, Agra 72, Lucknow 66]

4. A hollow right circular cylinder is filled with liquid and held with its axis horizontal. Find the resultant thrust on half the curved surface cut off by a vertical plane through the axis. **[Bh.U. 75H]**

   [*Ans.* $\frac{1}{2}\rho g r^2 l\sqrt{16+\pi^2}$ at $\tan^{-1}\frac{\pi}{4}$ with horizontal, where symbols have usual meaning]

5. A vessel in the shape of the segment of a sphere of radius a cut off by a plane at a distance c from the centre, rests with its base on a table. If the segment is filled with water, show that the resultant vertical thrust on the curved surface

will be upward force if $c < \frac{a}{2}$.

[I.A.S. 87]

6. A hollow sphere of radius a is just filled with water, find the resultant vertical thrusts on the two portions into which the surface is divided by a horizontal plane at a depth c below the centre.

[M.U. 79H]

[*Ans.* The thrust on the lower portion = $\frac{1}{3}\rho g\pi(a-c)(5a^2+5ac+2c^2)$ downwards and the thrust on the upper portion = $\frac{1}{3}\rho g\pi(a^3-3ac^2-2c^3)$ upwards]

7. The bottom and top ends of a frustum of a cone are respectively a and b $(a > b)$. If the frustum be filled with a liquid of density ρ, prove that the resultant vertical thrust on the curved surface of the frustum is

$$\frac{1}{3}\rho g\pi h(2a^2 - ab - b^2),$$

where h is the height of the frustum.

Also prove that; if the frustum is inverted, the resultant vertical thrust on the curved surface of the frustum of the cone is

$$\frac{1}{3}\rho g\pi h(a^2 + ab - 2b^2).$$

8. A hemispherical bowl is filled with water and inverted and placed with its plane base in contact with a horizontal table. Prove that the resultant vertical thrust on its surface is one-third the thrust on the table.

[B.U. 56A, R.U. 67A, 73, A.P.U. 75A, Delhi 53, M.U. 77A]

9. A cylindrical vessel full of water is held with its axis inclined at an angle of 45° to the vertical. Find the magnitudes of the pressures on the ends and show that the resultant pressure on the curved surface will be equal to the difference between the pressure on the ends.

[*Ans.* $\frac{1}{\sqrt{2}}\pi a^3\rho g$, $\frac{1}{\sqrt{2}}\pi a^2(h+a)\rho g$, where a, h being the radius and height of the cylinder]

10. A cone whose vertical angle is 2α, has its lowest generator horizontal and is filled with liquid. Prove that the resultant pressure on the curved surface is $\sqrt{1 + 15\sin^2\alpha}$ times the weight of the liquid.

[B.U. 69A, 80H, 52H, 54H, P.U. 76A, Vikram 65, Raj. 55, 65, Alld. 56, Agra 62, B.P.S.C. 83, R.U. 99H, C.U. 92H, 93H, 2000H, 03H]

11. A cylinder is filled with equal quantities of three liquids which do not mix. The sp. gravities of the liquids are s, $2s$ and $3s$, respectively. Prove that the whole pressures on the corresponding portions of the curved surface are in the ratio of $1 : 4 : 9$.

12. A hollow weightless hemisphere with a plane base is filled with water and hung by means of a string, one end of which is attached to a point of the rim of its base.

(a) Find the inclination to the horizontal of the resultant thrust on its curved surface.

[Agra 65, 68, Vikram 62, I.A.S. 52, 62, P.U. 54A, Luck. 79]

[*Ans.* $\tan^{-1}\frac{109}{96}$]

(b) Show that thrust on the plane base of the weight of contained liquid is $12 : \sqrt{73}$.

[P.U. 76H, B.P.S.C. 88]

13. A cone floats with its axis horizontal in a liquid of density double its own, find the pressure on its base

and prove that if θ be the inclination to the vertical of the resultant thrust on the curved surface, and θ the semi-vertical angle of the cone, then $\tan\theta = \frac{4}{\pi}\tan\alpha$.

[Lucknow 50, 65, 77, Raj. 66, 7?, 78, Agra 59, 69, Delhi 66, 69, Alld. 61, B.U. 62A, B.P.S.C. 86, Mith. 93H]

14. A solid cone is held with its axis in the surface of water. Find the resultant liquid pressure on the immersed portion of the curved surface in magnitude and direction.

[R.U. 66H, Bh.U. 72H]

[*Ans.* $\frac{1}{6}h^3\tan^2\alpha w\sqrt{\pi^2+16\tan^2\alpha}$ inclined to the vertical at $\tan^{-1}\left(\frac{4}{\pi}\tan\alpha\right)$, where h, α, w are the height, semi-vertical angle of the cone and weight of unit volume of water]

15. A right cone of semi-vertical angle α is just immersed with its slant side, of length l, in the surface of a liquid. Show that the resultant thrust on the curved surface will cut this slant side at a distance

$$\frac{3l}{4(1-3\sin^2\alpha)}$$

from the vertex, and find the magnitude of this thrust.

[P.U. 43H]

[*Ans.* $\frac{1}{3}\pi l^3\sin^2\alpha\cos\alpha\rho g\sqrt{1+3\sin^2\alpha}$]

16. The stem of a funnel has a radius a and its mouth has a radius b, the height of the stem is h and that of the sloped part is l. The mouth is placed on a horizontal plane, being greased so as to be water tight, if the funnel is then filled with water to the top of the stem, prove that the upward pressure on the funnel will be

$$\pi w(b-a)\left[h(b+a)+\tfrac{1}{3}l(a+2b)\right],$$

where w is the weight of the water.

17. An oblique cylinder standing on a horizontal plane, the generating lines making an angle α with the vertical, is filled with a liquid of weight w to a height h. Prove that the resultant thrust on the curved surface of the cylinder in a couple of moment is $\frac{1}{2}wh\tan\alpha$.

18. A vessel in the form of an oblique cylinder with its base horizontal, contains two liquids of densities $\rho, \sigma(\sigma > \rho)$ of equal depth h. Prove that the thrusts of the liquids on the curved surface are equivalent to a couple of moment

$$\frac{1}{2}wh\cot\alpha\frac{3\rho+\sigma}{\rho+\sigma},$$

where w is the total weight of the liquid, and α the inclination of the axis of the cylinder to the base.

[P.U. 55H]

19. A soid sphere of density ρ is placed at the bottom of a vessel which is horizontal and a liquid of density $\sigma(<\rho)$ is poured in so as just to cover up the sphere. The sphere is then cut along the vertical diametrical plane. Prove that the two parts will not separate if $\rho < 4\sigma$.

[P.U. 66H, Raj. 63, Agra 64, C.U. 48, Alld. 51, 64, Delhi 62, 64H]

20. A solid hemisphere of radius a is held completely in a heavy homogeneous liquid with its plane face inclined at an angle θ to the horizontal and with the highest point of the plane face in the free surface. Neglecting atmospheric pressure find the magnitude and the line of action of the thrust due to the liquid on the curved surface of the hemisphere. **[B.U. 56H]**

[*Ans.* $\pi a^3 \rho g \sqrt{\sin^2\alpha + \frac{4}{9} - \frac{4}{3}\sin\alpha\cos\alpha}$ acting at an angle $\tan^{-1}\frac{2-3\sin\alpha\cos\alpha}{3\sin\alpha}$ with the horizon]

21. A solid hemisphere is completely immersed in a fluid with the centre of its base at a given depth, if W be the weight of the fluid it displaces, P the resultant vertical thrust and Q the resultant horizontal thrust on the curved surface; prove that for all positions of the solid $(W - P)^2 + Q^2$ is constant. **[P.U. 58H]**

22. A solid circular cone of height h and vertical angle 2α made of uniform material, floats in water with its axis vertical and vertex downwards and a length h' of the axis immersed. The cone is bisected by a vertical plane through the axis and the two parts are hinged together at the vertex. Show that the two parts will remain in contact, if $h' > h\sin^2\alpha$ or $\tan^2\alpha < \frac{h'}{h-h'}$.

[P.U. 47H, M.U. 64H, Raj. 54, 56, Agra 70, Vikram 66, Luck. 64, 82, Sagar 65]

23. A hemispherical surface of radius a is immersed in a liquid of density ρ with its centre at depth h and its base inclined at an angle α to the horizontal. Find the resultant thrust on the curved surface and its direction.

[P.U. 63H, 67H, B.U. 62H, 68H, Bh.U. 67H, 70H, R.U. 71H, M.U. 68H, I.A.S. 91]

[*Ans.* $\pi a^2 w\rho$ $\sqrt{h^2 - \frac{4}{9}ah\cos\alpha + \frac{4}{9}a^2}$ inclined to the horizon at $\tan^{-1}\frac{2a-3h\cos\alpha}{3h\sin\alpha}$, where w is the weight of unit volume of the liquid.

24. A hemispherical bowl is filled with water, and two vertical planes are drawn through its central radius, cutting off a semi-line of the surface, if 2α be the angle between the planes, prove that the angle which the resultant pressure on the surface makes with verticals is $\tan^{-1}\left(\frac{\sin\alpha}{\alpha}\right)$]

[Agra 55, 65, Delhi 55H, Raj. 61, 64]

25. A hollow cone without weight, closed and filled with some liquid, is suspended from a point in the rim of its base, if θ be the angle which the direction of the resultant pressure makes with the vertical, then show that $\cot\theta = \frac{28\cot\alpha+\cot^3\alpha}{48}$, α being the semi-vertical angle of the cone.

[Lucknow 81, I.A.S. 96]

26. A spherical shell formed of two halves in contact along a vertical plane is filled with water, show that the resultant pressure on either half of the shell is $\frac{1}{4}\sqrt{13}$ of the total weight of the liquid.

[P.U. 63H, Luck. 77, 71, Raj. 75, Vik. 77, C.U. 94H]

27. A hemispherical bowl is filled with water. Find the horizontal thrust on one-half of the surface divided by a vertical diametral plane and show that it is $\frac{1}{\pi}$ of the magnitude of the resultant fluid thrust on the whole surface.

[Luck. 76, Raj. 75, Vik. 75, R.U. 74, P.U. 64A]

[*Ans.* $\frac{2}{3}\rho g r^3$, where r is the radius of the base of the hemisphere]

28. A hollow sphere of given radius is just filled with water, find the resultant vertical pressures on the two portions of the surface divided by a plane at a depth equal to half the radius below the centre.

[R.U. 63H]

[*Ans.* Upper part 0, lower part W, where W is the weight of total liquid]

CHAPTER 7

Equilibrium of Fluids in Given Field of Forces, Rotating Liquids

7.1 Introduction

If in a given region of space, every particle is subjected to a force which is proportional to the mass acted upon, then the region is called a *field of force*. The force per unit mass is called the *intensity of field*.

A line drawn in the field of force so that the tangent to it at any point, coincides with that of the resultant external forces of that point, is called a *line of force*.

For example, due to earth's gravitation a field of force is created. When gravity is the only external force, the lines of force are vertical or more strictly, they are directed towards the earth's centre.

7.2 Pressure Derivative in Terms of Force

[M.U. 91H]

Let a fluid be in equilibrium under the action of a given system of external forces. Let us take two pints P and Q situated along the given direction and in the fluid, where $PQ = \delta s$.

Join PQ and with PQ as axis we construct a cylinder, whose cross-section is small and a closed cuıve of any type whose area is α.

Fig. 7.1

Let ρ be the density of the fluid and F be the component of the external forces per unit mass of fluid in the cylinder along the direction PQ, then the external force on the fluid in the cylinder along the direction PQ is $\rho \alpha F \delta s$.

If p and $p + \delta p$ be the pressures at P and Q, then the fluid contained in the cylinder is in equilibrium under the following forces:

(*i*) $p\alpha$ along PQ; (*ii*) $(p + \delta p)\alpha$ along QP;
(*iii*) $S\rho\alpha\delta s$ along PQ.

Thus, resolving all the forces along PQ, we have

$$p\alpha + F\rho\alpha\delta s - (p + \delta p)\alpha = 0 \quad \text{or,} \quad F\rho\delta s = \delta p.$$

Now taking limit when Q tends to p, i.e., $\delta s \to 0$

$$\frac{dp}{ds} = \rho F. \tag{7.1}$$

This proves that the *rate of increase of the pressure in any direction is equal to the product of the density and the resolved part of the force in that direction.*

Corollary 7.2.1 *If* (x, y, z) *be the cartesian coordinates of the point* P *in the fluid and* X, Y, Z *be the component forces per unit mass in the increasing directions of* x, y *and* z, *respectively, then as in eqn.* (7.1) *we have*

$$\frac{\partial p}{\partial x} = \rho X, \quad \frac{\partial p}{\partial y} = \rho Y, \quad \frac{\partial p}{\partial z} = \rho Z, \tag{7.2}$$

where p *is the pressure at* $P(x, y, z)$ *and* ρ *is the density of the fluid.*

[B.B.A. B.U. 93H, M.U. 95H]

Also we have

$$\begin{aligned} dp &= \frac{\partial p}{\partial x} dx + \frac{\partial p}{\partial y} dy + \frac{\partial p}{\partial z} dz \\ &= \rho X, dx + \rho Y, dy + \rho Z, dz \\ dp &= \rho(Xdx + Ydy + Zdz) \quad \text{From (7.2).} \end{aligned} \tag{7.3}$$

This is known as *equation of pressure,* which determines the pressure p at a point.

Corollary 7.2.2 *If the fluid be homogeneous and incompressible,* ρ *is constant. Hence, from* (7.3), *we have*

$$Xdx + Ydy + Zdz = \frac{dp}{\rho} = d\left(\frac{p}{\rho}\right). \tag{7.4}$$

$\therefore$ $Xdx + Ydy + Zdz$ *must be a perfect differential, i.e., the force system must be conservative.*

Hence, a homogeneous liquid under the action of a system of forces is in equilibrium only if the system of forces belong to a conservative system.

Also $Xdx + Ydy + Zdz = -dV$, where V is the potential function.

Hence, from (7.4)

$$d\left(\frac{p}{\rho}\right) = -dV.$$

Integrating we get, $\frac{p}{\rho} = -V + C$, where C is constant of integration.
Also we have

$$Xdx + Ydy + Zdz = -\left(\frac{\partial V}{\partial x}\,dx\rho\frac{\partial V}{\partial y}\,dy\rho\frac{\partial V}{\partial z}\,dz\rho\right) = -dV.$$

Hence,

$$X = -\frac{\partial V}{\partial x}, \quad Y = -\frac{\partial V}{\partial y}, \quad Z = -\frac{\partial V}{\partial z}.$$

7.3 Equation of Pressure

[T.M.B.U. 92U, 93H, 99H, C.U. 2000H, 02H, R.U. 2000H, M.U. 95H, 01H, B.B.A.B.U. 93H]

A mass of fluid is at rest under the action of the given force to obtain the equation which determines the pressure at any point of the fluid.

Let us consider a point $P(x, y, z)$ in the mass of a fluid in equilibrium.

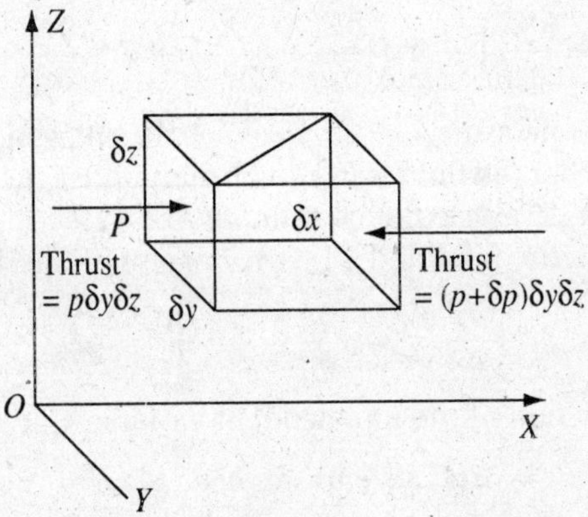

Fig. 7.2

We draw a small parallelepiped with P as a corner in the fluid whose faces are parallel to the coordinate planes and the edges be of lengths δx, δy and δz parallel to x, y and z axes, respectively.

Then the volume of the fluid in the parallelepiped = $\delta x\delta y\delta z$.

Let ρ be the mean density of the fluid in the parallelepiped and X, Y, Z the components of the external forces on the fluid per unit mass parallel to x, y, z axes, respectively. Let p and $p + \delta p$ be the pressure on faces parallel to the yz-plane.

Then by resolving forces, acting on the parallelepiped, along the x-axis, we have

$$p\delta y\delta z + \rho \times \delta x\delta y\delta z = (p + \delta p)\delta y\delta z \;\Rightarrow\; p + \rho \times \delta x = p + \delta p$$
$$\Rightarrow\; \rho \times \delta x = \delta p$$
$$\Rightarrow\; \frac{\delta p}{\delta x} = \rho X.$$

Similarly, resolving the forces parallel to the axes of y and z, we get for equilibrium

$$\frac{\delta p}{\delta y} = \rho Y \quad \text{and} \quad \frac{\delta p}{\delta z} = \rho Z.$$

Let $\delta x \to 0, \delta y \to 0, \delta z \to 0$, then the mean density ρ of the parallelepiped becomes the density at P.

$$\therefore \quad \frac{\partial p}{\partial x} = \rho X.$$

Similarly,

$$\frac{\partial p}{\partial y} = \rho Y \quad \text{and} \quad \frac{\partial p}{\partial z} = \rho Z.$$

Now,

$$dp = \frac{\partial p}{\partial x}dx + \frac{\partial p}{\partial y}dy + \frac{\partial p}{\partial z}dz = \rho(Xdx + Ydy + Zdz)$$

7.4 To Find the necessary and sufficient condition for equilibrium of a fluid under the action of a force whose components are X, Y, Z along the coordinate axes

The necessary and sufficient condition for equilibrium of a fluid under the action of a force whose components are X, Y, Z *along the coordinate axes is that*

$$X\left(\frac{\partial Y}{\partial z} - \frac{\partial Z}{\partial y}\right) + Y\left(\frac{\partial Z}{\partial x} - \frac{\partial X}{\partial z}\right) + Z\left(\frac{\partial X}{\partial y} - \frac{\partial Y}{\partial x}\right) = 0.$$

Proof. The condition is necessary [C.U. 94H, 99H, 03H, 84H, 85H, 86H, 89H]

Here we assume that the fluid is in equilibrium under the action of a force whose components are X, Y, Z along the coordinate axes. Hence, if p be the pressure at any point $P(x, y, z)$ in the fluid and ρ be the density of the liquid, then we have

$$\frac{\partial p}{\partial x} = \rho X, \frac{\partial p}{\partial y} = \rho Y, \frac{\partial p}{\partial z} = \rho Z. \tag{7.5}$$

Now as p is a function of the independent variable x, y, z, hence we have

$$\frac{\partial^2 p}{\partial y \partial z} = \frac{\partial^2 p}{\partial z \partial y}, \quad \frac{\partial^2 p}{\partial z \partial x} = \frac{\partial^2 p}{\partial x \partial z}, \quad \frac{\partial^2 p}{\partial x \partial y} = \frac{\partial^2 p}{\partial y \partial x}. \tag{7.6}$$

From (7.6), we have

$$\frac{\partial^2 p}{\partial y \partial z} = \frac{\partial^2 p}{\partial z \partial y}, \quad \Rightarrow \quad \frac{\partial}{\partial y}\left(\frac{\partial p}{\partial z}\right) = \frac{\partial}{\partial z}\left(\frac{\partial p}{\partial y}\right),$$

$$\Rightarrow \quad \frac{\partial}{\partial y}\left(\frac{\partial p}{\partial z}\right) - \frac{\partial}{\partial z}\left(\frac{\partial p}{\partial y}\right) = 0,$$

$$\Rightarrow \quad \frac{\partial}{\partial y}(\rho Z) - \frac{\partial}{\partial z}(\rho Y) = 0,$$

$$\Rightarrow \quad \frac{\partial}{\partial y}(\rho Z) = \frac{\partial}{\partial z}(\rho Y),$$

$$\Rightarrow \quad \rho\frac{\partial Z}{\partial y} + Z\frac{\partial \rho}{\partial y} = \rho\frac{\partial Y}{\partial z} + Y\frac{\partial \rho}{\partial z},$$

$$\Rightarrow \quad \rho\left(\frac{\partial Y}{\partial z} - \frac{\partial Z}{\partial y}\right) = Z\frac{\partial \rho}{\partial y} - Y\frac{\partial \rho}{\partial z}. \tag{7.7}$$

Similarly, we have

$$\rho\left(\frac{\partial Z}{\partial x}-\frac{\partial X}{\partial z}\right)=X\frac{\partial \rho}{\partial z}-Z\frac{\partial \rho}{\partial x} \tag{7.8}$$

$$\text{and}\quad \rho\left(\frac{\partial X}{\partial y}-\frac{\partial Y}{\partial x}\right)=Y\frac{\partial \rho}{\partial x}-X\frac{\partial \rho}{\partial y}. \tag{7.9}$$

Multiply (7.7), (7.8) and (7.9) by X, Y, Z and adding we get

$$X\left(\frac{\partial Y}{\partial z}-\frac{\partial Z}{\partial y}\right)+Y\left(\frac{\partial Z}{\partial x}-\frac{\partial X}{\partial z}\right)+Z\left(\frac{\partial X}{\partial y}-\frac{\partial Y}{\partial x}\right)=0, \tag{7.10}$$

which is the required necessary condition.

The condition is sufficient

Here we assume that

$$X\left(\frac{\partial Y}{\partial z}-\frac{\partial Z}{\partial y}\right)+Y\left(\frac{\partial Z}{\partial x}-\frac{\partial X}{\partial z}\right)+Z\left(\frac{\partial X}{\partial y}-\frac{\partial Y}{\partial x}\right)=0.$$

This shows that $Xds + Ydy + Zdz$ is an exact differential

$$\begin{aligned} \text{i.e.,}\quad \rho(Xdx + Ydy + Zdz) &= \text{an exact differential} \\ &= dp \\ &= \frac{\partial p}{\partial x}dx + \frac{\partial p}{\partial y}dy + \frac{\partial p}{\partial z}dz. \end{aligned}$$

Equating coefficients of dx, dy, dz on both the sides, we get

$$\frac{\partial p}{\partial x}=\rho X,\quad \frac{\partial p}{\partial y}=\rho Y,\quad \frac{\partial p}{\partial z}=\rho Z.$$

These are clearly the equations of equilibrium, hence the condition is sufficient also.

7.5 Surface of Equipressure

The surface passing through all the points of a fluid at which the pressure is the same is a surface of equipressure.

Also, for a liquid at rest, the pressure at any point is given by

$$dp = \rho(Xdx + Ydy + Zdz)$$

which on integration gives $p = \phi(x, y, z) = C$ (*say*).

At all points where p = constant we have

$$\phi(x, y, z) = C, \tag{7.11}$$

which is the equation of a surface.

Corresponding to the different values of the constant C we obtain different surfaces, at every point of these surfaces the pressure is same and thus we get from (7.11) the surface of equal pressures for different values of C.

The equation of the free surface or external surface is obtained by putting p equal to the external pressure to the fluid. If there is no external force, then $p = 0$ gives the equation of free surface as $\phi(x, y, z) = 0$.

7.6 Differential Equation of Lines of Force

Let X, Y, Z be the components of the resultant force acting on the fluid at any point (x, y, z) parallel to the coordinate axes. Then the direction cosines of the line of action of this resultant force are proportional to X, Y, Z.

Also from coordinate geometry we know that the direction cosines of the tangent to the curve at any point (x, y, z) on it are proportional to $\frac{dx}{ds}, \frac{dy}{ds}, \frac{dz}{dz}$.

For line of force these two directions are the same

$$\therefore \quad \frac{dx/ds}{X} = \frac{dy/ds}{Y} = \frac{dz/ds}{Z}.$$

Therefore, these are the differential equations of the lines of force.

7.7 To Prove that the surfaces of equal pressure are the surfaces intersecting orthogonally the lines of force

A surface of equal pressure is

$$p = \phi(x, y, z) = \text{constant} \tag{7.12}$$

The direction cosines of the normal at any point (x, y, z) to the surface (7.12) are proportional to

$$\frac{\partial \phi}{\partial x}, \quad \frac{\partial \phi}{\partial y}, \quad \frac{\partial \phi}{\partial z},$$

i.e., proportional to

$$\frac{\partial p}{\partial x}, \quad \frac{\partial p}{\partial y}, \quad \frac{\partial p}{\partial z},$$

as from (7.12)

$$\frac{\partial \phi}{\partial x} = \frac{\partial p}{\partial x}, \quad \frac{\partial \phi}{\partial y} = \frac{\partial p}{\partial y}, \quad \frac{\partial \phi}{\partial z} = \frac{\partial p}{\partial z},$$

i.e., proportional to ρX, ρY, ρZ, where X, Y, Z are the components of forces per unit mass acting at (x, y, z) in the increasing directions of x, y and z and ρ is the density of the fluid. Also

$$\frac{\partial p}{\partial x} = \rho X, \quad \frac{\partial p}{\partial y} = \rho Y \quad \text{and} \quad \frac{\partial p}{\partial z} = \rho Z.$$

Hence, the direction cosines of the normal at any point (x, y, z) to the surface (7.12) are proportional to X, Y, Z which are also direction ratios of lines of force.

Hence, lines of force are parallel to normals to the surface of equal pressure. Therefore, the surfaces of equal pressure are cut orthogonally by the lines of force.

7.8 In a conservative field of forces the surfaces of equipressure, equidensity and equipotential energy coincide

If X, Y, Z be the rectangular components of a force, then the expression $(X dx + Y dy + Z dz)$ represents the work done by the external forces on a unit mass in an infinitesimal displacement from the point (x, y, z).

When the force constitute a conservative system this expression represents the loss of potential energy and may be denoted by $-dV$, where V is the potential energy function for a unit mass in the given field.

Then,

$$\begin{aligned} dp &= \rho(Xdz + Ydy + Zdz) \\ dp &= \rho(-dv), \quad \text{or,} \quad \frac{dp}{\rho} = -dV. \end{aligned} \tag{7.13}$$

Since dV is an exact differential.

Hence, $\frac{dp}{\rho}$ must be an exact differential. $\therefore$ ρ is also a function of p.

Therefore when p is constant, ρ is constant.

Further from equation (7.13), as dp is an exact differential, ρdV is also an exact differential.

$\therefore$ ρ is also a function of V.

Hence, when ρ is constant, V is constant.

Thus, when any one of the three functions is constant, so are the other two.

Therefore *in a conservative field of forces the surfaces of equipressure, equidensity and equipotential energy coincide.*

[C.U. 92H, 96H, 2000H, 02H, 04H, M.U. 94H, 96H, 2000H, B.B.A.B.U. 92H, 94H, T.M.B.U. 93H, 96H]

7.9 Conversely, if the Surfaces of Equipressure coincide with those of Equidensity, the External Forces form a Conservative System

On a surface of equipressure, p is constant and according to the theorem, ρ is also constant.

Therefore, from the equation

$$dp = \rho(Xdx + Ydy + Zdz)$$

$Xdx + Ydy + Zdz$ must be an exact differential and may be denoted by $-dV$, indicating the loss of potential energy.

Hence, the forces constitute a conservative system. **[T.M.B.U. 92H]**

7.10 To Find the differential equations of the surfaces of equal pressure and equal density

Let the fluid be incompressible and let its density be a function of x, y, z.

Then the surfaces of equal pressure are given by p = constant or $dp = 0$

$$\begin{aligned} \text{or,} \quad & \rho(Xdx + Ydy + Zdz) = 0 \\ \text{or,} \quad & Xdx + Ydy + Zdz = 0 \end{aligned} \tag{7.14}$$

since $\rho \neq 0$.

Also the surfaces of equal density are given by ρ = constant or $d\rho = 0$.

$$\text{or,} \quad \frac{\partial \rho}{\partial x} dx + \frac{\partial \rho}{\partial y} dy + \frac{\partial \rho}{\partial z} dz = 0. \tag{7.15}$$

Since ρ is a non-zero constant.
Solving (7.14) and (7.15) by cross-multiplication we have

$$\frac{dx}{Y\frac{\partial\rho}{\partial z}-Z\frac{\partial\rho}{\partial y}}=\frac{dy}{Z\frac{\partial\rho}{\partial x}-X\frac{\partial\rho}{\partial z}}=\frac{dz}{X\frac{\partial\rho}{\partial y}-Y\frac{\partial\rho}{\partial x}}. \tag{7.16}$$

Also from the conditions of equilibrium

$$Z\frac{\partial\rho}{\partial y}-Y\frac{\partial\rho}{\partial z} = \rho\left(\frac{\partial Y}{\partial z}-\frac{\partial Z}{\partial y}\right),$$
$$X\frac{\partial\rho}{\partial z}-Z\frac{\partial\rho}{\partial x} = \rho\left(\frac{\partial Z}{\partial x}-\frac{\partial X}{\partial z}\right)$$
$$\text{and}\quad Y\frac{\partial\rho}{\partial x}-X\frac{\partial\rho}{\partial y} = \rho\left(\frac{\partial X}{\partial y}-\frac{\partial Y}{\partial x}\right).$$

Putting these values in (7.16), we get

$$\frac{dx}{\rho\left(\frac{\partial Y}{\partial z}-\frac{\partial Z}{\partial y}\right)}=\frac{dy}{\rho\left(\frac{\partial Z}{\partial x}-\frac{\partial X}{\partial z}\right)}=\frac{dz}{\rho\left(\frac{\partial X}{\partial y}-\frac{\partial Y}{\partial x}\right)}.$$

Hence, the required equations of the equal pressure and equal density are

$$\frac{dx}{\left(\frac{\partial Y}{\partial z}-\frac{\partial Z}{\partial y}\right)}=\frac{dy}{\left(\frac{\partial Z}{\partial x}-\frac{\partial X}{\partial z}\right)}=\frac{dz}{\left(\frac{\partial X}{\partial y}-\frac{\partial Y}{\partial x}\right)}.$$

7.11 Rotating Fluids: To determine the pressure at any point and the surfaces of equal pressure, when a mass of homogeneous liquid contained in a vessel resolves uniformly about a vertical axis

[T.M.B.U. 95H, 97H, 97H, R.U. 92H, C.U. 85H, M.U. 97H, 87H]

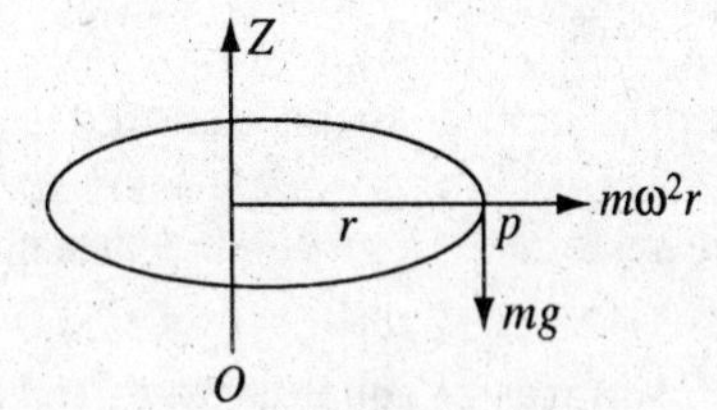

Fig. 7.3

Let a mass of liquid rotate with uniform angular velocity ω about a vertical axis OZ. Then at a point P at a distance r from the axis the acceleration of an element of the liquid is $\omega^2 r$ towards the axis. If the mass of this element be m, the resultant of the external forces acting on the element is $m\omega^2 r$ away from the axis. The only other force the element is subject to its weight mg acting vertically downwards.

Considering the equilibrium of a unit mass we get

$$dp = \rho(\omega^2 r dr - g dz). \tag{7.17}$$

Integrating, we get

$$p = \rho\frac{\omega^2 r^2}{2} - \rho g z + C, \tag{7.18}$$

where C is a constant.

Taking the origin at the vertex of the free surface $p = \pi$ when $z = 0$, $r = 0$.
Hence, from (7.18) we get $\pi = C$.
Therefore from (7.18), we have

$$p = \frac{\rho\omega^2 r^2}{2} - \rho g z + \pi. \tag{7.19}$$

This determines the pressure at any point of the revolving liquid.
Also eqn. (7.19) can be written as

$$\frac{\rho\omega^2 r^2}{2} = p + pgz - \pi, \quad \text{or,} \quad \frac{\rho^2 \omega r^2}{2} = (p - \pi) + \rho g z.$$

$$\therefore \quad r^2 = \frac{2g}{\omega^2} z + (p - \pi)\frac{2}{\rho\omega^2}. \tag{7.20}$$

On a surface of equipressure p = a constant so that $\frac{2(p-\pi)}{\rho\omega^2}$ is a constant quantity k (*say*).
Hence, (7.20) becomes

$$r^2 = \frac{2g}{\omega^2} z + K.$$

This equation represents a system of *paraboloid of revolution with the same latus rectums* $\frac{2g}{w^2}$. *This equation also represents the surface of equipressure.*

[P.U. 65H, 90H]

If the liquid is not homogeneous, ρ is not constant. Then from eqn. (7.17) dp being an exact differential, ρ is a function of $\frac{\omega^2 r^2}{2} - gZ$. Hence, the *same paraboloids of revolution are also the surfaces of equidensity.*

Illustrative Examples

▶ **Ex. 1.** Prove that if the forces per unit mass at (x, y, z) parallel to the axes are $y(a - z)$, $x(a - z)$, xy; the surfaces of equal pressure are hyperbolic paraboloid and the curves of equal pressure and density are rectangular hyperbolas.

[I.A.S. 76, 79, B.P.S.C. 83, C.U. 2000H]

Solution: Here, $X = y(a - z)$, $Y = x(a - z)$, $Z = xy$.

So that

$$\frac{\partial X}{\partial y} = a - z, \quad \frac{\partial X}{\partial z} = -y, \quad \frac{\partial Y}{\partial x} = a - z,$$

$$\frac{\partial Y}{\partial z} = -x, \quad \frac{\partial Z}{\partial x} = y \quad \text{and} \quad \frac{\partial Z}{\partial y} = x.$$

The condition of equilibrium is

$$X\left(\frac{\partial Y}{\partial z} - \frac{\partial Z}{\partial y}\right) + Y\left(\frac{\partial Z}{\partial x} - \frac{\partial X}{\partial z}\right) + Z\left(\frac{\partial X}{\partial y} - \frac{\partial Y}{\partial x}\right) = 0. \tag{1}$$

Now L.H.S. of (1)

$$= y(a-z)(-x-x) + x(a-z)(y+y) + xy[a-z-(a-z)]$$
$$= -2xy(a-z) + 2yx(a-z) = 0 = \text{R.H.S.}$$

For surface of equal pressure $dp = 0$, therefore, $X dx + Y dy + Z dz = 0$ as $\rho \neq 0$, or, $y(a-z)\,dx + x(a-z)\,dy + xy\,dz = 0$.

Dividing by $xy(a-z)$, the surfaces of equal pressure are given by

$$\frac{dx}{x} + \frac{dy}{y} + \frac{dz}{a-z} = 0.$$

Integrating, we get

$$\log x + \log y - \log(a-z) = \log c, \quad \text{or,} \quad \log \frac{xy}{a-z} = \log c.$$

$$\therefore \quad \frac{xy}{a-z} - c, \quad \text{or,} \quad xy = c(a-z)$$

which are hyperbolic paraboloid.

Also, the curves of equal pressure and equal density are given by

$$\frac{dx}{\dfrac{\partial Y}{\partial z} - \dfrac{\partial Z}{\partial y}} = \frac{dy}{\dfrac{\partial Z}{\partial x} - \dfrac{\partial X}{\partial z}} = \frac{dz}{\dfrac{\partial X}{\partial y} - \dfrac{\partial Y}{\partial x}},$$

$$\text{or,} \quad \frac{dx}{-x-x} = \frac{dy}{y+y} = \frac{dz}{(a-z)-(a-z)},$$

$$\text{or,} \quad \frac{dx}{-x} = \frac{dy}{y} = \frac{dz}{0}.$$

The last two equations give $dz = 0$, or, $z =$ constant.

Again, from the first two equations, we have

$$\frac{dx}{x} + \frac{dy}{y} = 0$$

$$\text{or,} \quad \log xy = \text{constant}$$

$$\text{or,} \quad xy = \text{constant}.$$

Then the curves of equal pressure and equal density are given by

$$xy = \text{constant}, \quad z = \text{constant},$$

which are clearly rectangular hyperbolas.

▶ **Ex. 2.** Show that the forces represented by

$$X = \mu(y^2 + yz + z^2), \; Y = \mu(z^2 + zx + x^2), \; Z = \mu(x^2 + xy + y^2)$$

will keep a mass of liquid at rest, if the density $\alpha \frac{1}{(\text{dist})^2}$ from the plane $x+y+z=0$; then the curves of equal pressure and density will be circles.

[Delhi 61, 69, Bombay 61, Agar 57, I.A.S. 72]

Solution: Let d be the distance of point (x, y, z) from the plane $x+y+z=0$, then

$$d = \frac{x+y+z}{\sqrt{3}}.$$

Let ρ be the density of the liquid, then $\rho \alpha \frac{1}{(\text{dist})^2}$

$$\text{or,} \quad \rho \alpha \frac{1}{d^2} \quad \text{or,} \quad \rho = \frac{\lambda}{(x+y+z)^2}, \tag{1}$$

where λ is a constant.

Now the given forces will keep the fluid in equilibrium if

$$\frac{\partial}{\partial y}(\rho Z) = \frac{\partial}{\partial z}(\rho Y), \frac{\partial}{\partial x}(\rho Z) = \frac{\partial}{\partial z}(\rho X), \frac{\partial}{\partial y}(\rho X) = \frac{\partial}{\partial x}(\rho Y).$$

Now,

$$\begin{aligned}
\frac{\partial}{\partial y}(\rho Z) &= \frac{\partial}{\partial y}\left[\frac{\lambda\mu(x^2+xy+y^2)}{(x+y+z)^2}\right] \\
&= \frac{\lambda\mu}{(x+y+z)^4}\left[(x+2y)(x+y+z)^2 - 2(x+y+z)(x^2+xy+y^2)\right] \\
&= \frac{\lambda\mu(x+y+z)}{(x+y+z)^4}\left[(x+2y)(x+y+z) - 2(x^2+xy+y^2)\right] \\
&= \frac{\lambda\mu}{(x+y+z)^3}[x^2+xy+xz+2xy+2y^2+2yz-2x^2-2xy-2y^2] \\
&= \frac{\lambda\mu}{(x+y+z)^3}(-x^2+2yz+xz+xy). \qquad (2)
\end{aligned}$$

Also,

$$\begin{aligned}
\frac{\partial}{\partial z}(\rho Y) &= \frac{\partial}{\partial z}\left[\frac{\lambda\mu(z^2+zx+x^2)}{(x+y+z)}\right] \\
&= \frac{\lambda\mu}{(x+y+z)^4}\left[(2z+x)(x+y+z)^2 - 2(x+y+z)(z^2+zx+x^2)\right] \\
&= \frac{\lambda\mu(x+y+z)}{(x+y+z)^4}\left[(2z+x)(x+y+z) - 2(z^2+zx+x^2)\right] \\
&= \frac{\lambda\mu}{(x+y+z)^3}[2zx+2zy+2z^2+x^2+xy+xz-2z^2-2zy-2x^2] \\
&= \frac{\lambda\mu}{(x+y+z)^3}(-x^2+zx+2yz+xy). \qquad (3)
\end{aligned}$$

Hence from (2) and (3), we get

$$\frac{\partial}{\partial y}(\rho Z) = \frac{\partial}{\partial z}(\rho Y)$$

Similarly, it can be shown that

$$\frac{\partial}{\partial x}(\rho Z) = \frac{\partial}{\partial z}(\rho X) \quad \text{and} \quad \frac{\partial}{\partial y}(\rho X) = \frac{\partial}{\partial x}(\rho Y).$$

Thus, the given forces will keep the mass of liquid at rest if ρ is given by (1).

Again, the curves of equal pressure and density are given by

$$\frac{dx}{\frac{\partial Y}{\partial z} - \frac{\partial Z}{\partial y}} = \frac{dy}{\frac{\partial Z}{\partial x} - \frac{\partial X}{\partial z}} = \frac{dz}{\frac{\partial X}{\partial y} - \frac{\partial Y}{\partial x}}$$

$$\frac{dx}{\mu[2z + x - (x + 2y)]} = \frac{dy}{\mu[2x + y - (y + 2z)} = \frac{dz}{\mu[2y + z - (z + 2x)]}$$

or, $$\frac{dx}{z - y} = \frac{dy}{x - z} = \frac{dz}{y - x} = \frac{dx + dy + dz}{0} = \frac{x\,dx + y\,dy + z\,dz}{0}.$$

Now, $dx + dy + dz = 0$ gives

$$x + y + z = A \tag{4}$$

and $x\,dx + y\,dy + z\,dz = 0$ gives

$$x^2 + y^2 + z^2 = B, \tag{5}$$

where A and B are constants.

Equations (4) and (5) together represent the curves of equal pressure and density. These curves being the curves of intersection of planes and spheres represent circles.

▶ **Ex. 3.** Fluid is contained within a circular tube of radius a in a vertical plane which can rotate about a vertical diameter. If the fluid subtend an angle θ at the centre, the least angular velocity to make the fluid divide into two parts is $\sqrt{\frac{g}{a}} \sec \frac{\theta}{4}$.

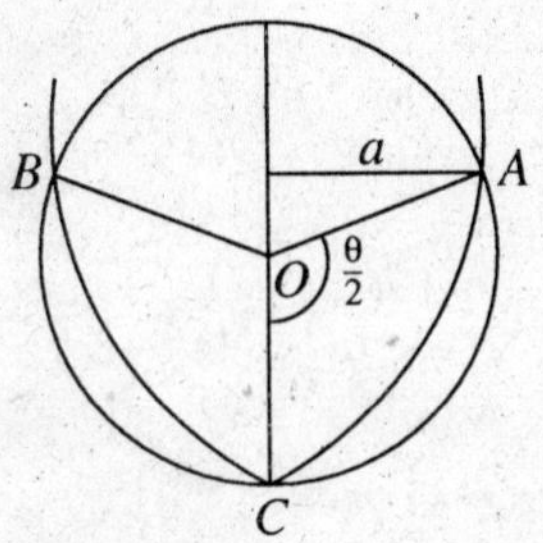

Fig. 7.4

Solution: Let the tube be ABC with centre O, radius a and let it rotate about the vertical OC. By symmetry, $\angle AOC = \angle BOC$.

Since the fluid subtends an angle θ at O, we have

$$\angle AOC = \angle BOC = \frac{\theta}{2}.$$

Let the least angular velocity to make the fluid divide (at the point C) be ω. The pressure at A is zero and it is clearly zero at A and B. Hence, the free surface passes through A, B, C. Taking C as the origin and CO as the axis of Z, the equation of the free surface is

$$r^2 = \frac{2g}{\omega^2} z. \tag{1}$$

For A, we have

$$r = a\sin\left(\pi - \frac{\theta}{2}\right), \quad z = a + a\cos\left(\pi - \frac{\theta}{2}\right)$$

$$\text{or,} \quad r = a\sin\frac{\theta}{2}, \quad z = a\left(1 - \cos\frac{\theta}{2}\right).$$

Since (1) passes through A.

$$a^2\sin^2\frac{\theta}{2} = \frac{2g}{\omega^2}\left(1 - \cos\frac{\theta}{2}\right), \quad \text{or,} \quad a\sin^2\frac{\theta}{2} = \frac{2g}{\omega^4}2\sin^2\frac{\theta}{4}$$

$$\text{or,} \quad \omega^2 = \frac{4g}{a}\frac{\sin^2\frac{\theta}{4}}{\left(2\sin\frac{\theta}{4}\cos\frac{\theta}{4}\right)^2}$$

$$\text{or,} \quad \omega^2 = \frac{g}{a\cos^2\frac{\theta}{4}}.$$

Hence,

$$\omega = \sqrt{\frac{g}{a}}\sec\frac{\theta}{4}.$$

▶ **Ex. 4.** A tank of heavy homogeneous liquid is moved horizontally with uniform acceleration f. Find the pressure at any point. **[Bih. 56H, 60H, R.U. 70H, Bhag. 70H]**

Solution: We take the axis of x along the acceleration and that of y vertically upwards. The reversed effective force per unit mass is $-f$ along the axis of x and the force of gravity per unit mass is $-g$ along the axis of y. Hence, the pressure equation is given by $dp = -\rho(f dx + g dy)$.

Integrating, we get

$$p = C - (fx + gy)\rho, \tag{1}$$

where C is a constant.

On the free surface $p = \pi$, so the equation of the free surface is $\pi = C - \rho(fx + gy)$.

If z be the distance of any point P from the free surface, then

$$Z = \frac{C - \pi - \rho(fx + gy)}{\rho\sqrt{f^2 + g^2}}, \quad \text{or,} \quad C - \pi - \rho(fx + gy) = Z\rho\sqrt{f^2 + g^2}$$

$$\text{or,} \quad C - \rho(fx + gy) = \pi + Z\rho\sqrt{f^2 + g^2}.$$

Thus, from (1), we get pressure at P

$$p = \pi + Z\rho\sqrt{f^2 + g^2}.$$

▶ **Ex. 5.** Show that the pressure at a small depth z below the surface of a sphere of water attracted to the centre of the sphere with a force producing an acceleration $\frac{\mu}{r^2}$ at a distance r is approximately $\pi + \rho g\left(z + \frac{z^2}{a}\right)$, where a is the radius of the sphere and g the attraction of a unit mass at the surface of the sphere.

[P.U. 56H, B.U. 55H, M.U. 64H]

Solution: We know that $\frac{dp}{dr} = \rho F$.
Force per unit mass at a point below the surface of the sphere $= -\frac{\mu}{r^2}$.
Here negative sign is taken since it is a case of attraction.

$$\therefore \quad dp = -\frac{\rho\mu}{r^2}\, dr.$$

Integrating, we get

$$p = \frac{\rho\mu}{r} + C. \tag{1}$$

At the free surface $p = \pi$ when $r = a$.
Hence, from (1)

$$\pi = \frac{\rho\mu}{a} + C. \tag{2}$$

Subtracting (2) from (1), we get

$$p - \pi = \rho\mu\left(\frac{1}{r} - \frac{1}{a}\right). \quad \therefore \quad p = \pi + \rho\mu\frac{1}{r}\left(-\frac{1}{a}\right).$$

$$\begin{aligned} p &= \pi + \rho\mu\left(\frac{1}{a-z} - \frac{1}{a}\right) \quad [\because r = a - z] \\ &= \pi + \rho\mu\frac{[a - a + z]}{(a-z)a} = \pi + \frac{\rho\mu z}{a(a-z)} \\ &= \pi + \frac{\rho\mu z}{a^2\left(1 - \frac{z}{a}\right)} = \pi + \frac{\rho\mu z}{a^2}\left(1 - \frac{z}{a}\right)^{-1} \\ &= \pi + \frac{\rho\mu z}{a^2}\left(1 + \frac{z}{a} + \frac{z^2}{a^2}\right). \end{aligned}$$

As z is very small, powers higher than second can be neglected, so that

$$p = \pi + \frac{\rho\mu z}{a^2}\left(1 + \frac{z}{a}\right) \text{ approximately.}$$

Now,

$$\begin{aligned} g &= \text{acceleration at the free surface} = \frac{\mu}{a^2} \\ p &= \pi + \rho z g\left(1 + \frac{z}{a}\right) = \pi + \rho g\left(z + \frac{z^2}{a}\right). \end{aligned}$$

▶ **Ex. 6.** A conical vessel, of height h and vertical angle 2α, contains water whose volume is one half that of the cone, if the vessel and the contained water revolve with uniform angular velocity ω, and no water overflows, show that ω must not be greater than

$$\sqrt{\frac{2g}{3h}} \cot\alpha.$$

[P.U. 51H, 62H, 78H, B.U. 58H, Bh.U. 65H, 67H, G.U. 04H]

Solution: When the liquid rotates, let V be the vertex of the paraboloid of free surface. If O be the centre of the cone and P a point on the rim, then

$$OP^2 = \frac{2g}{\omega^4}OV. \qquad (1)$$

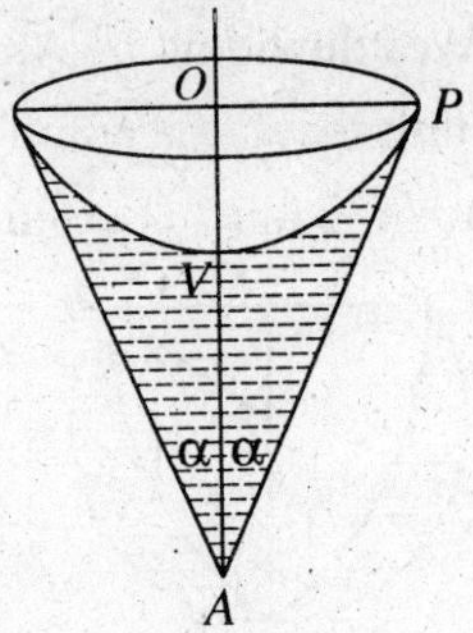

Fig. 7.5

In

$$\triangle OAP \tan\alpha = \frac{r}{h} = \frac{OP}{h}, \; OP = h\tan\alpha.$$

Hence from (1), we have

$$h^2\tan^2\alpha = \frac{2g}{\omega^2}OV. \quad \therefore \quad OV = \frac{\omega^2 h^2 \tan^2\alpha}{2g}. \qquad (2)$$

If no water is to overflow the volume of the empty space POV cannot be greater than half of the volume of the cone.

$$\text{or,} \quad \frac{1}{2}\pi h^2 \tan^2\alpha OV \not> \frac{1}{2}\frac{\pi}{3}h^3\tan^2\alpha,$$

$$\Rightarrow \quad OV \not> \frac{h}{3},$$

$$\Rightarrow \quad \frac{\omega^2 h^2\tan^2\alpha}{2g} \not> \frac{h}{3} \quad \text{From (2)}$$

$$\Rightarrow \quad \omega^2 \not> \frac{2g}{3h}\cot^2\alpha.$$

Hence,

$$\omega \not> \sqrt{\frac{2g}{3h}}\cot\alpha.$$

▶ **Ex. 7.** A hollow sphere of radius a, half filled with liquid is made to rotate with angular velocity ω about its vertical diameter. If the lowest of the sphere is just exposed, show that $2g = a\omega^2\left(2\sqrt[3]{4}\right)$.

[P.U. 62H, 65H, B.U. 62H, 64H, 69H, M.U. 65H, 66H, 69H, 73H, R.U. 67H, 71H, Bh.U. 69H, P.U. 95H]

Solution: When the sphere ABC half full of liquid rotates with angular velocity ω and the lowest point is just exposed, the vertex of the paraboloidal free surface is at A, the lowest point of the sphere.

If the height of the liquid be AK = h, volume of the portion ABC of the sphere

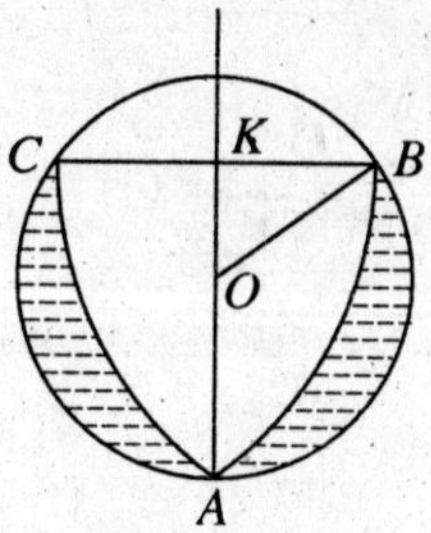

Fig. 7.6

$$\begin{aligned} V_1 &= \pi\int_0^h hy^2dz \\ &= \pi\int_0^h \{a^2-(a-z)^2\}\,dz \\ &= \pi\int_0^h (2az-z^2)\,dz \\ &= \pi\left[az^2-\frac{z^2}{3}\right]_0^h \\ &= \pi\left[ah^2-\frac{h^3}{3}\right] \\ &= \pi h^2\left(a-\frac{h}{3}\right). \end{aligned}$$

Volume V_2 of the liquid $= \frac{2}{3}\pi a^3$.

Volume V_3 of the empty space ABCK

$$= \frac{1}{2}\pi h KB^2 = \frac{1}{2}\pi h\left(\frac{2gh}{\omega^2}\right) = \frac{\pi g h^2}{\omega^2}$$

$$V_1 = V_2 + V_3$$

$$\pi h^2\left(a-\frac{h}{3}\right) = \frac{2}{3}\pi a^3 + \frac{\pi g h^2}{\omega^2}$$

$$\text{or,}\quad h^2\left(a-\frac{h}{3}-\frac{g}{\omega^2}\right) = \frac{2}{3}a^3. \tag{1}$$

Since,

$$KB^2 = \frac{2g}{\omega^2}AK. \tag{2}$$

In $\triangle OKB$,

$$OB^2 = OK^2 + KB^2,\quad \text{or,}\quad a^2 = (AK-OA)^2 + KB^2$$
$$\text{or,}\quad a^2 = (h-a)^2 + KB^2.$$

$$\therefore\quad KB^2 = a^2-(h-a)^2 = a^2-(h^2+a^2-2ah) = 2ah-h^2.$$

Therefore from (2), we have

$$2ah-h^2 = \frac{2g}{\omega^4}AK = \frac{2g}{\omega^2}h \quad \text{or,}\quad 2a-h = \frac{2g}{\omega^2}$$

$$\therefore\quad h = 2\left(a-\frac{g}{\omega^2}\right).$$

Substituting this value of h in eqn. (1) we get

$$4\left(a-\frac{g}{\omega^2}\right)^2\left(a-\frac{h}{3}-\frac{g}{\omega^2}\right)=\frac{2}{3}a^3,$$

$$\text{or,}\quad \frac{4(a\omega^2-g)^2}{\omega^4}\left[a-\frac{2}{3}\left(a-\frac{g}{\omega^2}\right)-\frac{g}{\omega^2}\right]=\frac{2}{3}a^3,$$

$$\text{or,}\quad \frac{2(a\omega^2-g)^2}{\omega^2}\left(\frac{a}{3}-\frac{1}{3}\frac{g}{\omega^2}\right)=\frac{a^3}{3},$$

$$\text{or,}\quad \frac{2(a\omega^2-g)^2(a\omega^2-g)}{3\omega^6}=\frac{a^3}{3},$$

$$\text{or,}\quad (a\omega^2-g)^3=\frac{a^3\omega^6}{2},$$

$$\text{or,}\quad a\omega^2-g=\frac{a\omega^2}{\sqrt[3]{2}},$$

$$\text{or,}\quad g=a\omega^2\left(1-\frac{1}{\sqrt[3]{4}}\right).$$

$$\therefore\quad 2g=a\omega^2\left(2-\sqrt[3]{4}\right).$$

▶ **Ex. 8.** A fluid is equilibrium under a given system of forces; if $\rho_1=\phi(x,y,z)$, $\rho_2=(\psi(x,y,z)$ be two possible values of the density at any point, show that the equations of the surfaces of equal pressure in either case are given by $\phi(x,y,z)+\lambda\psi(x,y,z)=0$, where λ is an arbitrary parameter. **[I.A.S. 77, Raj. 60, 66, 70]**

Solution: Let the given system of forces be X, Y, Z. If p and p′ are the pressures at a point x, y, z in two positions, we have

$$\begin{aligned} dp &= \rho_1(Xdx+Ydy+Zdz)\\ \text{and}\quad dp' &= \rho_2(Xdx+Ydy+Zdz). \end{aligned}$$

There give

$$\frac{dp}{\rho_1}=\frac{dp'}{\rho_2}\quad \text{or,}\quad dp=\frac{\rho_1}{\rho_2}\,dp'.$$

Since dp is a perfect differential.

So, $\frac{\rho_1}{\rho_2}\,dp'$ is also a perfect differential.

Hence, $\frac{\rho_1}{\rho_2}$ is some function of p'.

Let $\frac{\rho_1}{\rho_2}=f(p')=\text{constant}=-\lambda$ (*say*).

$$\therefore\quad \rho_1+\lambda\rho_2=0\quad \text{or,}\quad \phi(x,y,z)+\lambda\psi(x,y,z)=0.$$

▶ **Ex. 9.** A sphere of fluid, whose density ρ at a distance r from its centre varies as $\frac{\sin kr}{kr}$, is self attracting according to the law of gravitation. Prove that the pressure at a distance r from the centre is $\frac{2\pi\gamma}{k^2}(\rho^2-\rho_1^2)$, where ρ_1 is the surface density of the sphere and γ is the constant of gravitation. **[Part. 82H]**

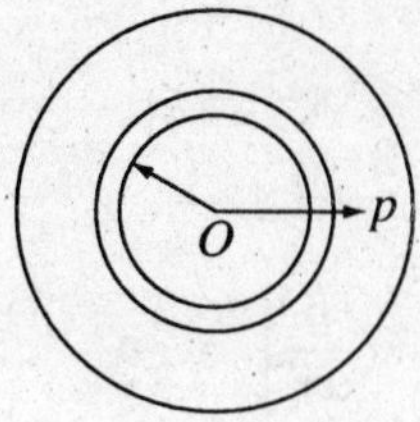

Fig. 7.7

Solution: Let O be the centre of the sphere. We consider a spherical shell of radius x and thickness δx. The attraction of this shell at an internal points is O and at an external point P is $\frac{\gamma M}{OP^2}$, where M is the mass of the shell

$$= \frac{\gamma \cdot 4\pi x^2 \rho \delta x}{r^2}, \quad \text{where } OP = r.$$

Given,

$$\rho a \frac{\sin kx}{kx}.$$

Hence,

$$\rho = \frac{\mu \sin kx}{kx}.$$

Hence, the attraction of the shell at P

$$= 4\pi\gamma\mu \frac{\sin kx x^2}{kxr^2} \delta x = \frac{4\pi\gamma\mu x \sin kx}{kr^2} \delta x.$$

Hence, the total attraction of the fluid at P

$$= \frac{4\pi\gamma\mu}{kr^2} \int_0^r x \sin kx \, dx = \frac{4\pi\gamma\mu}{kr^2} \left[-\frac{x \cos kx}{k} + \frac{\sin kx}{k^2} \right]_0^r$$

$$= \frac{4\pi\gamma\mu}{kr^2} \left[-\frac{r \cos kr}{k} + \frac{\sin kr}{k^2} - 0 \right] = \frac{4\pi\gamma\mu}{k^3 r^2} [\sin kr - kr \cos kr].$$

This is the attraction at P acting along OP.

Therefore, the pressure at P is given by

$$dp = -\rho \left[\frac{4\pi\gamma\mu}{k^3 r^2} (\sin kr - kr \cos kr) \right] dr. \tag{1}$$

Also,

$$\rho = \mu \frac{\sin kr}{kr}.$$

Hence,

$$d\rho = \frac{\mu}{kr^2} [rk \cos kr - \sin kr] \, dr.$$

Hence, (1) becomes

$$dp = \rho \frac{4\pi\gamma}{k^2} d\rho.$$

Integrating, we get

$$p = \frac{2\pi\gamma}{k^2} \rho^2 + C, \tag{2}$$

where C is a constant. But ρ_1 is surface density.

Hence, $p = 0$ when $\rho = \rho_1$.

So from (2), we get

$$0 = \frac{2\pi\gamma}{k^2}\rho_1^2 + C \quad \text{or,} \quad C = -\frac{2\pi\gamma}{k^2}\rho_1^2.$$

Hence, from (2) we get

$$p = \frac{2\pi\gamma}{k^2}(\rho^2 - \rho_1^2).$$

▶ **Ex. 10.** A right circular cylinder, open at the top, is filled with water and the whole of it revolves with angular velocity ω about the axis. If not more than half the water is spilt, prove that the thrust on the base is $\pi\rho g a^2 h\left(1 - \frac{a^2\omega^2}{4h}\right)$, where h is the height, a the radius of the base of the cylinder and ρ is the density of the water. **[P.U. 58H, Bhag. 63H, Bh.U. 73H, C.U. 97H]**

Solution: Let A be the vertex of the free surface, which is a paraboloid of revolution of latus rectum $\frac{2g}{\omega^2}$, and O the centre of the base.

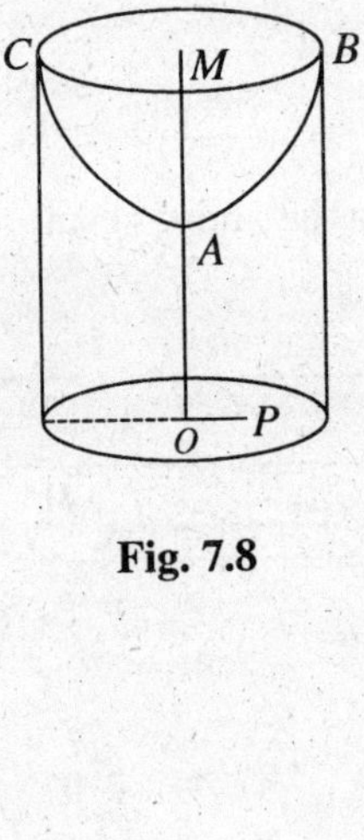

Fig. 7.8

Then pressure at any point is given by

$$p = \frac{\rho}{2}\omega^2 r^2 - g\rho z + k. \qquad (1)$$

At A, $r = 0$, $z = 0$, $p = 0$, then from (1)
$k = 0$

$$\therefore \quad p = \frac{1}{2}\rho\omega^2 r^2 - g\rho z. \qquad (2)$$

At B, we have $r = a$, $z = h - OA$, $p = 0$.

Hence from (2), we get

$$0 = \frac{\rho}{2}\omega^2 a^2 - \rho g(h - OA). \quad \therefore \quad \rho g OA = \rho g h - \frac{1}{2}\rho\omega^2 a^2. \qquad (3)$$

If P is a point on the base and p the pressure there, $OP = r$, then

$$\begin{aligned} p &= \frac{1}{2}\rho\omega^2 r^2 - \rho g(-OA) \\ &= \frac{1}{2}\rho\omega^2 r^2 + \rho g h - \frac{1}{2}\rho\omega^2 a^2 \quad \text{[From (3)]} \\ &= \rho g h - \frac{1}{2}\rho\omega^2(a^2 - r^2). \qquad (4) \end{aligned}$$

Hence, the required thrust

$$
\begin{aligned}
&= \int_0^a \int_0^{2\pi} pr d\theta dr = \int_0^a pr dr 2\pi \\
&= \int_0^a \pi\{2\rho gh - \rho\omega^2(a^2 - r^2)\} r dr \quad \text{[From (4)]} \\
&= \pi \left[2\rho gh \frac{r^2}{2} - \rho\omega^2 \left(a^2 \frac{r^2}{2} - \frac{r^4}{4} \right) \right]_0^a \\
&= \pi \left[\rho gha^2 - \rho\omega^2 \left(\frac{a^4}{2} - \frac{a^4}{4} \right) \right] \\
&= \pi \left[\rho gha^2 - \frac{\rho\omega^2}{4} a^4 \right] = \pi\rho ga^2 h \left[1 - \frac{a^4\omega^2}{4gha^2} \right] \\
&= \pi\rho ga^2 h \left[1 - \frac{a^2\omega^2}{4gh} \right].
\end{aligned}
$$

Also, volume of water spilt

$$
\begin{aligned}
&= \text{volume of the paraboloid} \\
&= \frac{1}{2} \times \text{area of BC} \times \text{MA} < \frac{1}{2} \times \text{area of BC} \times \text{MO} \\
&= \frac{1}{2} \times \text{volume of the cylinder.}
\end{aligned}
$$

Hence, for the above result not more than half the water is spilt.

Exercises 7(VI)

1. If

$$
\begin{aligned}
X &= y(y+z), \\
Y &= z(z+x), \\
Z &= y(y-x),
\end{aligned}
$$

show that surfaces of equal pressures are given by

$$y(x+z) = c(y+z),$$

and the curves of equal pressure and density are given by

$$y(x+z) = \text{constant},$$
$$y+z = \text{constant}.$$

2. If the components parallel to the axes of the forces acting on the element of fluid at (x, y, z) be proportional to

$$y^2 + 2\lambda yz + z^2,$$
$$z^2 + 2\mu zx + x^2,$$
$$x^2 + 2\gamma xy + y^2,$$

then show that if equilibrium be possible, we must have

$$2\lambda = 2\mu = 2\gamma = 1.$$

[Agra 79, L.N.M.U. 95H]

3. A liquid of given volume V is at rest under the forces

$$X = \frac{-\mu x}{a^2},\quad Y = \frac{-\mu y}{b^2},\quad Z = \frac{-\mu z}{c^2},$$

find the pressure at any point of the liquid and the equation of the surface of equal pressures.

[Agra 78, M.U. 95H, T.M.B.U. 91H, 04H]

[*Ans.* $p = -\frac{1}{2}\rho\mu\left(\frac{x^2}{a^2}+\frac{y^2}{b^2}+\frac{z^2}{c^2}\right) + \frac{\rho\mu}{2}\left(\frac{3V}{4\pi abc}\right)^{2/3}$. Equation of the surface of equal pressure is given by $\frac{x^2}{a^2}+\frac{y^2}{b^2}+\frac{z^2}{c^2}$ = constant.]

4. A liquid occupies a portion of a fine circular tube of radius a, subtending an angle $\pi + \theta$ at the centre. When the tube rotates uniformly about a vertical tangent the liquid just reaches the top of the vertical diameter, prove that the angular velocity is

$$\sqrt{\frac{g}{a\left(\tan\frac{\theta}{2} - \sin^2\frac{\theta}{2}\right)}}.$$ [P.U. 64H]

5. A circular cylinder of radius a floats in liquid at rest with its axis vertical and a length h unimmersed. Show that, if the cylinder is sufficiently long, it will float in equilibrium with its supper rim in the surface provided that the liquid is made to rotate with angular velocity $\frac{2\sqrt{gh}}{a}$.

6. When a cylinder open at the top and half-full of liquid rotates with angular velocity ω about its axis which is vertical, the liquid just reaches the upper rim. Prove that the velocity in order that $\frac{1}{n}$th of the liquid may remain in the cylinder is $\omega\sqrt{n}$.

7. A hollow cone, vertex upwards, is three-quarters full of water, and is set rotating about its axis which is vertical, with an angular velocity $\sqrt{\frac{8g}{3h}}\cot\alpha$, where α is the semivertical angle and h is the height of the cone. Show that the thrust on the base is to the weight of water in vessel as $10 : 3$.

[M.U. 93H]

8. A hemispherical shell of radius $2a$ containing water rotates with an angular velocity $\sqrt{\frac{2a}{3g}}$ about its axis which is vertical, a sphere of radius a rests on water with its lowest point in contact with the shell without pressing on it. The free surface of the fluid passes through the rim of the shell. Prove that density of sphere : density of water $= 3 : 8$.

[P.U. 53H, Bih. 57H, 62H, 68H]

9. Fluid is rotating in a cylinder of radius a, whose bottom is closed by a conical surface of semivertical angle α, the vertex being downwards. Prove that, the pressure at the surface of the cone is a minimum at a point distant $\frac{l}{2}\cot\alpha$, from the axis, where l is the latus rectum of the free surface provided $l < 2a\tan\alpha$.

10. A straight tube making an angle α with the vertical and filled with fluid of density ρ, rotates with uniform angular velocity ω about a vertical axis through its lower end which is closed. Prove that when the atmospheric pressure is Π, greatest length of the tube so

that no fluid may flow out is

$$\frac{\rho g \cos\alpha + \omega \sin\alpha\sqrt{2\Pi\rho}}{\omega^2\rho\sin^2\alpha}.$$

11. A closed vessel whose form is that of a right circular cylinder of radius a with plane ends, perpendicular to the axis of the cylinder, is nearly filled with liquid of density ρ, and placed with its axis vertical. A repelling centre of force ($\mu\times$ distance) resides at the centre of the base. Show that if h, the height of the cylinder be $\frac{g}{\mu}$, the resultant thrusts on the top and on the bottom will be in the ratio $a^2 : a^2 + 2h^2$.

12. A given volume of liquid is at rest on a fixed plane, under the action of a force, to a fixed point in the plane, varying as the distance. If the free surface be hemispherical, find the thrust on the plane.

[P.U. 66H, P.U. 62H]

[*Ans.* $\frac{1}{4}\pi\mu\rho a^4$]

13. A hollow sphere half-filled with water is moved with constant acceleration in a horizontal straight line. When the contents are in relative equilibrium the pressure at any point is $\rho z\sqrt{f^2+g^2}$, f being the acceleration and z the distance from the free surface, prove it.

14. A fine straight tube of length l, closed at both the ends and inclined to the vertical at an angle α; is just filled with liquid of density ρ. If the tube rotates with uniform angular velocity ω round a vertical axis through the lower end, prove that the pressure at the highest point is

$$\frac{1}{2}\rho\omega^2\sin^2\alpha\left(l - \frac{g\cos\alpha}{\omega^2\sin^2\alpha}\right)^2$$

or, zero according as $\omega^2 >$ or, $< \frac{g\cos\alpha}{l\sin^2\alpha}$.

[P.U. 51H, 67H, R.U. 65H, C.U. 91H, 01H]

15. A circular cylinder, radius a and height h; contains liquid to a height l. Show that if the cylinder and liquid rotate about their common axis, which is vertical, the angular velocity in order that the liquid may just not flow out, is

$$\frac{2}{a}\sqrt{g(h-l)} \quad \text{or,} \quad \frac{h}{a}\sqrt{\frac{g}{l}}$$

according as l is $>$ or $< \frac{h}{2}$.

[P.U. 47H, M.U. 63H, Raj. 70]

16. A cylinder containing water rotates round its vertical axis at such a speed that the cylinder is half empty and the bottom is just uncovered. Prove that volume of the liquid left in the cylinder if it rotates at a higher speed varies inversely as the square of the angular velocity. [B.U. 64H]

17. A hollow circular cylinder of radius r, height h and effective specific gravity σ, floats in water which is rotating round a vertical axis coincident with the axis of the cylinder; and the cylinder is open at its upper end. Show that the maximum angular velocity ω of the fluid so that the cylinder shall not sink is given by

$$r^2\omega^2 = 4gh(1-\sigma).$$

[Bih. 61H, 67H]

18. A sphere of radius a floats in a rotating fluid of four times its density, and the surface of the fluid meets it along a great circle. Find the angular velocity of rotation.

[B.U. 69H]

[*Ans.* $2\sqrt{\frac{g}{3a}}$]

19. A vessel, in the form of a paraboloid of revolution formed by the revolution of a parabola of latus rectum $4a$ about its axis, is filled to half its height with liquid, show that the greatest angular velocity with which it can revolve about its axis so that no liquid is spilt is $\frac{1}{4}\sqrt{\frac{6g}{a}}$. [C.U. 93H]

20. A given volume of heavy liquid is at rest under the action of a force to a fixed point, varying as the distance from that point. Find the surface of equipressure and the free surface.

[*Ans.* Surfaces of equipressure are sphere of centre O. The free surface is also a similar sphere]

21. Prove that, if the density of a liquid at rest under gravity varies as the square root of the pressure, the density increases uniformly with the depth. [M.U. 68H]

22. A semicircular tube is filled with water, and rotates about the diameter joining the two ends, which is vertical, where must a hole be made in the tube so that all the liquid may escape.

[R.U. 62H, M.U. 76H]

[*Ans.* $\frac{g}{\omega^2}$ below the centre]

23. A uniform semicircular closed tube of radius a is tightly filled with equal volumes of two liquids of densities ρ and σ which do not mix, and is rotated with angular velocity ω about a vertical radius making an angle α with the line of symmetry. Prove that the pressures at the two ends will be equal, if

$$\frac{\omega^2 a}{2g}(\sigma-\rho) = \frac{\sigma}{\cos\alpha+\sin\alpha} - \frac{\rho}{\cos\alpha-\sin\alpha}$$

the liquid of density σ being the lower of the two, and the convexity of the tube being downwards.

24. A cylinder of radius a is filled with water to a depth h. A fine vertical tube is at a distance $d(> a)$ from the axis of the cylinder and communicates with the bottom of the vessel. If the whole is made to rotate uniformly about the axis of the cylinder, show that the angular velocity w necessary to make the water rise in the tube to a height x above the bottom of the cylinder is given by

$$\omega^2\left(d^2-\frac{a^2}{2}\right) = 2g(x-h).$$

[Bhag. 62H]

25. A spherical shell of radius a is just filled with liquid of density ρ and the whole rotates with uniform angular velocity w about the vertical diameter. Find at what level the pressure on the shell is greatest and find the resultant thrusts on the upper and the lower hemispheres.

[*Ans.* Pressure greatest at a point $\frac{g}{\omega^2}$ below the centre of the sphere; $\frac{\pi a^3\rho}{12}(4g + 3a\omega^2)$ and $\frac{\pi\rho a^3}{12}(20g+3a\omega^2)$.]

26. If the liquid be heterogeneous and of density $\frac{\rho z}{a}$ at a depth z, find the pressure at a depth z, given that Π is the atmospheric pressure.

[P.U. 64H]

[*Ans.* $\Pi + \frac{\rho g z^2}{2a}$]

27. Find the surface of equipressure when a homogeneous liquid is subject to a constant force in a fixed direction and a force towards a fixed point varying as the distance. [B.U. 66H]

[*Ans.* $x^2 + y^2 + z^2 + 2az + b = 0$]

28. A mass of fluid rests upon a plane subject to a central attractive force $\frac{\mu}{\gamma^2}$ situated at a distance from the plane on the side opposite to that on which is the fluid, show that the pressure on the plane $= \frac{\pi\rho\mu(a-c)^2}{a}$. [C.U. 86H]

29. A tube of small section is in the form of three sides of a square, of which the middle one is horizontal; it is then filled with water and revolves about a vertical axis through the middle point of the horizontal side; prove that no liquid escapes unless $w > \sqrt{\frac{8g}{a}}$ and that, of this is so, the amount that escapes would fill a length of the tube equal to $a\sqrt{1 - \frac{8g}{aW^2}}$, where a is the side of the square.

If the tube revolves about a vertical side, prove that the amount of water that escapes would fill a length $\frac{w^2a^2}{2g}$ or $a + \sqrt{a^2 - \frac{2ag}{w^2}}$ according as $w >$ or $< \sqrt{\frac{2g}{a}}$. [C.U. 95H]

CHAPTER 8

Equilibrium of Floating Bodies

8.1 Introduction

According to Archimedes principle *When a body is partially or wholly immersed in a fluid at rest, the resultant thrust of the fluid on the solid is equal and opposite to the weight of the fluid displaced by the body and acts vertically upwards through the C.G. of the fluid displaced.*

The weight of the liquid displaced by the body is known as *the force of buoyancy* and the centre of gravity of this displaced liquid is called *the centre of buoyancy*.

[R.U. 95H]

If the weight of the body is greater than the weight of the liquid displaced, it will sink.

If the weight of the body is equal to the weight of the liquid displaced, it will sink. So, we state the *principle of floatation* which states that *The weight of a body floating freely in a fluid is equal to the weight of the fluid displaced by it.*

8.2 Conditions of Equilibrium of a body floating freely in a Homogeneous Liquid

[C.U. 93H, 97H, R.U. 91H, 93H, 96H]

When a body is freely floating in a homogeneous liquid, the only two forces acting on it are:

- Weight W of the body acting vertically downwards through G, the centre of gravity of the body.
- The force of buoyancy F acting vertically upwards through H, the centre of buoyancy, i.e., the centre of gravity of the liquid displaced.

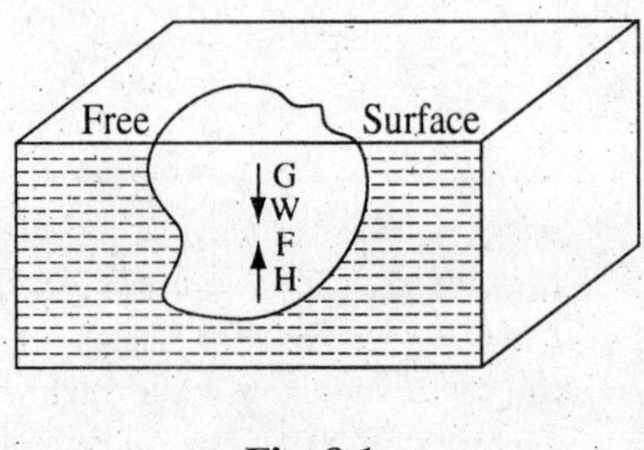

Fig. 8.1

For equilibrium these two forces must balance each other.

Thus, the conditions of equilibrium for a body floating freely in a homogeneous liquid are:

- The weight of the displaced liquid must be equal to the weight of the body.
- The centre of gravity of the body and the displaced liquid must be in the same vertical line.

8.3 Relations between the Volumes and Densities

We will establish a relation between the volume and the mean density of the floating body, and the volume of the body immersed and the density of the liquid.

Let V and ρ be the volume and mean density of a body floating in a liquid of density ρ' with a volume V' immersed inside the liquid.

Then, since the weight of the body = the weight of the liquid displaced

$$V\rho g = V'\rho' g, \text{ or, } V' = \frac{V\rho}{\rho'}. \tag{8.1}$$

Hence, of a solid of volume V and mean density ρ is floating in a liquid of density ρ', the volume immersed inside the liquid $\frac{V\rho}{\rho'}$.

Note 1. *For a floating body*

$$V' < V \text{ [From (8.1)]}$$
$$\textit{or,} \quad \rho < \rho'.$$

Hence, we conclude that a solid cannot float in liquid of less density than its own.

Note 2. *We define*

$$\textit{mean density} = \frac{\textit{mass of the body}}{\textit{volume of the body}}$$

which is generally different from the actual density. It is this fact which accounts for the ship to float. Though the density of iron is greater than that of sea-water yet the ship floats in sea-water because the volume occupied by the ship is much greater than the actual volume occupied by the iron, and hence the mean density of the ship is much less than the density of iron and also less than the density of sea-water.

8.4 Conditions of Equilibrium of a body floating in two or more liquids which do not mix

[R.U. 92H, 94H, L.N.M.U. 94H, 98H, T.M.B.U. 01H]

Let a body floats in two liquids, which do not mix. Let W be the weight of the body acting vertically downwards and and W_1, W_2 be the weights of the two displaced liquids. The forces of buoyancy, tnen are W_1 and W_2 acting vertically upwards. We, therefore, have for equilibrium

$$W = W_1 + W_2,$$

where W, W_1, W_2 are all acting vertically.

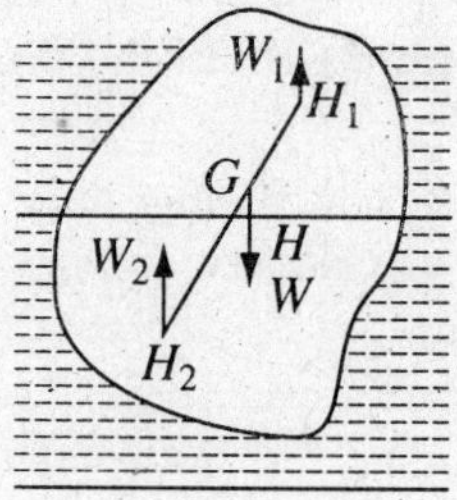

Fig. 8.2

Hence, the required conditions of equilibrium of a body floating in two liquids, are:

- The weight of the body must be equal to the sum of the weights of the two liquids displaced by the respective portion of the body.
- The centre of gravity of the body and the centre of gravity of the displaced liquids must be in the same vertical line.

Note 1. *The above conditions also include the case of a body floating partly in liquid and partly in air.*

Note 2. *Generalising the above result, we have*

$$W = \sum_{r=1}^{n} Wr.$$

Therefore, the above conditions are also true when the body is floating in more than two liquids.

8.5 Conditions of Equilibrium of a body floating in a Liquid

Case I. *When one point of the body is fixed.*

[M.U. 86H, 90H, R.U. 82, Mith. 82, 86H, B.U. 75, 81, 86H, B.B.A.B.U. 95H, M.U. 95H, R.U. 97H, T.M.B.U. 91H, 98H]

Let a string be attached to a point P of the body. Let W be the weight of the body, W' the weight of the liquid displaced by the body and T the tension in the string attached to the body.

The forces acting on the body are:

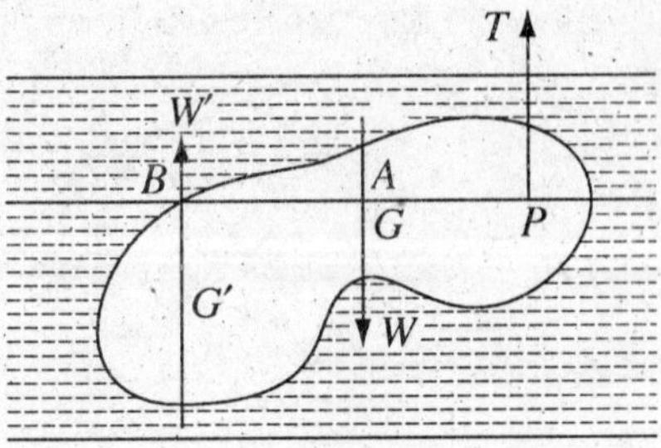

Fig. 8.3

- Weight W of the body acting vertically downwards through G, the C.G. of the body.
- Weight W' of the displaced liquid, i.e., force of buoyancy acting vertically upwards through G', the centre of buoyancy.
- Tension T of the string acting vertically upwards through the point of attachment P.

Let a horizontal line through P meet the line of action of W and W' in A and B, respectively.

Hence, for equilibrium

- The moments of W and W' about P balance each other, i.e., $W \cdot PA = W' \cdot PB$.
- P, G and G' must be in a vertical plane.
- $T + W' = W$ or $T = W - W'$.

If the body be lighter than the liquid

$$T = W' - W.$$

Note 1. *In case of a* body free to turn about a point *can be treated in a similar manner, simply by replacing the tension* T *of the string by a reaction* R *at the fixed point.*

Case II. *When two points of the body are fixed.*

Since two points of the body are fixed, it means that the straight line joining the fixed points is fixed. This means that the body is free to turn about a fixed line. Hence, for equilibrium *the moments of W and W′ about the line must be equal and opposite.*

8.6 Conditions of Equilibrium of a body immersed in a Liquid and supported by a string [M.U. 96H, R.U. 94H, 03H]

Let W be the weight of the body, W' is the weight of the liquid displaced by the body and T be the tension of the string attached to a point A of the body.

The forces acting on the body are:

- Weight W of the body immersed and acting vertically downwards through G, the C.G. of the body.
- Weight W' of the displaced liquid, i.e., force of buoyancy acting vertically upwards through H, the centre of buoyancy.
- Tension T of the string acting vertically upwards through the point of attachment P.

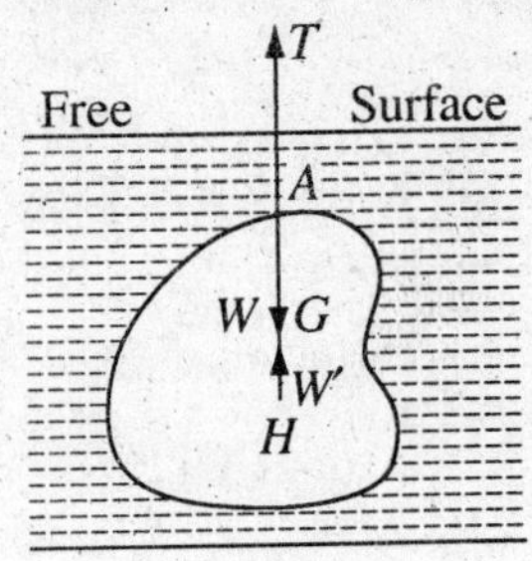

Fig. 8.4

Hence, for equilibrium

$$T + W' = W$$
$$\text{or,}\quad T = W - W'.$$

i.e., tension of the string = weight of the body − weight of the displaced liquid.

Note 1. *If the density of the body be lighter than the density of the liquid, then*

$$T = W' - W.$$

Note 2. *The lines of action of the weight W of the body, the weight W' of the displaced liquid and the tension T of the string must be in the same vertical plane.*

8.7 A body is Floating partly immersed in Liquid and with a String attached to a Point of it. To find the condition of equilibrium when

(*i*) the sp. gr. of the body > the sp. gr. of the liquid
and (*ii*) the sp. gr. of the body < the sp. gr. of the liquid

[M.U. 98H, 92H, 03H, 02H]

Case I. *The sp. gr. of the body > sp. gr. of the liquid.*

Let W be the weight of the body acting vertically downwards through G, the C.G of the body.

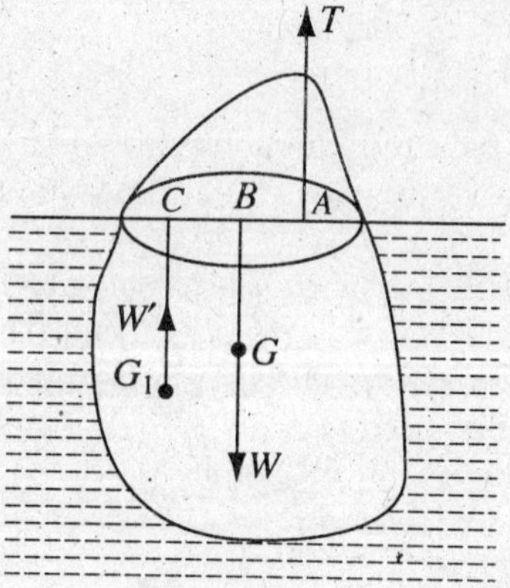

Fig. 8.5

Let W' be the force of buoyancy, which is equal to the weight of the liquid displaced and acts vertically upwards through G_1, the centre of buoyancy or the C.G of the liquid displaced.

$\therefore$ sp. gr. of the body > sp. gr. of the liquid and volume of the liquid displaced is also less than the volume of the body.

$$\therefore \quad W > W'.$$

Hence, for equilibrium, the tension T of the string should be acting vertically upwards. Let the string be attached to the body at P.

Let the vertical lines through P, G and G_1 meet the surface of the liquid in A, B and C, respectively.

Since the vertical forces acting on the body are T, W and W' and the body is in equilibrium. Hence, the algebraic sum of moments of all the forces about A is zero.

$$\therefore \quad W \cdot AB = W' \cdot AC.$$

Hence, the conditions of equilibrium in this case are:

(*i*) $W = T + W'$ and

(*ii*) $W \cdot AB = W' \cdot AC$.

Case II. *Sp. gr. of the body < sp. gr. of the liquid. As in Case I, let W and W′ be the weights of the body and the liquid displaced. Also in this case*

sp. gr. of the body < sp. gr. of the liquid

$\therefore$ The force of buoyancy W' will be greater than the weight of the body and the body will ascend.

In order to prevent the body from ascending a string will be attached at P and the body will be pulled downwards.

Hence, tension T in the string will be acting vertically downwards.

Hence, as in the last case, the conditions of equilibrium are:

(*i*) $W' = T + W$ and

(*ii*) $W \cdot AB = W' \cdot AC$.

8.8 Weighing a Body immersed in a Liquid

If a body immersed in a liquid, be weighed by a balance, the tension of the string supporting the body will not indicate the true weight of the body. The weight indicated is known as *apparent weight of the body*.

Hence,

Apparent weight of the body = True weight of the body − Weight of the liquid displaced by it.

8.9 Weighing in Air

As seen above, if the true weight of a body is required, the body should be weighed in vacuum which is very difficult to obtain. However, we can get the true weight as explained below.

To find the true weight of a body whose density is ρ when it is weighed by means of weights whose density is ρ′, σ being the density of air. **[T.M.B.U. 95H, 2000H, 04H]**

Let w be the true weight of the body and w be its corresponding apparent weight, which is equal to the sum of the weights used.

Let T be the tension of the string supporting the given body.

Then T = W− the weight of air displaced by the body

$$= W - \frac{W}{\rho}\sigma.$$

Also the tension T′ of the string supporting the weights is given by

$$T' = W - \text{weight of air displaced by } W$$

$$= w - \frac{w}{\rho'}\sigma.$$

For a correct balance the tensions T and T′ of the two strings must be equal

$$\therefore \quad W - \frac{W}{\rho}\sigma = w - \frac{w}{\rho'}\sigma,$$

$$\text{or,} \quad W\left(1 - \frac{\sigma}{\rho}\right) = W\left(1 - \frac{\sigma}{\rho'}\right).$$

$$\therefore \quad W = \frac{\left(1 - \frac{\sigma}{\rho'}\right)}{1 - \frac{\sigma}{\rho}} w$$

$$= w\left(1 - \frac{\sigma}{\rho'}\right)\left(1 - \frac{\sigma}{\rho}\right)^{-1}$$

$$= w\left(1 - \frac{\sigma}{\rho'}\right)\left(1 + \frac{\sigma}{\rho} + \frac{\sigma^2}{\rho^2} + \cdots\right).$$

As, the density of the air compared to that of the substance of the weights is very small, higher powers of $\frac{\sigma}{\rho}$ or $\frac{\sigma}{\rho'}$ may be neglected.

Thus,

$$W = W\left(1 - \frac{\sigma}{\rho'} + \frac{\sigma}{\rho}\right).$$

8.10 Principle of Potential Energy and Work Done

A mass of heavy liquid at rest possesses potential energy due to its position. When the liquid is without friction; change from one position to another takes place so slowly that the kinetic energy may be neglected, then loss of potential energy will be equal to the work done by the gravitational forces and again potential energy will be equal to the work done against the gravitational forces.

Thus, the principle of potential energy and work done can be applied to solve some problems of Hydrostatics in which the fluid changes its position very slowly.

Illustrative Examples

▶ **Ex. 1.** If the sp. gr. of air be S and W, W' be the weights of a body in air and water, respectively, prove that its weight in vacuum is

$$W + \frac{S}{1-S}(W - W')$$

[M.U. 73S, B.U. 62, P.U. 78A, 80A, T.M.B.U. 97H, 02H]

Solution: Let W_1 be the weight of the body in vacuum and S_1 be its sp. gr., then its volume $= \frac{W_1}{S_1}$

$$\begin{aligned} W &= \text{apparent weight in air} \\ &= \text{actual weight} - \text{weight of air displaced} \\ &= W_1 - \frac{W_1}{S_1}S \qquad (1) \end{aligned}$$

Also

$$\begin{aligned} W' &= \text{apparent weight of the body in water} \\ &= \text{true weight of the body} - (\text{weight of water displaced by it}) \\ W' &= W_1 - \frac{W_1}{S_1}\cdot 1 \qquad (2) \end{aligned}$$

Eliminating $\frac{W_1}{S_1}$ between (1) and (2), we get

$$W - W_1 = (W' - W_1)S$$

or, $W_1(1-S) = W - W'S$

or, $W_1(1-S) = W(1-S) + (W - W')S$

or, $W_1 = W + \frac{S}{1-S}(W - W').$

▶ **Ex. 2.** A solid displaces $\frac{1}{a}, \frac{1}{b}, \frac{1}{c}$ of its volume, respectively when it floats in three different liquids. Find what fraction of its volume it displaces when it floats in a mixture formed of equal volumes of the liquids.

[Bih. 56A, 76A, M.U. 63A, 69A, 78H, R.U. 65A, 76, B.P.S.C. 84, P.U. 80A, Raj. 74, Agra 83]

Also find the volume it displaces when if floats in a mixture formed of equal weight of the fluids. **[P.U. 64A, R.U. 65A, Bhag. 62A]**

Solution: Let ρ_1, ρ_2 and ρ_3 be the densities of the three liquids. Then the density of the mixture when equal volumes are mixed is

$$\sigma_1 = \frac{\rho_1 + \rho_2 + \rho_3}{3}. \tag{1}$$

Also the density of the mixture when equal weights are mixed is

$$\sigma_2 = \frac{3}{\frac{1}{\rho_1} + \frac{1}{\rho_2} + \frac{1}{\rho_3}} = \frac{3\rho_1\rho_2\rho_3}{\rho_2\rho_3 + \rho_1\rho_3 + \rho_1\rho_2}. \tag{2}$$

Let V be the volume and ρ be the density of the solid, then

$$\begin{aligned} V\rho &= \frac{V\rho_1}{a} = \frac{V\rho_2}{b} = \frac{V\rho_3}{c} \\ \text{or, } \rho &= \frac{\rho_1}{a} = \frac{\rho_2}{b} = \frac{\rho_3}{c}. \end{aligned} \tag{3}$$

If V_1 is the volume of the liquid immersed in a mixture of equal volumes of the fluids, then

$$\begin{aligned} V\rho &= V_1\sigma_1 \\ \therefore \quad V_1 &= \frac{V\rho}{\sigma_1} = \frac{3V\rho}{\rho_1 + \rho_2 + \rho_3} \quad \text{From (1)} \\ &= \frac{3V\rho}{\rho(a+b+c)} = \frac{3V}{a+b+c} \quad \text{From (3).} \end{aligned}$$

If V_2 is the volume of the solid immersed in a mixture of equal weights of the liquids, then

$$\begin{aligned} V\rho &= V_2\sigma_2 \\ V_2 &= \frac{V\rho}{\sigma_2} = \frac{V\rho(\rho_1\rho_2 + \rho_2\rho_3 + \rho_3\rho_1)}{3\rho_1\rho_2\rho_3} \quad \text{From (2)} \\ &= \frac{V\rho^3(ab + bc + ca)}{3\rho^3 abc} \quad \text{From (3)} \\ &= \frac{1}{3}\frac{V(ab + bc + ca)}{abc}. \end{aligned}$$

▶ **Ex. 3.** A thin rod of weight W is loaded at one end with a weight P of insignificant volume. If the rod floats in an inclined position with $\frac{1}{n}$th of its strength out of water, prove that $(n-1)P = W$.

[M.U. 65S, 74A, 68A, R.U. 65A, B.U. 62S, 59S, Bh.U 64A, 66S, Mith. U 80A, 82A, 88H, P.U. 59S, 65A, P.U. 63A, 65A, 72A, R.U. 80A, 83H, R.U. 91H, 93H, 97H, T.M.B.U. 2000H]

Solution: Let AB be the rod of length $2a$ floating with its length BC immersed in the liquid $AC = \frac{1}{n}2a$ is outside the liquid.

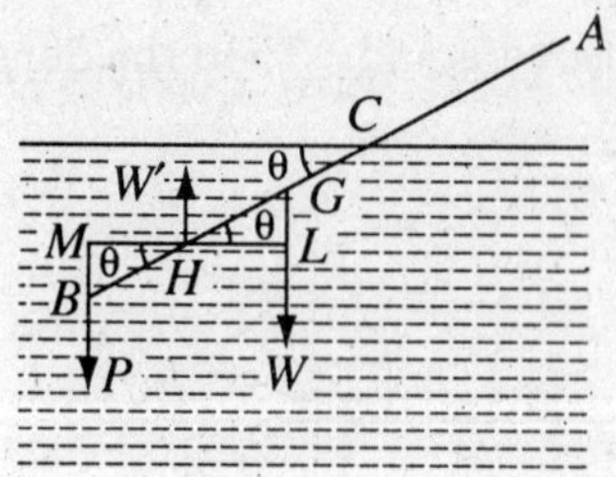

Fig. 8.6

Let G and H be the middle points of AB and BC. Let the load P is attached at the end B.

$$CB = AB - AC = 2a - \frac{2a}{n}$$
$$= 2a\left(1 - \frac{1}{n}\right) = \frac{2a(n-1)}{n}.$$

Also

$$AG = BG = a$$

and $$BH = \frac{1}{2}BC = \frac{1}{2}\frac{2a(n-1)}{n} = \frac{a(n-1)}{n}.$$

Also

$$HG = BG - BH$$
$$= a - \frac{a(n-1)}{n} = \frac{a[n-n+1]}{n} = \frac{a}{n}.$$

Now the forces acting on the rod are:

- The weight W of the rod, acting vertically downwards through G, the C.G. of the rod.
- The force of buoyancy acting vertically upwards through H, the centre of buoyancy, i.e., the C.G. of the rod CB.
- The weight P attached at the end B, acting vertically downwards.

Hence, for equilibrium, the moments of forces (1) and (3) about H must be equal, since the moment of the force (2) about H is zero. Therefore, if θ is the angle that rod makes with horizontal.

$$P \cdot BH\cos\theta = W \cdot HG\cos\theta$$
$$P \cdot a \cdot \frac{(n-1)}{n} = W\frac{a}{n},$$
or, $$P(n-1) = W.$$

► **Ex. 4.** A body of sp. gravity σ when weighted against a weight of sp. gravity ρ in water, the whole balance being immersed, appears to have a weight W. Show that its true weight is

$$W\left(\frac{\sigma}{\sigma-1}\right)\left(\frac{\rho-1}{\rho}\right)$$

[I.A.S. 78, B.U. 80A, Lucknow 74, 82, Vik. 80]

Solution: Let W′ be the true weight of the body.

Apparent weight of the body in water = true weight of the body – weight of the liquid displaced by the body.

$$= W' - \frac{W'}{\sigma g} 1 \cdot g \quad (\because \text{ sp. gravity of water is 1})$$

$$= W' - \frac{W'}{\sigma}.$$

Apparent weight of the weights in water = true weight of the weights – weight of the liquid displaced by weights.

$$= W - \frac{W}{\rho g} 1 \cdot g$$

$$= W\left(1 - \frac{1}{\rho}\right).$$

As, the weights balance the body when the whole system is wholly immersed in the water.

Hence, apparent weight of the body = apparent weight of the weights

$$W'\left(1 - \frac{1}{\sigma}\right) = W\left(1 - \frac{1}{\rho}\right)$$

$$\text{or,} \quad W' = W\left(\frac{\rho}{\rho - 1}\right)\left(\frac{\rho - 1}{\rho}\right).$$

▶ **Ex. 5.** A uniform rod of length $2l$ can turn freely about one end which is fixed at height $h (< 2l)$ above the surface of the liquid. If the densities of the rod and the liquid be ρ and σ show that the rod can rest either in a vertical position or inclined at an angle θ to the vertical such that

$$\cos\theta = \frac{h}{2l}\sqrt{\frac{\sigma}{\sigma - \rho}}$$

[I.A.S. 88, B.P.S.C. 82, P.U. 61A, 62H, R.U. 62A, 65S, 68A, 70A, 75S, 76A, 69H, M.U. 67A, Bih. 72A, Mith. 75A, Agra 75, 78, 82, Luck. 78, Raj. 75, 78, Vik. 65, 67, C.U. 82H, 90H, M.U. 92H]

Solution: Let the rod AB is fixed at the point A about which it is movable.

The portion BC of the rod is immersed in the liquid.

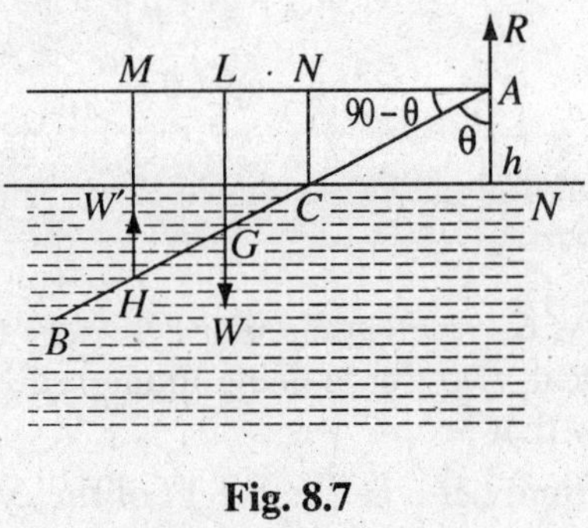

Fig. 8.7

Let the rod be inclined at an angle θ to the vertical in any position. From A we draw AN perpendicular to the free surface. Let S be the uniform cross-section of the rod. The condition $h < 2l$ implies that the rod is sub-merged in the liquid.

Forces acting on the rod are:

- The weight W of the rod acting vertically downwards through G, the C.G. of the rod.
- Force of buoyancy W' acting vertically upwards through H, the centre of buoyancy.
- Reaction R at the fixed point A.

Since there is equilibrium, taking moments about A, we have

$$\begin{aligned} W\cdot AL &= W'\cdot AM \\ W\cdot AG\sin\theta &= W'\cdot AH\sin\theta \\ \text{or,}\quad (W\cdot AG - W'AH)\sin\theta &= 0. \end{aligned}$$

If $\sin\theta = 0$, then $\theta = 0$.

Hence, the rod rests in a vertical position.

If $W\cdot AG - W'\cdot AH = 0$.

$$W\cdot AG = W'AH. \tag{1}$$

$$AG = \frac{AB}{2} = \frac{2l}{2} = l.$$

$$\begin{aligned} AH &= AB - BH \\ &= 2l - \tfrac{1}{2}BC \\ &= 2l - \tfrac{1}{2}(AB - AC) \\ &= 2l - \tfrac{1}{2}(2l - h\sec\theta) \\ &= \tfrac{1}{2}(2l + h\sec\theta). \end{aligned}$$

Also $W = (2lS)\rho g$ and $W' = S(BC)g\sigma = S(2l - h\sec\theta)g\sigma$.

Hence, (1) becomes

$$\begin{aligned} (2lS)\rho g\cdot l &= S(2l - h\sec\theta)g\sigma\tfrac{1}{2}(2l + h\sec\theta), \\ \text{or,}\quad 4l^2\rho &= (2l - h\sec\theta)(2l + h\sec\theta)\sigma, \\ \text{or,}\quad 4l^2\rho &= (4l^2 - h^2\sec^2\theta)\sigma, \\ \text{or,}\quad 4l^2(-\rho+\sigma) &= h^2\sigma\sec^2\theta, \\ \text{or,}\quad \cos^2\theta &= \frac{h^2\sigma}{4l^2(\sigma-\rho)}. \end{aligned}$$

$$\therefore\quad \cos\theta = \frac{h}{2l}\sqrt{\frac{\sigma}{\sigma-\rho}}.$$

▶ **Ex. 6.** A right circular cylinder of specific gravity ρ floats in water with its axis vertical, two-thirds being immersed in water. If σ be the specific gravity of the air, show that $3\rho = 2 + \sigma$. **[M.U. 66S, 72A, Bhag. 71A, 73S, Mith. 74A, M.U. 04H]**

Solution: Let h be the height of the cylinder and r be the radius of its base.

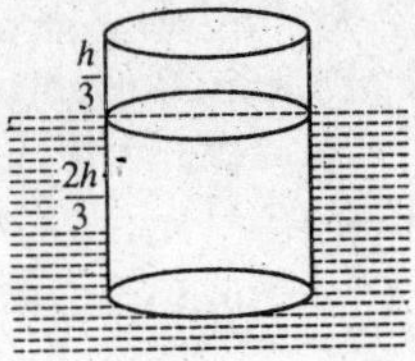

Fig. 8.8

Then the weight of the cylinder

$$= \pi r^2 h\rho g.$$

The weight of water displaced

$$= \pi r^2 \frac{2h}{3} g.$$

The weight of air displaced

$$= \pi r^2 \frac{h}{3} \sigma g.$$

For equilibrium,

weight of the cylinder = weight of air displaced + weight of water displaced,

$$\text{or,}\quad \pi r^2 h\rho g = \pi r^2 \frac{2}{3} hg + \pi r^2 \frac{h}{3} \sigma g,$$

$$\text{or,}\quad \rho = \frac{2}{3} + \frac{\sigma}{3},$$

$$\text{or,}\quad 3\rho = 2 + \sigma.$$

▶ **Ex. 7.** If W_1, W_2, W_3 be the apparent weights of a given body in fluids whose specific gravities are S_1, S_2, S_3, then

$$W_1(S_2 - S_3) + W_2(S_2 - S_1) + W_3(S_1 - S_2) = 0$$

[P.U. 52A, 61S, 82A, Bih. 66A, M.U. 65A, 68S, 82H, 70S, 72S, Luck. 76, 78, Raj. 75, Vik. 74, 75A, 68H, M.U. 03H]

Solution: Let V be the volume of the body and W be its true weight. Let ρ be the special gravity of the body.

Hence,

apparent weight of a body in a fluid = its true weight − weight of the fluid displaced by it

$$\therefore\ W_1 = W - VS_1\rho g. \qquad (1)$$

Similarly

$$W_2 = W - VS_2\rho g \qquad (2)$$

$$\text{and}\quad W_3 = W - VS_3\rho g \qquad (3)$$

Subtracting (2) from (1), we have $W_1 - W_2 = V\rho g(S_2 - S_1)$.

Similarly

$$W_2 - W_3 = V\rho g(S_3 - S_2)$$

$$W_3 - W_1 = V\rho g(S_1 - S_3).$$

Now,

$$\text{L.H.S.} = W_1(S_2 - S_3) + W_2(S_3 - S_1) + W_3(S_1 - S_2)$$

$$= \frac{1}{V\rho g}[W_1(W_3 - W_2) + W_2(W_1 - W_3) + W_3(W_2 - W_1)] = 0.$$

► **Ex. 8.** Two uniform straight rods of length $2a$, $2b$ and sp. gravities ρ, σ, respectively are joined together to form one straight rod. If it floats freely in water in an inclined position, the rod of length $2a$ and part of the other being immersed, prove that $a^2\rho(\rho-1)+2ab\sigma(\rho-1)+b^2\sigma(\sigma-1)=0$. **[I.A.S. 89]**

Solution: Let two uniform straitht rods AB and BC be joined together to form one straight rod AC. Let S be the breadth of the rod.

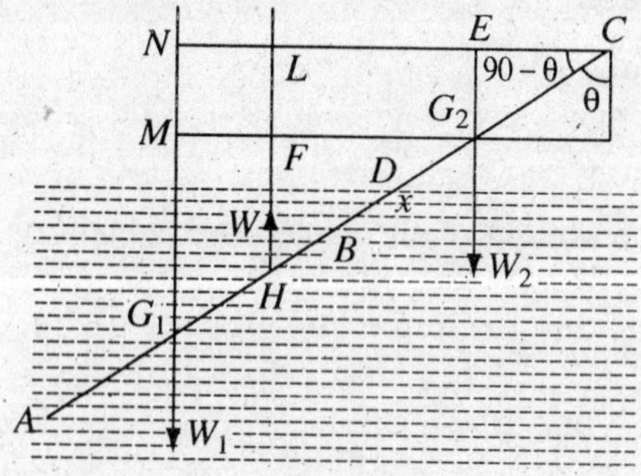

Fig. 8.9

Given $AB = 2a$, $BC = 2b$.

Let $BD = x$

$$AD = AB + BD = 2a + x$$
$$DC = BC - BD = 2b - x.$$

Now, the weight W_1 of the rod $AB = (2aS)\rho g$

The weight W_2 of the rod $BC = (2bS)\sigma g$

The weight W of the water displaced $= (ADS)g = (2a+x)Sg.$

For equilibrium,

$$W = W_1 + W_2,$$
$$\text{or,} \quad (2a+x)gS = 2gS(\rho a + b\sigma),$$
$$\text{or,} \quad 2a + x = 2(\rho a + b\sigma),$$
$$\text{or,} \quad x = 2a(\rho - 1) + 2b\sigma.$$

Again, taking moments about C, we get

$$W_1 CN + W_2 G_2 M = W \cdot CL,$$
$$\text{or,} \quad W_1 \cdot G_1 C \sin\theta + W_2 \cdot G_2 C \sin\theta = W \cdot CH \sin\theta.$$

But $\sin\theta \neq 0$

$$\therefore \quad W_1 G_1 C + W_2 \cdot G_2 C = W \cdot CH \qquad (1)$$

Now,

$$\begin{aligned} G_1C &= G_1B + BC = a + 2b \\ G_2C &= \frac{1}{2}BC = \frac{1}{2}2b = b \\ HC &= HD + DC = \frac{1}{2}AD + (2b - x) = \frac{1}{2}(2a + x) + (2b - x). \end{aligned}$$

Hence, from (1) we get

$$\begin{aligned} & W_1(a+2b) + W_2 b = W\left[\frac{1}{2}(2a+x) + (2b-x))\right], \\ \Rightarrow\ & W_1(a+2b) + W_2 b = (W_1 + W_2)\left[(a+2b) - \frac{x}{2}\right], \\ \Rightarrow\ & W_1(a+2b) + W_2 b = (W_1 + W_2)(a+2b) - \frac{x}{2}(W_1 + W_2), \\ \Rightarrow\ & W_2(a+b) = \frac{x}{2}(W_1 + W_2). \qquad (2) \end{aligned}$$

Putting the values of W_1, W_2 and x in (2), we get

$$\begin{aligned} & (2bs)\sigma g(a+b) = \frac{x}{2}[(2aS)\rho g + (2bs)\sigma g] \\ \Rightarrow\ & b\sigma(a+b) = \frac{x}{2}[a\rho + b\sigma] \\ \Rightarrow\ & 2b\sigma(a+b) = x(a\rho + b\sigma) \\ \Rightarrow\ & 2b\sigma(a+b) = [2a(\rho-1) + 2b\sigma](a\rho + b\sigma) \\ \Rightarrow\ & b\sigma a + b^2\sigma = a^2\rho(\rho-1) + ab\rho\sigma + ba\sigma(\rho-1) + b^2\sigma^2 \\ \Rightarrow\ & 2ab\sigma\rho - 2ab\sigma + b^2\sigma^2 - b^2\sigma + a^2\rho(\rho-1) = 0 \\ \Rightarrow\ & 2ab\sigma(\rho-1) + b^2\sigma(\sigma-1) + a^2\rho(\rho-1) = 0. \end{aligned}$$

▶ **Ex. 9.** Show that a right circular cone of density ρ and semi-vertical angle α can float vertex downwards in a liquid of density σ with one generator vertical and the base just clear of the liuqid of $\rho = \sigma(\cos 2\alpha)^{3/2}$.

[I.A.S. 73, P.U. 60H, 65H, Bih. 60H, 69H, R.U. 68H, 70H, Bhag. 72H, M.U. 93H, 04H]

Solution: Let h be the height of the cone and A be the centre of its base. The generator OB of the right circular cone of density ρ is vertical and the base CB is just outside the liquid of density σ. Let CD be the surface of floatation. From D and M we draw lines DE and LN parallel to BC.

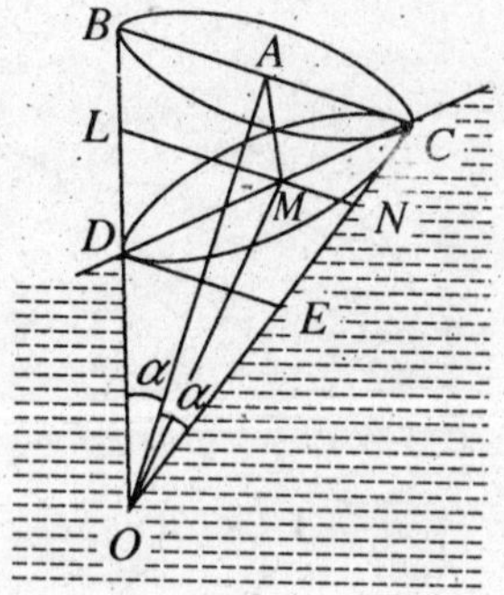

Fig. 8.10

In $\triangle BCD$, since M is the middle point of CD and LM is parallel to BC, hence, $LM = \frac{1}{2}BC$.

Similarly in $\triangle CDE, MN = \frac{1}{2}DE$.

From A, we draw AM perpendicular to CD. So M is the middle point of CD as AM is parallel to BD and both are vertical.

Now, the section of the cone by the surface of floatation is an ellipse with centre M, the middle point of CD and CD as major axis and let the lengths of its axes be $2a$ and $2b$.

In $\triangle OAC$

$$\cos\alpha = \frac{OA}{OC}, \text{or}, OC = h\sec\alpha \quad (OA = h).$$

From $\triangle ODC$

$$\begin{aligned} \sin 2\alpha &= \frac{DC}{OC} \\ \text{or}, CD &= OC\sin 2\alpha \\ &= h\sec\alpha 2\sin\alpha\cos\alpha = 2h\sin\alpha. \end{aligned}$$

Hence, $2a = 2h\sin\alpha \Rightarrow a = h\sin\alpha$.

Also $\cos 2\alpha = \frac{OD}{OC}$

$$OD = OC\cos 2\alpha = h\sec\alpha\cos 2\alpha$$

By the property of cone

$$b^2 = LM\cdot MN = \frac{1}{2}BC\cdot\frac{1}{2}DE$$

In $\triangle OAB$

$$\begin{aligned} \tan\alpha &= \frac{AB}{OA} = \frac{BC}{2h}, \quad BC = 2h\tan\alpha \\ \cos\alpha &= \frac{OA}{OB} = \frac{h}{OB}, \quad OB = h\sec\alpha. \end{aligned}$$

Since the triangles ODE and OBC are similar,

$$\begin{aligned} \therefore \quad \frac{DE}{BC} &= \frac{OD}{OB} \\ DE &= \frac{OD}{OB}BC = \frac{h\sec\alpha\cos 2\alpha}{h\sec\alpha}\cdot 2h\tan\alpha \\ &= 2h\cos 2\alpha\tan\alpha. \end{aligned}$$

Hence,

$$b^2 = \frac{1}{4} 2h \tan\alpha \cdot 2h \cos 2\alpha \tan\alpha = h^2 \tan^2\alpha \cos 2\alpha$$

$$b = h \tan\alpha \sqrt{\cos 2\alpha}.$$

Now, the weight of the cone $= \frac{1}{3}\pi\rho g h^3 \tan^2\alpha$

The weight of the liquid displaced by the cone

$$= \text{volume OCD} \times g\sigma$$

$$= \left(\frac{1}{3}\text{area of the ellipse} \times \text{OD}\right) g\sigma$$

$$= \left(\frac{1}{3}\pi abh \sec\alpha \cos 2\alpha\right) g\sigma$$

$$= \frac{1}{3}\pi h \sin\alpha h \tan\alpha \sqrt{\cos 2\alpha} h \sec\alpha \cos 2\alpha g\sigma$$

$$= \frac{1}{3}\pi h^3 \tan^2\alpha (\cos 2\alpha)^{3/2} g\sigma.$$

For equilibrium, the weight of the cone = the weight of the displaced liquid

$$\text{or,} \quad \frac{1}{3}\pi\rho g h^3 \tan^2\alpha = \frac{1}{3}\pi h^3 \tan^2\alpha (\cos 2\alpha)^{3/2} g\sigma$$

$$\text{or,} \quad \rho = \sigma(\cos 2\alpha)^{3/2}.$$

► **Ex. 10.** A ship sailing out of the sea into a river sinks through a distance b feet, on unloading a cargo of weight P, the ship rises through c feet, show that the weight of the ship after unloading is

$$\left[\frac{b\sigma}{c(\sigma - \rho)} - 1\right] P,$$

where σ and ρ are the sp. gr. of sea-water and river water, respectively. The cross-section of the ship near the level of water may be assumed to be uniform.

Solution: Let A be the cross-section of the ship near water and V the original volume below the sea-water.

Let W be the weight of the ship without load.

Hence, for equilibrium in sea-water, weight of the ship before unloading = weight of the sea-water displaced.

$$\text{or,} \quad W + P = V\sigma \tag{1}$$

when it goes into river

$$W + \rho = (V + bA)\rho \tag{2}$$

Again, for equilibrium after discharging cargo

$$W = [V + (b - c)A]\rho \tag{3}$$

Subtracting (3) from (2), we get

$$P = \rho[V + bA - \{V + (b - c)A\}]$$

$$P = \rho c A, \text{ or, } A = \frac{P}{\rho c} \quad (4)$$

Also from (1) and (2), we have

$$V\sigma = (V + bA)\rho$$

$$\text{or, } bA\rho = V(\sigma - \rho)$$

$$\text{or, } \frac{bP}{c} = V(\sigma - \rho) \quad \text{From (4)}$$

$$\text{or, } V = \frac{bP}{c(\sigma - \rho)}.$$

Therefore, from (1), we have

$$W = V\sigma - P = \frac{bP\sigma}{c(\sigma - \rho)} - P = P\left[\frac{b\sigma}{c(\sigma - \rho)} - 1\right]$$

► **Ex. 11.** A vertical cylinder of density ρ; floats in two liquids, the density of the upper liquid being ρ_1 and that of the lower ρ_2; if the length h, of the cylinder be n times the depth of the upper liquid, prove that the depth of immersion of the upper face of the cylinder is $\frac{h}{n} - \frac{h(\rho_2 - \rho)}{\rho_2 - \rho_1}$ provided that $\rho < \rho_2$ and $\rho_2 - \left(\frac{\rho_2 - \rho_1}{n}\right)$.

[R.U. 72H]

Solution: Let A be the area of the cross-section of the cylinder.

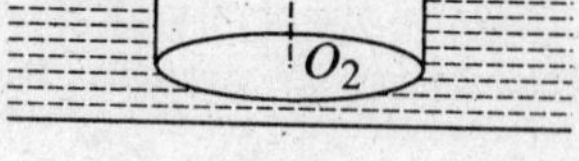

Fig. 8.11

The depth of the lighter liquid is $\frac{h}{n}$.

Then weight of the cylinder = $(Ah)\rho g$.

Let a length x of the axis of the cylinder be immersed in lighter liquid of density ρ_1, then the length of the axis immersed in heavier liquid of density ρ_2 is $(h - x)$.

For equilibrium,

the weight of the cylinder
= weight of the liquid of density ρ_1 displaced
+ weight of the liquid of density ρ_2 displaced,

$$\text{or, } (Ah)\rho g = (Ax)\rho_1 g + A(h - x)\rho_2 g$$

$$\text{or, } h\rho = x\rho_1 + (h - x)\rho_2$$

$$\text{or, } h(\rho - \rho_2) = x(\rho_1 - \rho_2).$$

$$\therefore \quad x = \frac{h(\rho_2 - \rho)}{\rho_2 - \rho^2}.$$

Hence, the required depth of O_1, the centre of the upper end

$$= z = \frac{h}{n} - x = \frac{h}{n} - \frac{h(\rho_2 - \rho)}{\rho_2 - \rho_1}$$

Since $\rho < \rho_2$ otherwise the cylinder will sink entirely in the liquid of the density ρ_2 and no part of it will be in the lighter liquid of density ρ_1.

Also of z is positive, then

$$\frac{h}{n} > \frac{h(\rho_2 - \rho)}{\rho_2 - \rho_1},$$

$$\text{or,} \quad \rho_2 - \rho_1 > n(\rho_2 - \rho),$$

$$\text{or,} \quad \rho n > n\rho_2 - \rho_2 + \rho_1,$$

$$\text{or,} \quad \rho > \rho_2 - \frac{(\rho_2 - \rho_1)}{n}.$$

▶ **Ex. 12.** A uniform rod, six feet long, can move about a fulcrum which is above the surface of some water. In the position of equilibrium four feet of the rod are immersed, prove that its sp. gravity is $\frac{8}{9}$. **[M.U. 64A, 66A, 70H, R.U. 66A]**

Solution: Let AB be the given uniform rod.

Given AB = 6 ft.

The immersed length BC = 4ft.

Hence, the length out of the water

= AC = AB − BC = 6 − 4 = 2 ft.

Let the rod makes an angle θ with the horizontal. Let α and 6 be the area of cross-section and sp. gravity of the rod.

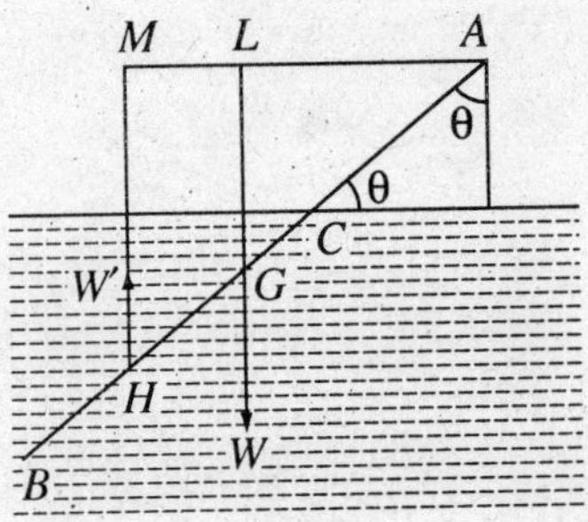

Fig. 8.12

The rod AB is in equilibrium under the action of following forces:

- The weight $W = 6\sigma\rho g$ of the rod acting vertically downwards through G, the C.G. of the rod AB.
- The force of buoyancy $W' = 4\alpha \times 1 \times g$ acting vertically upwards through H, the centre of buoyancy.
- The reaction at the fulcrum at A. Taking moment of these forces about A, we

get

$$W \cdot AL = W' \cdot AM$$
$$W \cdot AG \cos\theta = 4\alpha g AH \cos\theta$$
$$6\alpha\rho g AG = 4\alpha g AH, \text{ since } \cos\theta \neq 0,$$
$$\text{or, } 3\rho AG = 2AH,$$
$$\text{or, } 3\rho \cdot 3 = 2(AB - BH) = 2(6-2) = 8,$$
$$\text{or, } \rho = \frac{8}{9}.$$

► **Ex. 13.** A uniform prism whose cross-section is an isosceles triangle of vertical angle 2α, floats freely in a liquid with its base just immersed, one edge being in the surface, show that the ratio of its density to that of the liquid is $2\sin^2\alpha$.

[P.U. 67H, 75H, Bhag. 67H, Bih. 59H, B.P.S.C 82]

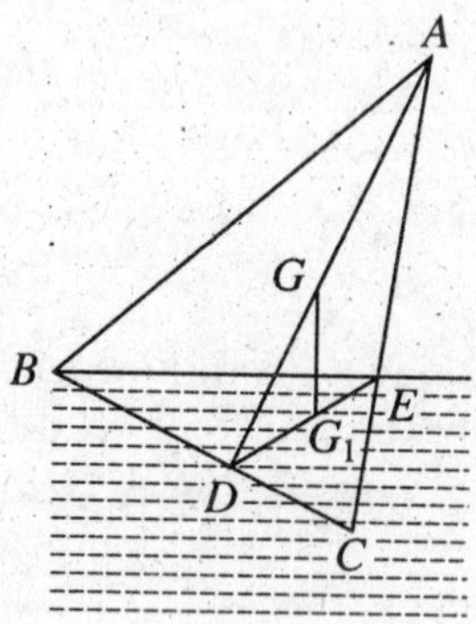

Fig. 8.13

Solution: Let ρ and σ be the densities of the prism and the liquid, respectively. Let h be the length of the prism and let ABC be its cross-section through G, its centre of gravity. Only two forces are responsible for its equilibrium, its weight acting downwards at G and the force of buoyancy acting upward at G_1, the centre of buoyancy. Let the free surface be BE.

Now, $EG : G_1D = 2 : 1$. Also $AG : GD = 2 : 1$

$$\therefore \quad \frac{EG_1}{G_1D} = \frac{AG}{GD}.$$

Hence, GG_1 is parallel to AC.

The two forces responsible for equilibrium are both vertical, so GG_1 is vertical, so AC is vertical.

For equilibrium, the two forces are equal in magnitude.

$$\therefore \quad (\Delta ABC \times h)\rho g = (\Delta BCE \times h) g\sigma$$

$$\therefore \quad \frac{\rho}{\sigma} = \frac{\Delta BCE}{\Delta ABC} = \frac{\frac{1}{2}BEEC}{\frac{1}{2}BEAC} = \frac{EC}{AC} = \frac{EC}{BE}\frac{BE}{AC}$$
$$= \cot C \cdot \frac{BE}{AB} (AB = AC)$$
$$= \cot(90-\alpha)\sin 2\alpha \quad [\because C = 90 - \angle CAD = 90 - \alpha]$$
$$= \tan\alpha 2\sin\alpha\cos\alpha = 2\sin^2\alpha.$$

▶ **Ex. 14.** A rod of density ρ and length a is freely movable about one end fixed at depth c below the surface of a liquid of density σ, show that the rod may rest in a position inclined to the vertical provided that $1 < \frac{\sigma}{\rho} < \frac{a^2}{c^2}$. **[Bhag. 63H, Luck 75]**

Solution: The rod cannot remain completely immersed in an inclined position, for in that case the C.G. of the rod and centre of buoyancy will coincide, which is impossible in an inclined position. Hence, a portion of the rod must remain out. For this it is necessary that the density of the rod must be less than the density of the fluid

i.e., $\rho < \sigma$. (1)

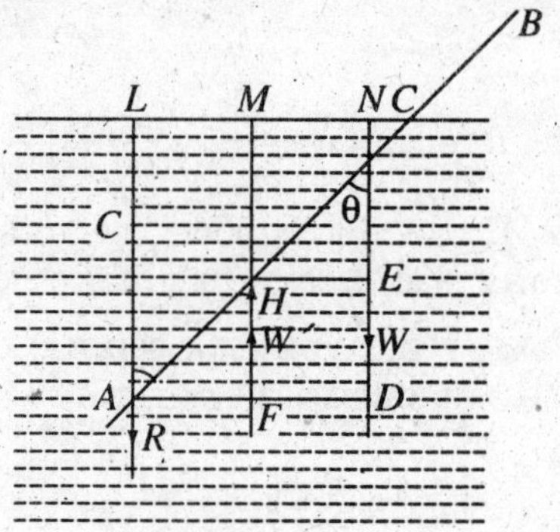

Fig. 8.14

Let ACB be the position of the rod with AC immersed and the point A fixed at a depth c. Let the rod be inclined at an angle θ to the vertical.

If G is the centre of gravity of the rod and H the centre of buoyancy, then

$$AG = \frac{a}{2}.$$

In $\triangle ACL$

$$\begin{aligned} \cos\theta &= \frac{AL}{AC} = \frac{c}{AC} \\ AC &= c\sec\theta \\ AH &= \frac{1}{2}AC = \frac{c}{2}\sec\theta. \end{aligned}$$

Let S be the cross-section of the rod. The forces acting on the rod are:

- Its weight $W = aS\rho g$ acting vertically downwards through G.
- The force of buoyancy $W' = c\sec\theta\cdot Sg\sigma$ acting vertically upwards through H and
- The reaction at A.

Taking moments of these forces about A, we get

$$W\cdot AD = W'AF$$

$$W\frac{a}{2}\sin\theta = W'AH\sin\theta,$$

$$\Rightarrow \quad aS\rho g\frac{a}{2} = c\sec\theta g\sigma S\frac{c}{2}\sec\theta \text{ since } \sin\theta \neq 0$$

$$\Rightarrow \quad \rho a^2 = c^2\sec^2\theta\cdot\sigma,$$

$$\Rightarrow \quad \cos^2\theta = \frac{c^2\sigma}{a^2\rho}.$$

In order that this equation may give a real value of θ, we must have

$$\cos^2\theta < 1,$$
$$\Rightarrow \quad \frac{c^2\sigma}{a^2\rho} < 1 \Rightarrow \frac{\sigma}{\rho} < \frac{a^2}{c^2}.$$

Hence, from (1), we get $1 < \frac{\sigma}{\rho} < \frac{a^2}{c^2}$.

▶ **Ex. 15.** A thin hollow cone with a base floats completely immersed in water, wherever placed, show that the vertical angle is $2\sin^{-1}\frac{1}{3}$.

[Bhag. 70H, P.U. 66H, 75A, 80H, B.P.S.C. 88, Delhi 64, Agra 60, 65, R.U. 81H, M.U. 94H, 98H, L.N.M.U. 94H, T.M.B.U. 92H, 94H, 99H]

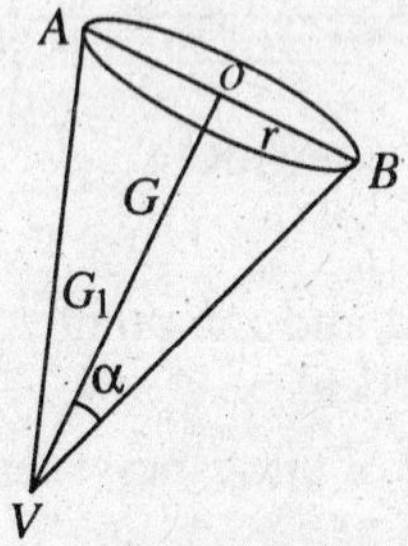

Fig. 8.15

Solution: Since the cone can float in any position, the centre of gravity of the cone must coincide with the centre of gravity of the displaced liquid which is the same as that of a solid cone. If h be the length of the axis and r the radius of the base, then the C.G. of the base is at its centre O and the C.G. of the curved surface of the cone is at a distance $\frac{h}{3}$ from O. The C.G. of the contained liquid is at G, where $OG = \frac{h}{4}$.

The mass of the base is $\pi r^2\sigma$ and that of the curved surface is $\pi r l\sigma$, where σ is the mass per unit area. The resultant C.G. must be at G, so that

$$\frac{h}{4} = OG = \frac{\pi r l\sigma\cdot\frac{h}{3} + \pi r^2\sigma\cdot 0}{\pi r l\sigma + \pi r^2\sigma},$$

[where r, h and l are the radius of the base, the height and slant side of the cone]

$$\Rightarrow \quad \frac{h}{4} = \frac{\frac{lh}{3}}{l+r} = \frac{lh}{3(l+r)}$$
$$\Rightarrow \quad 3l + 3r = 4l$$
$$\Rightarrow \quad 3r = l.$$

In △OVB

$$\sin\alpha = \frac{OB}{VB} = \frac{r}{l} = \frac{r}{3r} = \frac{1}{3}$$
$$\therefore \quad \alpha = \sin^{-1}\left(\frac{1}{3}\right).$$

Hence, the vertical angle of the cone $= 2\alpha = 2\sin^{-1}\left(\frac{1}{3}\right)$.

► **Ex. 16.** A cylindrical piece of wood of length l and sectional area α is floating with its axis vertical and in a cylindrical vessel of sectional area A which contains water. Prove that the work done in slowly pressing down the wood until it is just completely immersed is

$$\frac{1}{2}g\alpha l^2\rho\left(1-\frac{\alpha}{A}\right)\left(1-\frac{\sigma}{\rho}\right)^2,$$

where ρ and σ are the densities of water and the wood. **[I.A.S. 74, Raj. 62]**

Solution: When pressing the wood slowly downwards, let in any position, x be the depth immersed and y be the height of the bottom of the vessel.

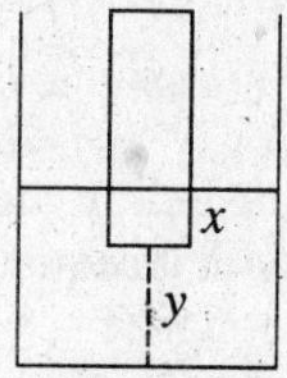

Fig. 8.16

$$\begin{aligned} \therefore \quad \text{External force} &= \text{upward force} \\ &= \text{weight of the liquid displaced} - \text{weight of the wood} \\ &= x\alpha\rho g - l\alpha\sigma g = \alpha g(x\rho - l\sigma) \\ &= P \text{ (Let).} \end{aligned}$$

Also $Ay + (A-\alpha)x$ = volume of water = constant.

Differentiating it, we get $dy = -\frac{A-\alpha}{A}dx$.

Due to its own weight, the length $\frac{l\sigma}{\rho}$ of the wood remains immersed in the liquid, so external force is applied to further immerse it.

Hence, the total work done on completely immersing the wood

$$\begin{aligned} &= -\int P\,dy \\ &= -\int_{\frac{l\sigma}{\rho}}^{l} \alpha g(x\rho - l\sigma)\frac{A-\alpha}{A}dx \\ &= \frac{(A-\alpha)}{A}\alpha g\left[\frac{x^2}{2}\rho - l\sigma x\right]_{\frac{l\sigma}{\rho}}^{l} \\ &= \alpha g\frac{(A-\alpha)}{A}\left[\frac{\rho}{2}\left(l^2 - \frac{l^2\sigma^2}{\rho^2}\right) - l\sigma\left(l - \frac{l\sigma}{\rho}\right)\right] \\ &= \alpha g\left(1-\frac{\alpha}{A}\right)l^2\left[\frac{\rho}{2}\left(1-\frac{\sigma^2}{\rho^2}\right) - \sigma\left(1-\frac{\sigma}{\rho}\right)\right] \\ &= \frac{\alpha g}{2}l^2\left(1-\frac{\alpha}{A}\right)\left[\rho\left(1-\frac{\sigma^2}{\rho^2}\right) - 2\sigma\left(1-\frac{\sigma}{\rho}\right)\right] \end{aligned}$$

$$
\begin{aligned}
&= \frac{\alpha g}{2} l^2 \left(1 - \frac{\alpha}{A}\right) \left[\frac{(\rho^2 - \sigma^2) - 2\sigma(\rho - \sigma)}{\rho}\right] \\
&= \frac{1}{2} g\alpha l^2 \left(1 - \frac{\alpha}{A}\right) \frac{(\rho - \sigma)}{\rho} [\rho + \sigma - 2\sigma] \\
&= \frac{1}{2} g l \alpha^2 \left(1 - \frac{\alpha}{A}\right) \left(\frac{\rho - \sigma}{\rho}\right)^2 \\
&= \frac{1}{2} g l \alpha^2 \rho \left(1 - \frac{\alpha}{A}\right) \left(\frac{\rho - \sigma}{\rho}\right)^2 \\
&= \frac{1}{2} g l \alpha^2 \rho \left(1 - \frac{\alpha}{A}\right) \left(1 - \frac{\sigma}{\rho}\right)^2.
\end{aligned}
$$

▶ **Ex. 17.** A life belt in the form of an anchor-ring generated by a circle of radius a floats in water with its equatorial flame horizontal, show that π, the depth immersed is given by the equation

$$z = a(1 - \cos\beta), \; 2\pi s = (2\beta - \sin 2\beta),$$

where s is sp. gravity of the material of the belt. **[I.A.S. 78, Agra 65]**

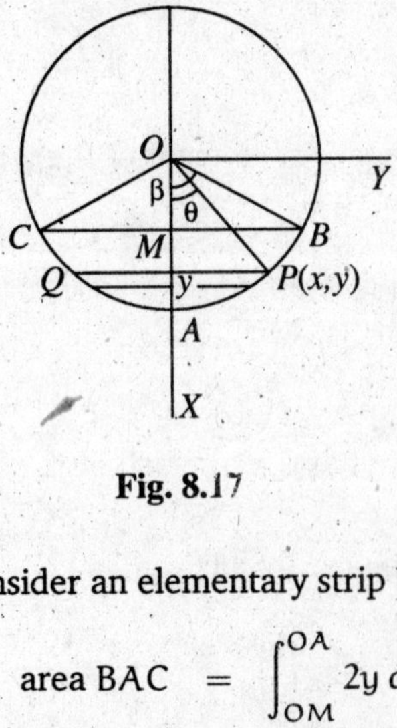

Fig. 8.17

Solution: We take the meridian section of the ring which is a vertical circle of radius a. Let BC be the line of floatation on this section, so that depth immersed is $AM = z$.

Also let $\angle BOA = \angle COA = \beta$ so that

$$
\begin{aligned}
z &= MA = OA - OM = a - a\cos\beta \\
z &= a(1 - \cos\beta).
\end{aligned}
$$

We consider an elementary strip PQ of breadth δx, then

$$
\begin{aligned}
\text{area BAC} &= \int_{OM}^{OA} 2y\, dx \\
&= 2\int_{a\cos\beta}^{a} \sqrt{a^2 - x^2}\, dx \quad [\because x^2 + y^2 = a^2] \\
&= 2\left[\frac{1}{2}x\sqrt{a^2 - x^2} + \frac{a^2}{2}\sin^{-1}\frac{x}{a}\right]_{a\cos\beta}^{a} \\
&= [0 + a^2 \sin^{-1} 1 - \{a\cos\beta\, a\sin\beta + a^2\sin^{-1}(\cos\beta)\}] \\
&= a^2\left[\frac{\pi}{2} - \sin\beta\cos\beta - \left(\frac{\pi}{2} - \beta\right)\right] \\
&= a^2[\beta - \sin\beta\cos\beta] \\
&= \frac{a^2}{2}[2\beta - \sin 2\beta].
\end{aligned}
$$

By condition of floatation

weight of the body = weight of the liquid displaced

$$\pi a^2\sigma = \pi a^2(2\beta - \sin 2\beta)\rho$$

$$\text{or,}\quad 2\pi\frac{\sigma}{\rho} = 2\beta - \sin 2\beta$$

$$\text{or,}\quad 2\pi s = 2\beta - \sin 2\beta,$$

where s = specific gravity = $\frac{\sigma}{\rho}$.

▶ **Ex. 18.** A sphere of radius a and mass M is loaded so that its centre of gravity G is at a distance c from its centre O, and is suspended by a string attached to a point P of its surface, GP subtending an angle θ at O. The sphere is partly immersed in a liquid of density ρ, and the tension in the string is $M'g$. Show that the depth h of O below the surface of the liquid is given by

$$M - M' = \frac{1}{3}\pi\rho(a+h)^2(2a-h)$$

and the inclination of GO to the vertical is

$$\tan^{-1}\frac{M'a\sin\theta}{Mc - M'a\cos\theta}.$$

Solution: The forces acting on the sphere are:

- The weight Mg of the sphere acting at the centre O vertically downwards.
- The weight $M'g$ acting vertically upwards at P.
- The force of buoyancy ρgV acting vertically upwards at G', the centre of buoyancy.

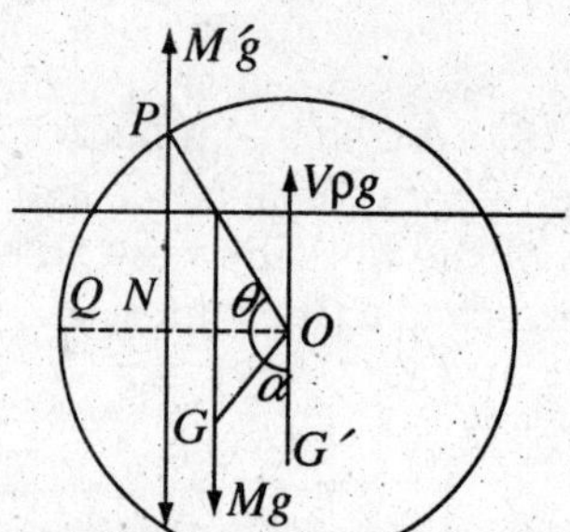

Fig. 8.18

It is clear that the force of buoyancy ρgV will pass through O.

Now,

$$\begin{aligned} V &= \frac{2}{3}\pi a^3 + \int_0^h \pi(a^2 - x^2)\,dx \\ &= \frac{2}{3}\pi a^3 + \pi\left(a^2 h - \frac{h^3}{3}\right) \\ &= \frac{\pi}{3}[2a^3 + 3a^2 h - h^3] \\ &= \frac{\pi}{3}[2a^3 - a^2 h + 4a^2 h - 2ah^2 + 2ah^2 - h^3] \\ &= \frac{\pi}{3}[a^2(2a - h) + 2ha(2a - h)^2 + h^2(2a - h)] \\ &= \frac{\pi}{3}(2a - h)(a^2 + 2ah + h^2) \\ &= \frac{\pi}{3}(2a - h)(h + a)^2. \end{aligned}$$

For equilibrium,

$$\begin{aligned} Mg - M'g &= \rho g \frac{\pi}{3}(2a - h)(a + h)^2 \\ \text{or,} M - M' &= \frac{1}{3}\pi\rho(2a - h)(a + h)^2. \end{aligned}$$

Also for equilibrium

$$\text{Moment of } Mg \text{ about } O = \text{Moment of } M'g \text{ about } O$$

$$Mgc\sin\alpha = M'ga\sin(\alpha + \theta), \text{ where } \angle GOG' = \alpha \text{ and } \angle POG = \theta$$

$$\begin{aligned} &\Rightarrow \quad Mc\sin\alpha = M'a(\sin\alpha\cos\theta + \cos\alpha\sin\theta) \\ &\Rightarrow \quad \sin\alpha(Mc - M'a\cos\theta) = M'a\cos\alpha\sin\theta \\ &\Rightarrow \quad \tan\alpha = \frac{M'a\sin\theta}{Mc - M'a\cos\theta}. \end{aligned}$$

▶ **Ex. 19.** A cylindrical bucket with water in it balances a mass M over a pulley. A piece of cork; of mass m and sp. gr. ρ is then tied to the middle point of the bottom of the bucket so as to be totally immersed. Prove that the tension of the string attached to the cork is

$$\frac{2Mmg}{(2M + m)}\left(\frac{1}{\sigma} - 1\right)$$

Also show that the pressure at the lowest point of the curved surface of the bucket will be $\gtrless$ than it was originally, according as the volume of the cork $\gtrless \frac{m}{2M}$ times the volume of the water.

Solution: Since mass M over the pulley balances the bucket with water, hence, the mass of bucket with water is M, which becomes $(M + m)$ when a cork of mass m is tied to the bottom of the bucket.

Let in this case the acceleration of the system be f, then for the bucket with cork we have the equation of motion

$$(M+m)f = (M+m)g - T \quad (1)$$

And for the mass M,

$$Mf = T - mg \quad (2)$$

where T is the tension in the string round the pulley.

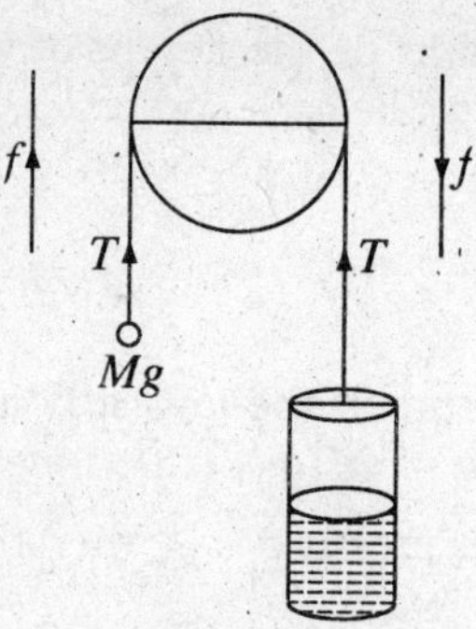

Fig. 8.19

Adding (1) and (2), we have

$$\begin{aligned} (2M+m)f &= mg \\ f &= \frac{mg}{2M+m} \end{aligned} \quad (3)$$

The forces acting on the cork are:

- Its weight mg acting vertically downwards.
- The force of buoyancy R acting vertically upwards.
- The tension T of the string acting vertically downwards.

$$\therefore \quad mf = mg + T - R$$

Now R, which is equivalent to the resultant thrust of the water displaced, whose mass is $\frac{m}{6}$ is equal to the weight of the water displaced by the resultant thrust caused by the acceleration.

$$\therefore \quad R = \frac{m}{\sigma}(g-f)$$

Hence,

$$\begin{aligned} mf &= mg + T - \frac{m}{\sigma}(g-f) \\ T &= m(f-g) \Rightarrow \frac{m}{\sigma}(g-f) \\ &= m(f-g)\left(1-\frac{1}{\sigma}\right) \\ &= m\left[\frac{m}{2M+m}g - g\right]\left(1-\frac{1}{\sigma}\right) \\ &= mg\left[\frac{m-2M-m}{2M+m}\right]\left(1-\frac{1}{\sigma}\right) \\ &= \frac{2Mmg}{2M+m}\left(\frac{1}{\sigma}-1\right). \end{aligned}$$

Let h and h' be the depths of water initially and afterwards. Let V_1 and V_2 be the volumes of the cork and water, respectively.

Then

$$\frac{h'}{h} = \frac{V_1 + V_2}{V_2}.$$

The pressure at the lowest point of the curved surface initially $= m'gh$ and while in motion $= m'(g - f)h'$.

The latter pressure is $\gtrless$, then the former, according as

$$m'(g - f)h' \gtrless m'gh$$

$$\text{or,} \quad \left(g - \frac{mg}{2M + m}\right)\frac{V_1 + V_2}{V_2}h \gtrless gh$$

$$\text{or,} \quad \frac{2M}{2M + m} \cdot \frac{V_1 + V_2}{V_2} \gtrless 1$$

$$\text{or,} \quad 2M(V_1 + V_2) \gtrless (2M + m)V_2$$

$$\text{or,} \quad 2MV_1 \gtrless V_2 m$$

$$\text{or,} \quad \frac{V_1}{V_2} \gtrless \frac{m}{2M}.$$

▶ **Ex. 20.** A uniform rod of length $2a$, floats partly immersed in a liquid being supported by a string fastened to one of its ends. If density of the liquid be $\frac{4}{3}$ times that of the rod, prove that the rod will rest with half its length out of the liquid, find also the tension of the string. **[M.U. 82H, R.U. 81]**

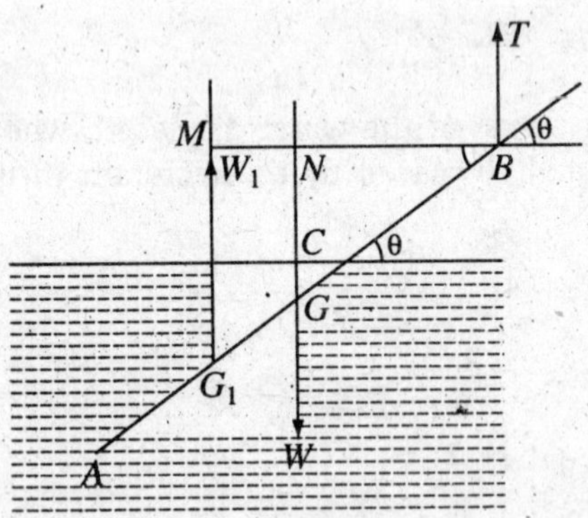

Fig. 8.20

Solution: Let AB be the rod with a length AC immersed in the liquid and supported by a string at the ends B. Let σ be the density of the rod, then the density of the liquid is $\frac{4}{3}\sigma$.

Let $AC = 2x$. Let G_1 and G be the centres of gravity of the liquid displaced and the rod, respectively, then $CG_1 = x$ and $BG = a$. Let α be the area of cross-section of the rod.

Now,

$$\begin{aligned} BG_1 &= BC + CG_1 = (AB - AC) + x \\ &= (2a - 2x) + x = 2a - x. \end{aligned}$$

The forces acting on the rod are:

- Its weight $W = 2a\alpha\sigma g$, acting vertically downwards through G.
- The force of buoyancy $W_1 = 2x\alpha\frac{4}{3}\sigma g$, acting vertically upwards through G_1.
- The tension T at B.

Let the rod AB makes an angle θ with by horizon, the taking moment of all the forces about B, we have

$$
\begin{aligned}
& W_1 \cdot BM = W \cdot BN, \\
\Rightarrow\ & W_1 BG_1 \cos\theta = W \cdot BG \cos\theta, \\
\Rightarrow\ & \frac{8x\alpha}{3}\sigma g(2a - x) = 2a\alpha\sigma g a, \\
\Rightarrow\ & \frac{4x}{3}(2a - x) = a^2, \\
\Rightarrow\ & 8ax - 4x^2 = 3a^2, \\
\Rightarrow\ & 4x^2 - 8ax + 3a^2 = 0, \\
\Rightarrow\ & 4x^2 - 6ax - 2ax + 3a^2 = 0, \\
\Rightarrow\ & 2x(2x - 3a) - a(2x - 3a) = 0, \\
\Rightarrow\ & (2x - 3a)(2x - a), \\
& 2x - 3a \neq 0.
\end{aligned}
$$

$$\therefore\ 2x = a, \text{or}, AC = \frac{1}{2}AB.$$

So, C is the middle point of AB. Hence, the rod is half immersed in the liquid.

Also for equilibrium,

$$
\begin{aligned}
T &= W - W_1 \\
&= 2a\alpha\sigma g - 2x\alpha\frac{4}{3}\sigma g \\
&= 2a\alpha\sigma g - 2\frac{a}{2}\alpha\frac{4}{3}\sigma g \\
&= 2a\alpha\sigma g - \frac{4a\alpha}{3}\sigma g \\
&= \frac{2a}{3}\alpha g\sigma = \frac{1}{3}W \\
&= \frac{1}{3}\text{ weight of the rod.}
\end{aligned}
$$

▶ **Ex. 21.** A sphere of density σ floats just immersed in three liquids. The densities of the liquids in descending order are $\rho, 4\rho, 9\rho$ and the thickness of the two upper liquids are each one-third of the sphere. Prove that $27\sigma = 122\rho$. **[B.B.A.U. 94H]**

Solution: Let a be the radius of the sphere.

∴ Weight of the sphere $= \frac{4}{3}\pi a^3 \sigma g$.

The volume of the uppermost liquid displaced

$$
\begin{aligned}
&= \text{the volume of the lowermost liquid displaced} \\
&= \frac{1}{3}\pi(h_1 - h_2)\{3a^2 - (h_1^2 + h_1 h_2 + h_2^2)\},
\end{aligned}
$$

where h_1, h_2 are the distances of the bounding planes of the zone from the centre, so that here $h_1 = a, h_2 = \frac{a}{3}$.

$$= \frac{1}{3}\pi\left(a - \frac{a}{3}\right)\left\{3a^2 - \left(a^2 + \frac{a^2}{3} + \frac{a^2}{9}\right)\right\}$$

$$= \frac{2\pi}{3}\frac{a}{3}\cdot a^2\left\{3 - \left(\frac{9+3+1}{9}\right)\right\}$$

$$= \frac{2\pi}{9}a^3\left(3 - \frac{13}{9}\right) = \frac{28\pi a^3}{81}.^1$$

The volume of the middle liquid displaced

$$= \text{(volume of the sphere)}$$
$$-2\text{ (volume displaced by uppermost liquid)}$$
$$= \frac{4\pi a^3}{3} - \frac{56\pi a^3}{81} = \frac{52\pi a^3}{81}.$$

For equilibrium, the weight of the sphere = sum of the weights of the displaced liquids

$$\Rightarrow \frac{4\pi a^3}{3}\sigma g = \frac{28\pi a^3}{81}\rho g + \frac{52\pi a^3}{81}4\rho g + \frac{28\pi a^3}{81}9\rho g$$

$$\Rightarrow 27\sigma = 7\rho + 52\rho + 63\rho$$

$$\Rightarrow 27\sigma = 122\rho \quad \text{Proved.}$$

▶ **Ex. 22.** A cone of given weight and volume floats. With its vertex downward, prove that the surface of the cone in contact with the liquid is least when its vertical angle is $2\tan^{-1}\left(\frac{1}{\sqrt{2}}\right)$. **[L.N.M.U. 95H, C.U. 98H]**

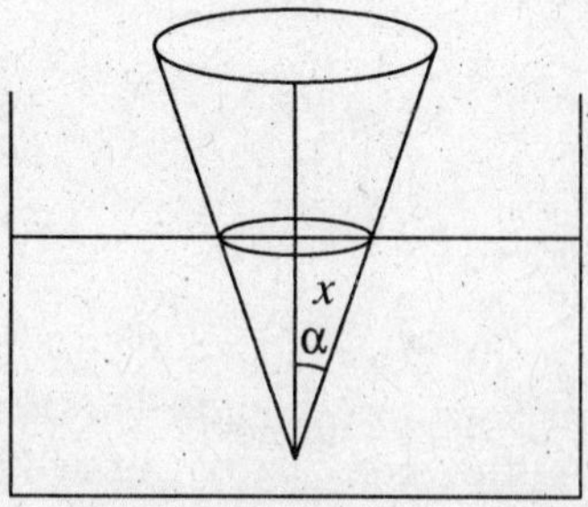

Fig. 8.21

Solution: Let α be the semi-vertical angle and h the height of the cone immersed to a length x of its axis in the fluid. Let W be the constant weight and V the constant volume, then

$$W = \frac{1}{3}\pi h^3 \tan^2 \alpha\rho g \qquad (1)$$

Also

$$V = \frac{1}{3}\pi h^3 \tan^2 \alpha \qquad (2)$$

Thus, ρ is also constant and we have

$$h^3 \tan^2 \alpha = \frac{3V}{\pi g} = \lambda \text{ (say)} \qquad (3)$$

The cone is floating in equilibrium.

$\therefore$ weight of the cone = weight of the displaced liquid

$$\Rightarrow \quad \frac{1}{3}\pi h^3 \tan\alpha\rho g = \frac{1}{3}\pi x^3 \tan^2\alpha\sigma g$$

$$\Rightarrow \quad h^3\rho = x^3\sigma. \tag{4}$$

If S denotes the surface in contact with water, then

$$\begin{aligned} S &= \pi(x\tan\alpha)(x\sec\alpha) \quad (\because\ S = \pi r l) \\ &= \pi x^2 \tan\alpha \sec\alpha \\ &= \pi\left(\frac{h^3\rho}{\sigma}\right)^{2/3} \tan\alpha\sec\alpha \quad \text{From (4)} \\ &= \pi\left[\frac{\lambda}{\tan^2\alpha}\frac{\rho}{\sigma}\right]^{2/3} \tan\alpha\sec\alpha \\ &= \pi\left(\frac{\lambda\rho}{\sigma}\right)^{2/3} \frac{\sec\alpha}{(\tan\alpha)^{1/3}} \\ &= \frac{k\sec\alpha}{(\tan\alpha)^{1/3}}, \text{ where } k = \pi\left(\frac{\lambda\rho}{6}\right)^{2/3} \end{aligned}$$

$$\therefore \quad \frac{ds}{d\alpha} = \frac{k}{(\tan\alpha)^{2/3}}\left[\sec\alpha\tan\alpha(\tan\alpha)^{1/3} - \frac{1}{3}(\tan\alpha)^{-2/3}\sec^3\alpha\right].$$

If S is max or min. then $\frac{ds}{d\alpha} = 0$

$$\begin{aligned} \Rightarrow \quad & \sec\alpha(\tan\alpha)^{4/3} = \frac{1}{3}\frac{\sec^3\alpha}{(\tan\alpha)^{2/3}} \\ \Rightarrow \quad & 3\tan^2\alpha = \sec^2\alpha = 1 + \tan^2\alpha \\ \Rightarrow \quad & 2\tan^2\alpha = 1 \\ \Rightarrow \quad & \tan\alpha = \frac{1}{\sqrt{2}} \\ \Rightarrow \quad & \alpha = \tan^{-1}\left(\frac{1}{\sqrt{2}}\right) \end{aligned}$$

$$\therefore \quad \text{Vertical angle} = 2\alpha = 2\tan^{-1}\left(\frac{1}{\sqrt{2}}\right) \quad \text{Proved.}$$

▶ **Ex. 23.** If the height of a right circular cone be equal to the diameter of the base, it will float, with its slant side horizontal, in any liquid of greater density. Prove this. **[B.B.A.B.U. 95H, T.M.B.U. 98H]**

Solution: The cone will float with its slant side horizontal, if in this position the C.G. of the cone and the C.G. of the displaced liquid are in the same vertical line.

The free surface is a parabola $A'O'B'$, since density of the liquid is greater than that of the cone, some portion of the cone will remain outside.

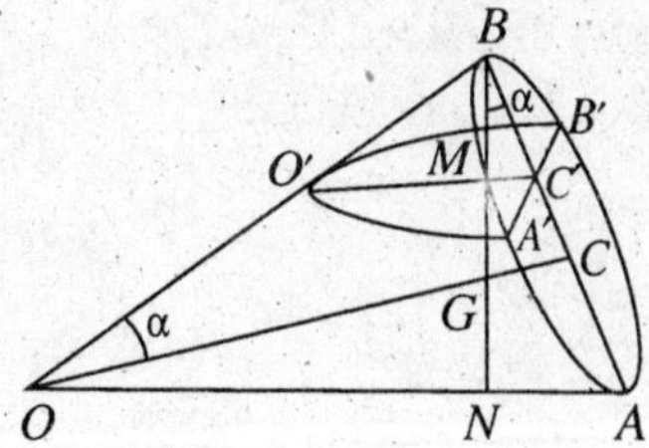

Fig. 8.22

Now, let h be the height and α the semi-vertical angle of the cone so that

$$OC = AB = h.$$

We draw a line BM perpendicular to the free surface cutting OC at G. Since OA is horizontal, BMG is vertical.

If $\angle BOC = \alpha$, then

$$\tan\alpha = \frac{BC}{OC} = \frac{\frac{1}{2}h}{h} = \frac{1}{2}, \quad \text{so that } \angle ABN = \alpha.$$

In $\triangle BGC$,

$$\tan\alpha = \frac{GC}{BC}$$

$$\Rightarrow \quad GC = BC\tan\alpha = \frac{h}{2}\cdot\frac{1}{2} = \frac{h}{4}.$$

This G is the C.G. of the cone.

We now want to show that C.G. of the displaced liquid or of the portion outside the liquid also lies on the vertical line BN.

The section of the cone by free surface (a plane parallel to a generator) is a parabola $O'B'C'$. Referred to O' as origin let its equation be

$$y^2 = 4ax.$$

Distance of its C.G. from O'

$$= \frac{\int_O^{O'B'} xy\,dx}{\int_O^{O'B'} y\,dx} = \frac{\int_O^{O'B'} x2\sqrt{ax}\,dx}{\int_O^{O'B'} 2\sqrt{ax}\,dx}$$

$$= \frac{2\sqrt{a}\int_O^{O'B'} x^{3/2}\,dx}{2\sqrt{a}\int_O^{O'B'} x^{1/2}\,dx} = \frac{\frac{2}{5}(O'B')^{5/2}}{\frac{2}{3}(O'B')^{3/2}} = \frac{3}{5}O'B' \qquad (1)$$

Also BN cuts $O'C'$ at M, then

$$\angle BO'M = 2\alpha$$

$$\therefore \quad O'M = BM\cot 2\alpha = \frac{BM(1-\tan^2\alpha)}{2\tan\alpha}$$

$$= \frac{BM\left(1-\frac{1}{4}\right)}{2\times\frac{1}{2}} \qquad \left(\because \tan\alpha = \frac{1}{2}\right)$$

$$= \frac{3}{4}BM$$

and $$MC' = BM \tan \alpha = \frac{1}{2}BM$$

so that $$\frac{O'M}{C'M} = \frac{\frac{3}{4}BM}{\frac{1}{2}BM} = \frac{3}{2}$$

$$\Rightarrow \frac{C'M}{O'M} = \frac{2}{3}$$

$$\Rightarrow \frac{C'M}{O'M} + 1 = \frac{5}{3}$$

$$\Rightarrow \frac{O'M + C'M}{O'M} = \frac{5}{3}$$

$$\Rightarrow \frac{3}{5}O'C' = O'M \quad (2)$$

Comparing (1) and (2), we find that the C.G. of the section coincides with M. Similarly C.G. of other sections will also lie on the vertical line AM.

Hence, the C.G. of the displaced liquid and that of the cone lie in the same vertical line. Thus, the equilibrium will exist in this position. This proves the result.

▶ **Ex. 24.** A triangular prism floats with its edge horizontal, find the positions of equilibrium. **[L.N.M.U. 95H]**

Solution: We take the vertical section OAB through the centre of gravity G of the prism. EF is the line of floatation. $\angle AOB = \theta$H is the C.G. of the immersed portion.

Let

$$OA = 2a, \quad OB = 2b,$$
$$OE = 2x, \quad OF = 2y.$$

For floating in equilibrium, we have

- Weight of the prism = weight of the liquid displaced

$$\Rightarrow \frac{1}{2}2a2b \sin\theta g\sigma = \frac{1}{2}2x \cdot 2y \sin\theta \rho g$$

$$\Rightarrow xy = \frac{\sigma}{\rho}ab = c^2 \quad \text{(say)} \quad (1)$$

- The line GH must be vertical and must therefore be perpendicular to the surface EF.

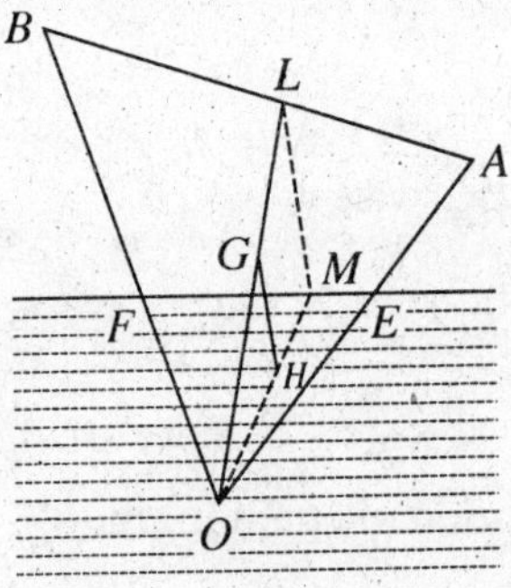

Fig. 8.23

Let us take OA and OB as the oblique axes of x and y. Then the coordinates of various points can be written as

$$A(2a, 0),\ B(0, 2b),\ E(2x, 0),\ F(0, 2y),\ G\left(\frac{2a}{3}, \frac{2b}{3}\right) \text{ and } H\left(\frac{2x}{3}, \frac{2y}{3}\right)$$

$$\therefore \text{ gradient of EF} = \frac{2y - 0}{0 - 2x} = -\frac{y}{x} = m_1 \quad \text{(say)}$$

$$\text{and gradient of GH} = \frac{\frac{2}{3}b - \frac{2}{3}y}{\frac{2}{3}a - \frac{2}{3}x} = \frac{b - y}{a - x} = m_2 \quad \text{(say)}.$$

These lines are at right angles, the condition for which is

$$1 + (m_1 + m_2)\cos\theta + m_1 m_2 = 0$$

$$\Rightarrow\ 1 + \left(-\frac{y}{x} + \frac{b-y}{a-x}\right)\cos\theta - \frac{y}{x}\frac{b-y}{a-x} = 0$$

$$\Rightarrow\ x(x-a) + [x(y-b) - y(x-a)]\cos\theta - y(y-b) = 0 \quad (2)$$

$$\Rightarrow\ x^2 - ax + x\left[x\left(\frac{c^2}{x} - b\right) - \frac{c^2}{x}(x-a)\right]\cos\theta - \frac{c^2}{x}\left(\frac{c^2}{x} - b\right) = 0, \text{ From (1)}$$

$$\Rightarrow\ x^4 - ax^3 + [x^2(c^2 - bx) - c^2x(x-a)]\cos\theta - c^2(c^2 - bx) = 0$$

$$\Rightarrow\ x^4 - x^3(a + b\cos\theta) + xc^2(a\cos\theta + b) - c^4 = 0 \quad (3)$$

This equation has one negative root by Descarte's rule of signs. Discarding this negative root of the remaining three roots either all are real or one real and two imaginary. Thus, at least there is one real position of equilibrium.

If the prism is isosceles, then putting $a = b$, the eqn. (3) becomes

$$x^4 - ax^3(1 + \cos\theta) + ac^2x(1 + \cos\theta) - c^4 = 0$$

$$\Rightarrow\ (x^2 - c^2)\left(x^2 - 2ac\cos^2\frac{\theta}{2} + c^2\right) = 0$$

$$\Rightarrow\ x = \pm c \text{ and } x = a\cos^2\frac{\theta}{2} \pm \sqrt{a^2\cos^4\frac{\theta}{2} - c^2}.$$

But $x = -c$ is inadmissible. Thus, of $a^2\cos^4\frac{\theta}{2} > c^2$, then we will have three real roots, otherwise one.

But in this case, $\frac{\sigma}{\rho}a^2 = c^2$, or, $\frac{\sigma}{\rho} = \frac{c^2}{a^2}$

Thus, of $\cos^4\frac{\theta}{2} > \frac{\sigma}{\rho}$ there are three positions of equilibrium, otherwise there is only one position of equilibrium.

Exercises

1. A body of specific gravity ρ floats half immersed in a liquid but is three-quarters immersed in a mixture of equal volumes of the liquid and water. Neglecting the atmospheric pressure, find ρ. [*Ans.* 1.5]

2. A semi-circular cylinder floats in water with its axis fixed in the surface of water. If the cylinder be movable about the fixed axis and if the density be half of that of water, show that it will be in equilibrium in any position.

[P.U. 61A, M.U. 63H]

3. A thin uniform rod of weight W has a particle of weight w attached to one end. It is floating in an inclined position in water with this end immersed. Prove that the length of the rod above water is $\frac{w}{w+W}$ times its whole length and that the sp. gravity of the rod is $\frac{W^2}{(w+W)^2}$.

[R.U. 63H, Vikram 63, Delhi 59]

4. A rod of small section and of density ρ has a small portion of metal of weight $\frac{1}{n}$th that of the rod attached to one extremity, prove that the rod will float at any inclination in a liquid of density σ if $(n+1)^2\rho = n^2\sigma$.

[P.U. 54A, 62A, 69A, 74A, R.U. 66A, 74A; Bhag. 69A, Bih. 71A, Luck. 74, 81, Raj. 76, R.U. 95H]

5. A right circular cone of density ρ, floats just immersed with its vertex downwards in a vessel containing two liquids of densities σ_1 and σ_2, show that the plane of separation of the two liquids cut off from the axis of a cone fraction $\left(\frac{\rho-\sigma_2}{\sigma_1-\sigma_2}\right)^{1/3}$ of its length.

[P.U. 63A, 70A, Bhag. 65A, Vik. 61, 65, Delhi 60, Raj 63, IAS 98]

6. A body floating in water has volumes V_1, V_2, V_3 above the surface when the densities of the surrounding are respectively ρ_1, ρ_2, ρ_3, prove that

$$\frac{\rho_2-\rho_3}{V_1}+\frac{\rho_3-\rho_1}{V_2}+\frac{\rho_1-\rho_2}{V_3}=0.$$

[Bih. 60S, 65S, 67S, R.U. 63A, M.U. 69S, 71A, Bhag. 66A, 68H, Agra 83, Vik. 78, Luck 75, 79, 82, M.U. 97H, 99H, R.U. 98H, 96H, M.U. 02H]

7. A triangular lamina ABC of density ρ floats in a liquid of density σ with its plane vertical; the angle B being in the surface of the liquid and the angle A not immersed. Show that

$$\frac{\rho}{\sigma}=\frac{\sin A\cos C}{\sin B}=\frac{a^2+b^2-c^2}{2b^2}$$

a, b, c being the lengths of the sides of the triangle.

[Bih. 58A, R.U. 71H, 73A, 76H, Luck. 83, Raj. 77, Mith. 93H, C.U. 04H]

8. A thin uniform wooden rod AB is in equilibrium in an inclined position with one end AC immersed in a bowl of water and one point D supported on the edge of the bowl. Show that the specific gravity σ of the wood is

$$\frac{AC}{BC}\cdot\frac{2AD-AC}{2AD-AB}$$

and that the greatest fraction of the length of the rod which can immersed is $(1-\sqrt{1-\sigma})$.

9. A semi-circular lamina has one of the ends of its diameter smoothly hinged to a fixed point above the surface of a liquid and floats with its plane vertical and its diameter half immersed. If the inclination of the diameter to the horizon is $\frac{\pi}{4}$, prove that the ratio of the density of the liquid to that of the lamina is $\frac{4(3\pi-4)}{9\pi-8}$.

[P.U. 52H, 57H, B.U. 57H, C.U. 94H]

10. A solid displaces $\frac{1}{2}, \frac{1}{3}$ and $\frac{1}{4}$ of its volume, respectively when it floats in three different liquids; find the volume it displaces when it floats in a mixture formed of equal volumes of the fluids.

[*Ans.* $\frac{1}{3}$ of its volume]

[M.U. 78A, P.U. 65S, 70A, 77A, Mith. U. 76A]

11. A block of wood floats in a liquid with $\frac{4}{5}$th of its volume immersed. In another liquid it floats with $\frac{2}{3}$rds of its volume immersed. If the liquids be mixed together in equal quantities by weight, what fraction of the volume of the wood would now be immersed?

[*Ans.* $\frac{11}{15}$]

12. A cylinder of wood (sp. gr. $\frac{3}{4}$) of length h floats with its axis vertical in water and oil (sp. gr. $\frac{1}{2}$), the length of the solid in contact with oil being $a(< \frac{h}{2})$. Find how much of the wood is above the liquid. Also find to what additional depth must oil be added so as to cover the cylinder.

[*Ans.* (i) $\frac{h-2a}{4}$; (ii) $\frac{h}{2} - a$.]

[I.A.S. 97]

13. A uniform rod capable of turning about one of its ends, which is out of water, rests inclined to the vertical. If the sp. gr. is $\frac{5}{9}$, find the length immersed.

[*Ans.* $\frac{1}{3}$ of length of rod]

[R.U. 67A, Bih. 76A]

14. A thin rod of sp. gr. $\frac{3}{4}$ and of length 4 ft, floats partly immersed in water, being supported by a string fastened to one end of the rod. How much of the rod is immersed. If the upper end of the rod be 1 foot above the surface of the water, prove that the rod is inclined at an angle 30° to the horizontal.

[*Ans.* 2 ft]

[B.U. 66S, P.U. 62S, 66, 71A]

15. A body floats in liquid of sp. gr. S with as much of its volume out of it as would be immersed in a second liquid of specific gravity S' if it floats in that liquid. Find the sp. gravity of the body. [*Ans.* $\frac{SS'}{S+S'}$]

[Agra. 78, 81, Bih. 57]

16. Two solids are each weighed in succession in three homogeneous liquids of different densities, if the weights of one are W_1, W_2 and W_3 and these of the other are w_1, w_2 and w_3, prove that

$$W_1(w_2 - w_3) + W_2(w_3 - w_1) + W_3(w_1 - w_2) = 0.$$

[Bih. 82, Mith. 81, Agra 74, Luck. 81, Vik. 80, 79, I.A.S. 99]

17. If a body floats in a liquid with volumes V_1, V_2 and V_3 above the surface when the barometric weights are h_1, h_2 and h_3, prove that

$$h_1V_1(V_2 - V_3) + h_2V_2(V_3 - V_1) + h_3V_3(V_1 - V_2) = 0.$$

[M.U. 74A, ů', Bih. 69A, P.U. 79A]

18. A, B, C are balls of equal weight. A balances B and C when all are suspended in a liquid of density σ, B balances C and A in a liquid of density σ_2, C balances A, B in a liquid of density σ_3. Find the sp. gravity of A, B and C.

[*Ans.* $\frac{2\sigma_1\sigma_2}{\sigma_1+\sigma_2}, \frac{2\sigma_2\sigma_3}{\sigma_2+\sigma_3}, \frac{2\sigma_3\sigma_1}{\sigma_3+\sigma_1}$]

19. A triangular lamina ABC, of which AB and AC are equal, floats in water with BC vertical, and three-quarters of its length immersed, being kept in equilibrium in this position by means of a string fastened to A and the bottom of the vessel. Find the sp. gravity of the lamina and show that the tension of the string is $\frac{1}{27}$ of the weight of the lamina. [*Ans.* $\frac{27}{32}$]

20. A square lamina is placed vertically in a liquid of double its density, prove that it can rest only with an edge of a diagonal vertical.

[L.N.M.U. 96H, P.U. 51H, 58H, 78H, Bih. 64H, T.M.B.U. 93h, 01H, M.U. 80H, 01H]

21. An equilateral triangle ABC, of weight W and sp. gravity σ is movable about a hinge at A, and is in equilibrium when the angle C is immersed in water and the side AB is horizontal. If it is then turned about A in its own plane until the whole of the side BC is in water and horizontal, prove that the action of the hinge in this position is

$$\frac{2(1-\sqrt{\sigma})}{\sqrt{\sigma}}W.$$

22. A solid cube made of uniform materials can turn freely about one edge, which is fixed in the surface of water, prove that if the cube rests with the face which is not immersed inclined at an angle 30° to the horizon, the density of the cube is to that of the water in the ratio $25 - 7\sqrt{3} : 18$.

[P.U. 53H, 63H, 66H, Bih. 51H, 67H]

23. A right circular cone has a plane base in the form of an ellipse, the cone floats on a fluid with its longest generator horizontal, if 2α be the vertical angle of the cone, and β the angle between the plane base and the shortest generating line, show that

$$\cot\beta = \frac{1}{5}\operatorname{cosec}4\alpha - \cot 4\alpha.$$

24. A cone of semi-vertical angle α and density ρ has its vertex fixed above the surface of a liquid of density σ; if its axis be inclined to the vertical at an angle θ and it floats with one point of its base in the surface of the liquid, prove that

$$\rho = \sigma\left\{1 - \frac{\cos\theta\cos^{3/2}(\theta+\alpha)}{\cos\alpha\cos^{3/2}(\theta-\alpha)}\right\}.$$

[M.U. 91H]

25. A solid uniform hemisphere of radius a and density σ can turn about O, the centre of its base. The point O is fixed in the surface of a liquid of density ρ. Show that for a certain value of $\frac{\rho}{\sigma}$, the hemisphere will rest in any position.

If the hemisphere is completely immersed with its centre fixed and kept at rest with its base vertical by means of a couple, determine, for any values of σ and ρ, there reaction of the centre and moment of the couple.

[*Ans.* $\frac{2}{3}\pi a^3(\sigma-\rho)g$, $\frac{1}{4}\pi a^4(\sigma-\rho)g$]

26. A sphere of radius a and sp. gravity $\frac{1}{2}$, is held completely immersed at the bottom of a circular cylinder of radius b, which is filled with water to a depth d. The sphere is set free and takes up its position of equilibrium, show that the potential energy lost is

$$W\left[d - \frac{11}{8}a - \frac{1}{3}\frac{a^3}{b^2}\right],$$

where W is the weight of the sphere. [P.U. 55H, 57H, B.U. 68H]

27. A uniform cube of specific gravity S floats in water with two of its faces vertical and one specified edge above the water, the other three horizontal edges being immersed. Show that if S lies between $\frac{23}{32}$ and $\frac{3}{4}$, there are three positions of equilibrium.

[Bih. 65H, I.A.S. 75]

28. To one end of a thin rod floating horizontally in water is attached string by which the rod is slowly raised from the water. Show that until the vertical position is reached, the tension of the string remains constant; and that the work done is then

$$\frac{Wl(1-S)}{1+\sqrt{1-S}},$$

where l, W, S are the length, weight and sp. gravity of the rod.

29. A wooden sphere of radius r is held just immersed in a cylindrical vessel of radius R containing water, and is allowed to rise gently as to be completely out of water, prove that the loss of potential energy of the water is

$$Wr \cdot \frac{3R^2 - 2r^2}{3R^2}.$$

W being the weight of the water displaced by the sphere.

[P.U. 60H, 65H]

30. A block of stone lies in a tank, with its top in the surface of water. Prove that the work done in lifting the stone slowly just clear of water is $\left(W - \frac{W'}{2}\right) l \left(1 - \frac{A}{B}\right)$, where W is the weight of the block, W' that of the displaced water, A the area of the section of the block, B that of the tank and l the depth of the block. [P.U. 56H, 58H]

31. A ship sailing from sea into a river sinks a inches and on discharging x tons of her cargo rises b inches. If sea water is one-fortieth heavier than river water, prove that weight of the ship is $\frac{41ax}{b}$ tons. Assume the sides of ship to be vertical at water level. [I.A.S. 80]

32. Show that a homogeneous body in the shape of a right circular cone can float in a liquid of twice its own density with its axis horizontal. [T.M.B.U. 94H, 02H]

CHAPTER 9

The Stability of Equilibrium of Floating Bodies

9.1 Introduction

If a body, floating partially immersed in a liquid, be given a small displacement from its position of equilibrium and left to move under the action of the forces acting on it. The three possibilities are there, either the forces acting on it tend to bring the body to its original position, or they tend to take the body further away from the original position of equilibrium. In the first case equilibrium is said to be *stable*, while in the second case equilibrium is said to be *unstable*. Thirdly, the body may remain at rest in the displaced position, in this case equilibrium is said to be *neutral*.

9.2 Types of Displacement

A small displacement may be of translation type or rotation type. Here we shall consider equilibriums for the different types of displacements.

9.2.1 Vertical Displacement

Let the body be pushed vertically downwards without rotation. In this case, the force of buoyancy increases, but weight and force of buoyancy (the only two forces acting) act in the same vertical line, and so there will be a resultant upward force on the body. This force will restore the body to its original position, i.e., the body will remain in *stable* equilibrium.

In the same manner, if the body is pushed up without rotation, then the resultant force will be acting downwards and this will restore the body to its original position. Hence, again the equilibrium is *stable*.

9.2.2 Horizontal Displacement

Let the body be pushed horizontally without rotation. In this case, the force of buoyancy will not change and the weight and upthrust will be acting in the same vertical line and so the equilibrium will remain undisturbed. Hence, for horizontal displacement equilibrium is *neutral*.

9.2.3 Rotational or Angular Displacement

Let W be the weight of a floating body, G its centre of gravity and H the centre of buoyancy. Let the body be given a small rotational displacement about a horizontal axis. Then the position of the centre of buoyancy will change, let it be H_1. Since the mass of the liquid displaced remains unaltered, the force of buoyancy will be the same as before, and is equal to the weight W of the body.

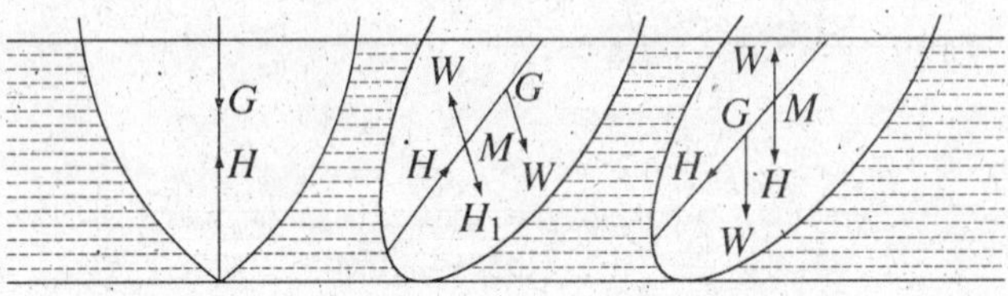

Fig. 9.1 **Fig. 9.2** **Fig. 9.3**

In the original position the force of buoyancy, which is equal to W, acts vertically upwards through H but in the displaced position it acts through H_1.

If the displacement takes place in the plane of symmetry of the body, then let the vertical through the new centre of buoyancy H_1, meet the line HG at M. This point M is called the *metacentre* and GM is called the *metacentric Height* of the floating body.

Hence, we define a metacentre as, [C.U. 93H, 95H, 97H]

> *Let a body floating in equilibrium, receive a small rotational displacement such that the quantity of liquid displaced remains unaltered. If the displacement takes place in the plane of the symmetry of the body, the vertical through the new centre of buoyancy meets the line joining the centre of gravity of the body and the original centre of buoyancy in a point called the Metacentre. The distance between the C.G. of the body and metacentre is called metacentric height.*

The forces acting on the body in displaced position are:

- The weight W of the body acting vertically downwards through G, the C.G. of the body.
- The force of buoyancy W acting vertically upwards through the H_1, the new centre of buoyancy.

These two forces constitute a couple which is, generally, called the *restoring couple*.

1. *When the metacentre M is above G*

   In fig. (2) M is above G, it is clear that the restoring couple has a tendency to turn the body in anti-clockwise direction and thus, to restore the body to its original position and hence, the equilibrium in this case is *stable*.

2. *When the metacentre M is below G*

   In fig.(3) M is below G, the tendency of the restoring couple is to rotate the body in clockwise direction and thus, to recede the body further away from its original position and so the equilibrium in this case is *unstable*.

3. When the metacentre M coincides with G. In this case, GH, will be vertical and so conditions of equilibrium are satisfied. The body will remain at rest in the displaced position and so the equilibrium is *neutral*.

Thus, we see that equilibrium of a floating body is stable, unstable or neutral according as the metacentre is above, below or coincident with the centre of gravity of the body.

In otherwords it can be expressed as the equilibrium is stable, neutral or unstable according as $HM >, =, < HG$.

9.3 Definitions

1. **Surface of Buoyancy:** If a body floating in a homogeneous liquid be supposed to take in turn every possible position for which by volume of the liquid displaced remains constant, the locus of the centre of buoyancy is called the *Surface of Buoyancy*.

2. **Plane of Floatation:** If a body floats partially immersed in a liquid, in any position, the section of the body made by the plane of the surface of the liquid is known as the corresponding *Plane of Floatation*.

3. **Surface of Floatation:** If a body be floating in such a way that the volume of the liquid displaced throughout remain the same, then the envelope of the planes of floatation is called *Surface of Floatation*.

4. **Curve of Buoyancy:** If the rotation of a floating body, with the volume of the liquid remaining constant, takes place in the plane of symmetry, the centres of buoyancy lie on a curve in the plane of symmetry, known as *Curve of Buoyancy*.

5. **Curves of Floatation:** The principal normal sections through the corresponding points of the surface of floatation are called the *Curves of Floatation*.

9.4 Theorem on Stability

To show that the equilibrium is stable or unstable according as the metacentre is above or below the centre of gravity of the body. **[L.N.M.U. 95H]**

Let W be the weight of the body, then after displacement also from the position of equilibrium, it acts vertically downwards through G.

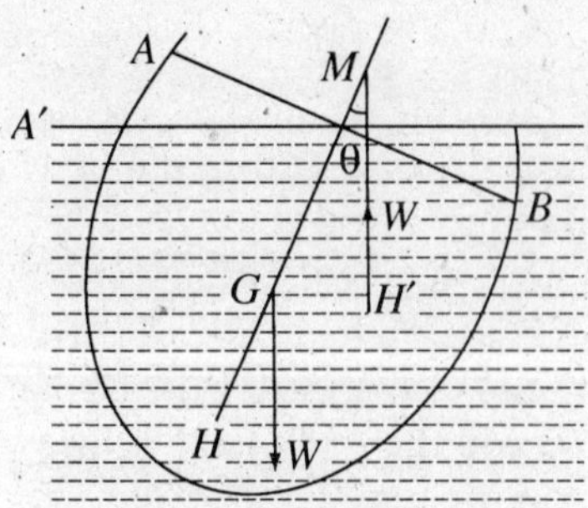

Fig. 9.4

Since the weight of the liquid displaced remains unchanged (weight of the body, i.e., W) the upthrust now acts vertically upwards through the new centre of buoyancy H′.

These two forces form a couple.

Moment of this couple

$$= W \cdot \mathrm{GM} \sin\theta, \text{ where } \angle \mathrm{GMH'} = \theta$$

$$= W(\mathrm{HM} - \mathrm{HG}) \sin\theta.$$

If the moment of this couple is positive, then the body will tend to come back to its original position.

So the equilibrium will be stable or unstable according as $\mathrm{HM} - \mathrm{HG} \gtrless 0$ as $\sin\theta$ is +ve for small θ or according as $\mathrm{HM} \gtrless \mathrm{HG}$, i.e., according as M is above or below G.

9.5 The tangent plane at any point of the surface of buoyancy is parallel to the corresponding plane of floatation

Let the plane of floatation be AOB. Let it be turned through a small angle. Let its new position be A′OB′.

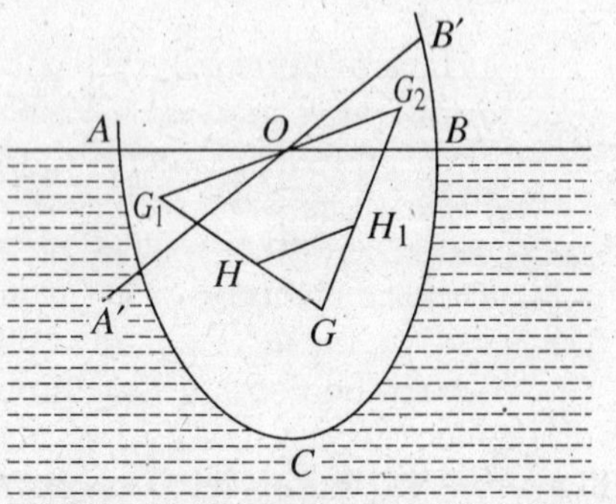

Fig. 9.5

Let H and H_1 be the centres of buoyancy corresponding to the planes of floatation AOB and A′OB′.

Since the volume of the liquid displaced in the displaced position is equal to the volume of the liquid displaced in the original position.

Hence, volume ACB = volume A′CB′.

$$\therefore \text{ volume of the wedge AOA}' = \text{volume of the wedge BOB}'. \tag{9.1}$$

Let G_1, G_2 and G be the centre of gravity of AOA′, BOB′ and A′CB, respectively. Join G_1G_2, G_1G and G_2G.

Now the centre of buoyancy H (i.e., C.G. of the displaced liquid ACB) must lie on GG_1 such that

$$\frac{\mathrm{G_1H}}{\mathrm{HG}} = \frac{\text{Volume of A}'\text{CB}}{\text{Volume of AOA}'}. \tag{9.2}$$

Similarly the centre of buoyancy H_1 must lie on G_2G, such that

$$\frac{\mathrm{G_2H_1}}{\mathrm{H_1G}} = \frac{\text{Volume of A}'\text{CB}}{\text{Volume of BOB}'}. \tag{9.3}$$

Hence, from (9.1), (9.2) and (9.3), we get

$$\frac{G_1H}{HG} = \frac{G_2H_1}{H_1G}.$$

Hence, HH_1 is parallel to G_1G_2 and therefore, when the displacement is very small HH_1 is parallel to AB the surface of floatation.

Similarly it can be proved that the line joining any point on the surface of buoyancy near H with H_1 is parallel to the plane of floatation AB.

Hence, *the tangent plane of any point on the surface of buoyancy is parallel to the corresponding position of the plane of floatation.*

9.6 To Find the Position of Metacentre

Let AB be the plane of floatation in the position of equilibrium. H is the metacentre in the position of equilibrium. G is the C.G. of the body. In the position of equilibrium, the axes OX, OY lie in the plane of floatation and axis OZ is vertically downwards as shown in the figure.

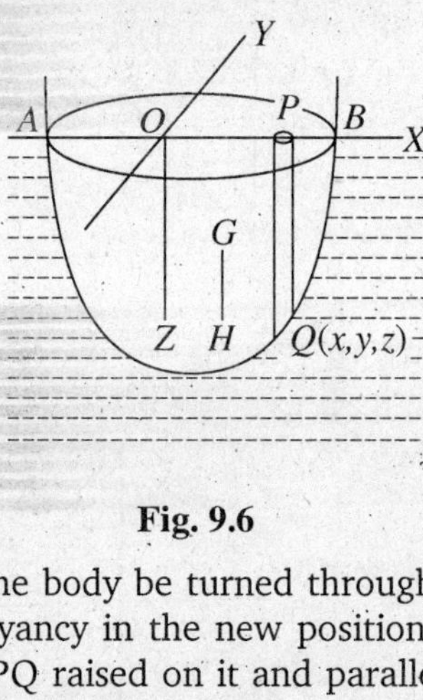

Fig. 9.6

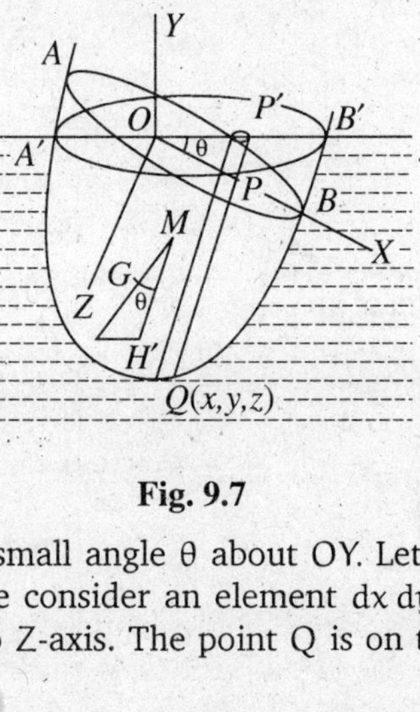

Fig. 9.7

Now, let the body be turned through a small angle θ about OY. Let H' be the centre of buoyancy in the new position. We consider an element $dx\,dy$ at P and a filament $P'PQ$ raised on it and parallel to Z-axis. The point Q is on the surface whose coordinates are $(x, y; z)$.

Now, $OP = x$, $PQ = Z$, $\angle P'OP = \theta$.

In $\triangle POP'$

$$\tan\theta = \frac{PP'}{OP} = \frac{PP'}{x}$$
$$PP' = x\tan\theta.$$

Hence, $P'Q = P'P + PQ = x\tan\theta + z = x\theta + z$, as θ is very small.

Therefore, volume of the element $P'Q' = (x\theta + z)\,dx\,dy$.

Hence, volume immersed

$$= \iint (x\theta + z)\,dx\,dy$$
$$= \iint x\theta dx\,dy + \iint z\,dx\,dy,$$

where the integration is extended over the new plane of floatation.

In the original position the volume immersed $= \iint z\,dx\,dy$, where the integration is extended over the plane of floatation.

If the volume immersed is to be constant, then

$$\iint x\theta dx\,dy + \iint z\,dx\,dy = \iint z\,dx\,dy$$

$$\therefore \quad \iint x\,dx\,dy = 0, \quad \text{as } \theta \neq 0,$$

i.e., the Yaxis, about which the plane turns through the centroid of the surface section AOB.

Thus, the volume of liquid displaced by a body under going small displacements remains constant if the axis about which the plane turns passes through the centroid of the plane of floatation.

Let coordinates of H and H_1 be $(\bar{x}, \bar{y}, \bar{z})$ and $(\bar{x}_1, \bar{y}_1, \bar{z}_1)$, respectively.

Then,

$$\left.\begin{aligned} \bar{x} &= \frac{1}{V}\iint xz\,dx\,dy \\ \bar{y} &= \frac{1}{V}\iint yz\,dx\,dy \\ \bar{z} &= \frac{1}{V}\iint \frac{z^2}{2}\,dx\,dy \end{aligned}\right\}. \tag{9.4}$$

Since the coordinates of C.G. of PQ are $(x, y, \frac{z}{2})$

Coordinates of C.G. of PP′ are $\left[x, y, \frac{1}{2}(z - x\theta)\right]$. Hence,

$$\left.\begin{aligned} \bar{x}_1 &= \frac{1}{V}\iint x(z + x\theta)\,dx\,dy \\ \bar{y}_1 &= \frac{1}{V}\iint y(z + x\theta)\,dx\,dy \\ \bar{z}_1 &= \frac{1}{V}\iint \frac{1}{2}(z - x\theta)(z + x\theta)\,dx\,dy \\ &= \frac{1}{2V}\iint (z^2 - x^2\theta^2)\,dx\,dy\,. \end{aligned}\right\}. \tag{9.5}$$

Where V is the volume of the portion of the body immersed in liquid.

Since θ is very small, θ^2 can be neglected and hence,

$$\bar{z} = \bar{z}_1. \tag{9.6}$$

Hence, HH′ is parallel to the plane AOB, i.e., plane of floatation. Also H, H′ are neighbouring point. Hence, *the tangent plane at any point on the surface of buoyancy is parallel to the corresponding position of the plane of floatation.*

M, the metacentre, the intersection of GH and vertical through H′ will exist if the above two lines are coplanar. These lines are both parallel to the XZ plane. Hence, the two lines will intersect if

$$\bar{y} = \bar{y}_1, \tag{9.7}$$

$$\text{or,} \iint yz\, dx\, dy = \iint y(z + x\theta)\, dx\, dy$$

$$\therefore \quad \iint xy\, dx\, dy = 0$$

The implies the vanishing of the product of inertia of the area of the plane of floatation with regard to the horizontal axes. This is possible only when they are the principal axes of the surface section at some point.

Hence, a *metacentre exists of the axis of rotation is also a principle axis of the plane of floatation at some point.* **[C.U. 91H]**

On using (9.3) and (9.4), we get

$$\begin{aligned} HH' &= \bar{x}_1 - \bar{x} \\ &= \frac{1}{V}\iint x(z + x\theta)\, dx\, dy - \frac{1}{V}\iint xz\, dx\, dy = \frac{\theta}{V}\iint x^2 dx\, dy. \end{aligned}$$

But $\iint x^2 dx\, dy$ being the moment of inertia of the plane of floatation AOB about the axis of rotation, i.e., y-axis may be denoted by AK^2, where A is the area of AOB and K its radius of gyration about the axis of rotation.

Hence,

$$HH' = \frac{AK^2}{V}\theta.$$

Also in $\Delta HH'M$

$$\begin{aligned} \sin\theta &= \frac{HH'}{HM} \\ HH' &= HM \sin\theta = HM\theta, \text{ since } \theta \text{ is small.} \end{aligned}$$

Hence,

$$HM\theta = \frac{AK^2}{V}\theta \quad \text{but } \theta \neq 0.$$

Therefore,

$$HM = \frac{Ak^2}{V}. \tag{9.8}$$

This gives the height of metacentre. **[C.U. 92H, 95H, 98H, 99H, 01H, 03H]**

Note 1. *Equation (9.8) shows that the height of the metacentre above the centre of buoyancy in the original position depends on area of plane of floatation and the constant volume of the body immersed.*

Note 2. *If ρ be the density of the liquid in which the body is floating, then*

$$W = \rho g V.$$

As θ is small, the restoring couple $= W \cdot GM \cdot \theta$.

To find the work done in this angular moment let a further displacement of $d\theta$ be assumed.

Then the work done in this infinitesimal displacement $d\theta = W \cdot GM \cdot \theta\, d\theta$.

Therefore, work done in producing a displacement θ

$$
\begin{aligned}
&= \int_0^\theta W\cdot GM\cdot \theta\, d\theta \\
&= \frac{1}{2} W\cdot GM\cdot \theta^2 \\
&= \frac{\rho g V}{2}\theta^2 GM \\
&= \frac{\rho g V}{2}(HM - HG)\theta^2 \\
&= \frac{\rho g}{2}\left(\frac{AK^2}{V} - HG\right) V\cdot \theta^2 \\
&= \frac{\rho g}{2}(AK^2 - V\cdot HG)\theta^2.
\end{aligned}
$$

Note 3. *Metacentre is a point, where two consecutive normals to the surface of buoyancy meet, so in fact it is the* centre of curvature of the surface of buoyancy.

9.7 The distance between the centre of gravity of a floating body, in equilibrium and the centre of buoyancy is a minimum or maximum according as the equilibrium is stable or unstable.

Let G be the centre of gravity, H, H′ be the centres of buoyancy in the position of equilibrium and in a slightly displaced position, respectively.

The metacentre M lies on HG.

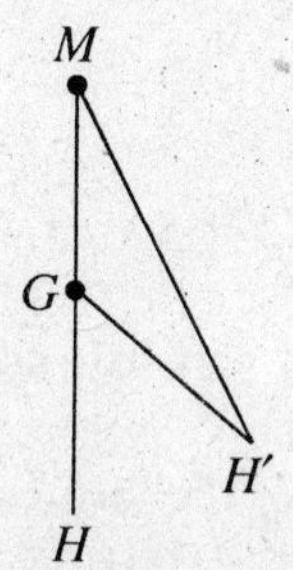

Fig. 9.8

If the equilibrium is stable G lies between H and M.

Then,

$$H'G + GM > H'M. \tag{9.9}$$

If H, H′ be sufficiently near, then HM = H′M.

Hence, from (9.9)

$$H'G + GM > HM = HG + GM,$$

or, $\quad H'G > HG.$

Therefore, HG is minimum.
Hence, in the case of stable equilibrium, HG is minimum.
If M lies between G and H, then we can similarly show that HG > H′G.
Therefore, HG is maximum.
Hence, in the case of unstable equilibrium; HG is maximum.

9.8 A solid of uniform density σ floats partly immersed in a homogeneous liquid of density ρ. Show that as solid of the same size and shape and of density $\rho - \sigma$ can float inverted with the same plane of floatation in the same liquid, and that if the equilibrium is stable in the first case, it is also stable in the second case.

[I.A.S. 74]

When a body of density σ floats in a liquid of density ρ, let V and V′ be the volumes immersed and unimmersed. If H, H′ be the centroids of V and V′ and G be the centre of gravity of the body, HH′ must pass through G. But HG is vertical.

Hence, H′G is also vertical and it is so when the body is inverted.

Again,

$$\frac{V}{\sigma} = \frac{V+V'}{\rho} = \frac{V'}{\rho-\sigma},$$

or, $V'\rho = (V+V')(\rho-\sigma)$

i.e., weight of liquid displaced = weight of the body of density $(\rho - \sigma)$.

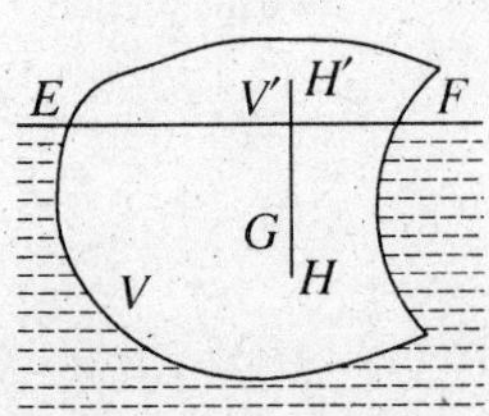

Fig. 9.9

Therefore, in the inverted position the body will float freely.

Now, of the equilibrium is stable in first case; M is above G

$$\text{i.e., } HM > HG \text{ or, } \frac{AK^2}{V} > HG \text{ or, } AK^2 > V \cdot HG. \tag{9.10}$$

But

$$V \cdot HG = V' \cdot H'G,$$

$$\text{or, } AK^2 > V'H'G \quad \text{From (9.10)}$$

$$\therefore \quad \frac{AK^2}{V'} > H'G,$$

i.e., in the inverted position the metacentre lies above the C.G.

Hence, the equilibrium is stable.

Illustrative Examples

▶ **Ex. 1.** A cone whose vertical angle is 60° floats in water, its axis vertical and vertex downwards, show that its metacentre lies in the plane of floatation and that its equilibrium will be stable of its sp. gravity is greater than $\frac{27}{64}$.

[P.U. 61H, 62H, R.U. 62H, M.U. 66H, 69H, Bhag. 70H, Raj. 61, Agra 61, Cal 84H, 90H]

Solution: Let h be the height of the cone and let length h′ of its axis be immersed in water.

The surface of floatation is a circle of radius $= h' \tan 30 = \dfrac{h'}{\sqrt{3}}$.

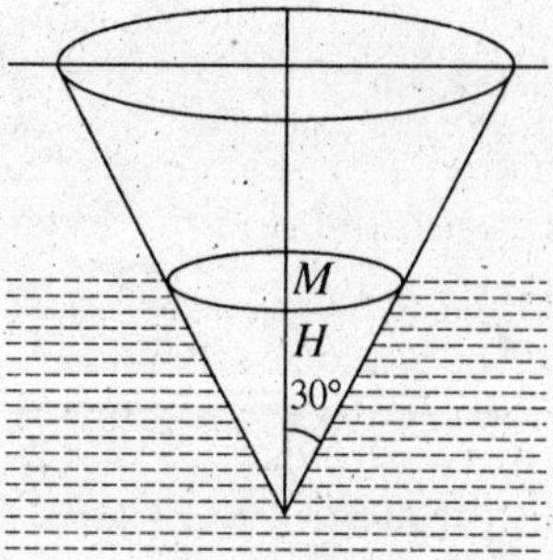

Fig. 9.10

$$\begin{aligned} AK^2 &= \text{M.I. of the plane of floatation about a principal axis, i.e., diameter} \\ &= \pi\left(\frac{h'}{\sqrt{3}}\right)^2 \frac{1}{4}\left(\frac{h'}{\sqrt{3}}\right)^2 \\ &= \frac{1}{36}\pi h'^4 \\ \text{and } V &= \text{volume immersed} \\ &= \frac{1}{3}\pi h'^3 \tan^2 30 = \frac{1}{3}\pi h'^3 \frac{1}{3} \\ &= \frac{1}{9}\pi h'^3 \\ HM &= \frac{AK^2}{V} = \frac{\pi h'^4}{36} \times \frac{9}{\pi h'^3} = \frac{h'}{4}. \end{aligned}$$

Also

$$OH = \frac{3}{4}h'.$$

Now,

$$OM = OH + HM = \frac{3}{4}h' + \frac{h'}{4} = h'.$$

Hence, metacentre lies on the plane of floatation

Also

$$OG = \frac{3}{4}h.$$

So that

$$HG = OG - OH = \frac{3}{4}h - \frac{3}{4}h' = \frac{3}{4}(h - h').$$

For stable equilibrium, $HM > HG$

$$\frac{h'}{4} > \frac{3}{4}(h - h'), \quad \text{or,} \quad 4h' > 3h,$$

$$\text{or,} \quad \frac{h'}{h} > \frac{3}{4}. \tag{1}$$

Also for floating, weight of the cone = weight of the liquid displaced

$$\text{or, } \frac{1}{3}\pi h^3 \tan^2 30^\circ \sigma g = \frac{1}{3}\pi h'^3 \tan^2 30^\circ \rho g$$

where σ and ρ are the densities of cone and water

$$\text{or,} \quad \frac{\sigma}{\rho} = \frac{h'^3}{h^3} > \frac{27}{64} \quad \text{From (1)}$$

▶ **Ex. 2.** A solid and homogeneous body consists of a cylinder joined to a hemisphere on the same base and floats with the hemispherical portion partly immersed in water. Find the greatest height of the cylinder consistent with stability.

[I.A.S. 73]

Solution: Let a cylinder of height h be joined to a hemisphere of radius r on the same whose centre is C.

Let ρ be the density of the body.

Hence, the weight of the hemisphere = $\frac{2}{3}\pi r^3 \rho g$ which act at its C.G. G_1.

The weight of the cylindrical portion $\pi r^2 h\rho g$ acts through G_2, its centre of gravity. Let O be the lowest point of the body.

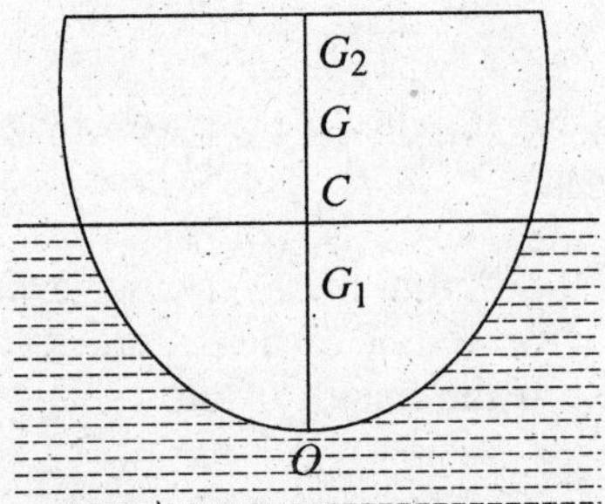

Fig. 9.11

Now,

$$OG_1 = OC - CG_1 = r - \frac{3r}{8} = \frac{5r}{8}$$

and $$OG_2 = OC + CG_2 = r + \frac{h}{2}.$$

Therefore, the centre of gravity G of the whole body is given by

$$OG = \frac{\frac{2}{3}\pi r^3 \rho g \cdot \frac{5r}{8} + \pi r^2 h\rho g\left(r + \frac{h}{2}\right)}{\frac{2}{3}\pi r^3 \rho g + \pi r^2 h \rho g}$$

$$OG = \frac{\frac{5r^4}{12} + r^2 h\left(r + \frac{h}{2}\right)}{\frac{2}{3}r^3 + r^2 h}$$

$$= \frac{3\left[\frac{5r^2}{12} + h\left(r + \frac{h}{2}\right)\right]}{2r + 3h}$$

$$= \frac{\frac{5r^2}{4} + 3h\left(r + \frac{h}{2}\right)}{2r + 3h}.$$

The metacentre is at C.

For stable equilibrium the centre of gravity cannot be above the metacentre

$$OG \not> OC$$

$$\Rightarrow \frac{\frac{5r^2}{4} + 3h\left(r + \frac{h}{2}\right)}{2r + 3h} \not> r,$$

$$\Rightarrow \frac{5r^2}{4} + 3h\left(r + \frac{h}{2}\right) \not> r(2r + 3h),$$

$$\Rightarrow \frac{5r^2}{4} + \frac{3h^2}{2} \not> 2r^2,$$

$$\Rightarrow \quad \frac{3h^2}{2} \not> 2r^2 - \frac{5r^2}{4},$$
$$\Rightarrow \quad \frac{3h^2}{2} \not> \frac{3r^2}{4},$$
$$\Rightarrow \quad h^2 \not> \frac{r^2}{2},$$
$$\Rightarrow \quad h \not> \frac{r}{\sqrt{2}}.$$

Hence, for stability the greatest height of the cylinder is $\frac{1}{\sqrt{2}}$ times the radius of the hemisphere.

► **Ex. 3.** A thin metal circular cylinder contains water to a depth h and floats in water with its axis vertical, immersed to a depth h'. Show that the vertical position is stable if the height of the centre of gravity of the cylinder above its base is lies than $\frac{1}{2}(h + h')$. **[I.A.S. 76, B.P.S.C. 86, Vik. 69, Delhi 60, Agra 62]**

Solution: Let O be the centre of the base, G is the C.G. of the cylinder. Also let H, H′ be the C.G.s of the contained and displaced water in the equilibrium position of the cylinder.

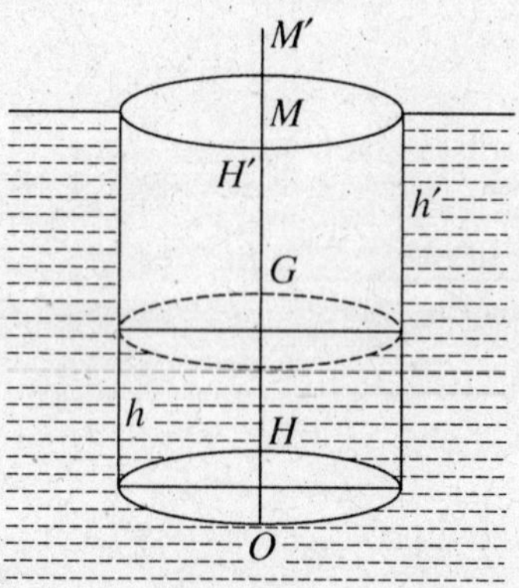

Fig. 9.12

Let M and M′ be the metacentres of the contained and the displaced liquid.

Let x be the height above the base G and r be the radius of the circular base of the cylinder.

Now,

$$HM = \frac{AK^2}{V} = \frac{\pi r^2 \frac{r^2}{4}}{\pi r^2 h} = \frac{r^2}{4h}$$

Similarly

$$H'M' = \frac{r^2}{4h'}.$$

The upward force of buoyancy acting at $M' = \pi r^2 h' g$, and the weight of contained water acting at $M = \pi r^2 h g$.

Now,

$$\begin{aligned} GM &= OM - OG = OH + HM - OG \\ &= \frac{h'}{2} + \frac{r^2}{4h} - x \ (OG = x). \end{aligned}$$

Similarly

$$GM' = \frac{h'}{2} + \frac{r^2}{4h'} - x.$$

For stability,

$$\pi r^2 h' g \cdot GM^2 > \pi r^2 g GM$$

$$h'\left(\frac{h'}{2}+\frac{r^2}{4h'}-x\right) > h\left(\frac{h}{2}+\frac{r^2}{4h}-x\right),$$

$$\text{or,}\quad \frac{1}{2}(h'^2-h^2) > x(h'-h),$$

$$\text{or,}\quad \frac{1}{2}(h'+h)(h'-h) > x(h'-h),$$

$$\text{or,}\quad x < \frac{h+h'}{2}.$$

▸ **Ex. 4.** If a solid homogeneous right circular cone floats with the axis vertical and vertex downwards, show that the floating is stable if $\frac{\sigma}{\rho} > \cos^6\alpha$, where α and ρ are the densities of the solid and the liquid, respectively and σ is the semi-vertical angle of the cone.

[P.U. 78H, 80H, 58H, 61H, 63H, R.U. 63H, M.U. 64H, 76H, 77H, C.U. 80H, 87H, L.N.M.U. 95H]

Solution: Let h be the height and α the semi-vertical angle of the cone. Let the length h' of the axis of the cone be immersed.

For equilibrium, we have

weight of the cone

= weight of the liquid displaced.

$$\text{or,}\quad \frac{1}{3}\pi h^2\tan^2\alpha\sigma g = \frac{1}{3}\pi h'^2\tan^2\alpha\rho g,$$

$$\text{or,}\quad \frac{h^3}{h'^3} = \frac{\rho}{\sigma}. \qquad (1)$$

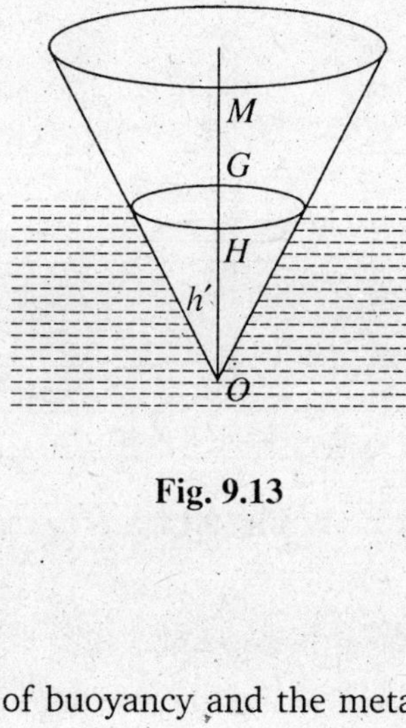

Fig. 9.13

Let G, H and M be the centre of gravity, centre of buoyancy and the metacentre, respectively.

Also

$$OH = \frac{3}{4}h' \quad\text{and}\quad OG = \frac{3}{4}h.$$

So that

$$HG = OG - OH = \frac{3}{4}(h-h').$$

Again,

$$HM = \frac{AK^2}{V} = \frac{\pi h'^2\tan^2\alpha\left(\frac{1}{4}h'^2\tan^2\alpha\right)}{\frac{1}{3}\pi h'^3\tan^2\alpha} = \frac{3}{4}h'\tan^2\alpha.$$

The equilibrium will be stable if HM > HG

$$\Rightarrow \quad \frac{3}{4}h'\tan^2\alpha > \frac{3}{4}(h - h'),$$
$$\Rightarrow \quad h'\sec^2\alpha > h,$$
$$\Rightarrow \quad \frac{h'}{h} > \cos^2\alpha,$$
$$\Rightarrow \quad \frac{h'^3}{h^3} > \cos^6\alpha,$$
$$\Rightarrow \quad \frac{\rho}{\sigma} > \cos^6\alpha \quad \text{From (1).}$$

▶ **Ex. 5.** A cone of density ρ whose height is h and the radius of whose base is a floating with its axis vertical and vertex upwards in liquid of density σ. Prove that the equilibrium is stable of $\rho < \sigma(1 - \cos^6\alpha)$.

[Bih. 63H, R.U. 65H, Bh. 65H, M.U. 67H, 77H, 81H]

Solution: Let h be the height and α be the semi-vertical angle of the cone. Let h' be the length of the axis of the cone immersed in the liquid.

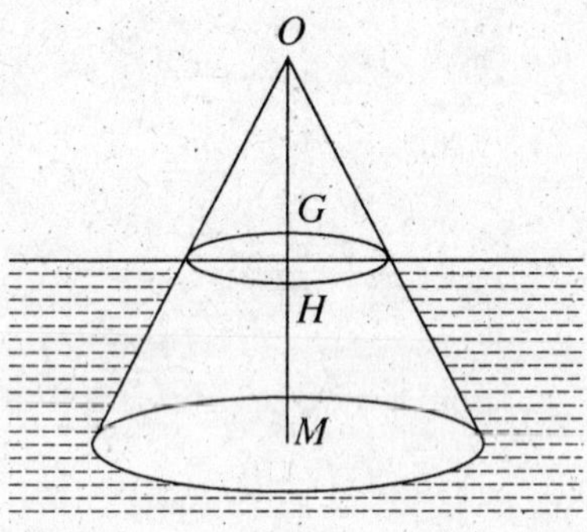

Fig. 9.14

For equilibrium

weight of the cone
= weight of the liquid displaced

$$\frac{1}{3}\pi h^3\tan^2\alpha\rho g = \frac{\sigma g}{3}(\pi h^3\tan^2\alpha - \pi h'^3\tan^2\alpha),$$

$$\text{or,} \quad \rho h^3 = \sigma(h^3 - h'^3),$$

$$\text{or,} \quad \frac{\rho}{\sigma} = 1 - \frac{h'^3}{h^3},$$

$$\text{or,} \quad \frac{h'^3}{h^3} = 1 - \frac{\rho}{\sigma}. \qquad (1)$$

Let G, H and M be the centre of gravity, centre of buoyancy and the metacentre, respectively.

Then,

$$OG = \frac{3h}{4}$$

$$\text{and} \quad OH = \frac{\frac{3h}{4}h^3 - \frac{3h'}{4}h'^3}{h^3 - h'^3} = \frac{3(h^4 - h'^4)}{4(h^3 - h'^3)}.$$

Also

$$HM = \frac{AK^2}{V} = \frac{\pi h'^2\tan^2\alpha\frac{h'^2\tan^2\alpha}{4}}{\frac{\pi}{3}(h^3 - h'^3)\tan^2\alpha} = \frac{3h'^4\tan^2\alpha}{4(h^3 - h'^3)}.$$

Now,

$$HG = OH - OG = \frac{3}{4}\left(\frac{h^4 - h'^4}{h^3 - h'^3} - h\right)$$

$$= \frac{3}{4}\left(\frac{h^4 - h'^4 - h^4 + hh'^3}{h^3 - h'^3}\right) = \frac{3}{4}\frac{h'^3(h - h')}{h^3 - h'^3}.$$

The equilibrium will be stable if HM > HG

$$\Rightarrow \frac{3h'^4 \tan^2\alpha}{4(h^3 - h'^3)} > \frac{3h'^3(h - h')}{4(h^3 - h'^3)},$$

$$\Rightarrow h'(\tan^2\alpha > h - h',$$

$$\Rightarrow h' \sec^2\alpha > h,$$

$$\Rightarrow \frac{h'}{h} > \cos^2\alpha,$$

$$\Rightarrow \frac{h'^3}{h^3} > \cos^6\alpha,$$

$$\Rightarrow 1 - \frac{\rho}{\sigma} > \cos^6\alpha \quad \text{From (1)},$$

$$\Rightarrow 1 - \cos^6\alpha > \frac{\rho}{\sigma},$$

$$\therefore \rho < \sigma(1 - \cos^6\alpha).$$

▶ **Ex. 6.** A homogeneous circular cylinder of length h, radius a and sp. gravity s floats in water. Prove that the position with the axis vertical is stable if

$$\frac{a^2}{h^2} > 2s(1 - s).$$

[I.A.S. 82, Bih. 62H, 71H, R.U. 68H, C.U. 88H, 99H, 02H, M.U. 71H, 73H]

Solution: Let x be the length of axis of the cylinder immersed in, then for equilibrium, weight of the cylinder = weight of the liquid displaced, $gs\pi a^2 h = g\pi a^2 x$

$$x = hs. \qquad (1)$$

Let O, G, H and M be the centre of base, the centre of gravity, the centre of buoyancy and the metacentre, respectively, then we have

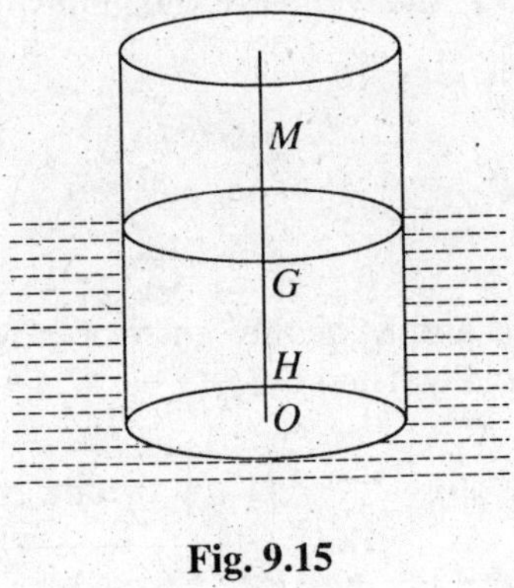

Fig. 9.15

$$OG = \frac{h}{2}, \quad OH = \frac{x}{2} = \frac{hs}{2}, \quad \text{From (1)}$$

$$HM = \frac{AK^2}{V} = \frac{\pi a^2 \frac{a^2}{4}}{\pi a^2 x} = \frac{a^2}{4x} = \frac{a^2}{4hs}.$$

For stability of equilibrium, $HM > HG$

$$\frac{a^2}{4hs} > OG - OH = \frac{h}{2} - \frac{hs}{2}, \quad \text{or,} \quad \frac{a^2}{4hs} > \frac{h}{2}(1-s)$$
$$\text{or,} \quad a^2 > 2h^2 s(1-s)$$
$$\text{or,} \quad \frac{a^2}{h^2} > 2s(1-s).$$

► **Ex. 7.** Show that when a uniform hemisphere of density ρ and radius a floats with its plane base immersed in a homogeneous liquid of density σ, the equilibrium is stable and the metacentric height is $\frac{3a}{8\rho}(\sigma - \rho)$.

[I.A.S. 75, Bih. 61H, 68H, Bhag. 63H, P.U. 66H]

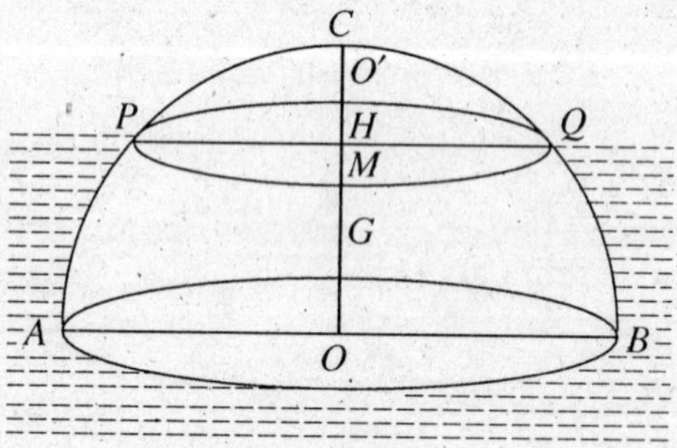

Fig. 9.16

Solution: Let the plane of floatation cut the hemisphere ABC in PQ, the centre of the section being O′. Let the centre of the base be O and $OO' = h$.

By the formula

$$\frac{\pi}{3}(h_1 - h_2)[3r^2 - (h_1^2 + h_1 h_2 + h_2^2)]$$

the volume of the displaced liquid

$$= \frac{\pi}{3}(h-0)[3a^2 - (h^2 + 0 + 0)]$$
$$= \frac{\pi h}{3}(3a^2 - h^2). \qquad (1)$$

For the equilibrium,

the weight of the hemisphere = weight of the displaced liquid

$$\text{or,} \quad \frac{2}{3}\pi a^3 \rho g = \frac{\pi}{3} h(3a^2 - h)\sigma g,$$
$$\text{or,} \quad 2a^3\rho = h(3a^2 - h)\sigma. \qquad (2)$$

Let G, H and M be the centre of gravity, the centre of buoyancy and the metacentre, respectively. Then $OG = \frac{3}{8}a$.

$$OH = \frac{\int_0^h \pi r^2 dz \cdot z}{\int_0^h \pi r^2 dz} = \frac{\pi \int_0^h (a^2 - z^2) z\, dz}{\frac{\pi h}{3}(3a^2 - h^2)} \quad \text{From (1)}$$
$$= \frac{3\left[a^2 \frac{z^2}{2} - \frac{z^4}{4}\right]_0^h}{h(3a^2 - h^2)} = \frac{\frac{3}{4}(2a^2h^2 - h^2)}{h(3a^2 - h^2)}$$
$$= \frac{\frac{3}{4}h(2a^2 - h^2)}{4(3a^2 - h^2)}.$$

Also

$$
\begin{aligned}
HM &= \frac{AK^2}{V} = \frac{\pi(a^2-h^2)\frac{1}{4}(a^2-h^2)}{\frac{\pi h}{3}(3a^2-h^2)} \\
OM &= OH + HM \\
&= \frac{3h(2a^2-h^2)}{4(3a^2-h^2)} + \frac{3(a^2-h^2)^2}{4h(3a^2-h^2)} \\
&= \frac{3h^2(2a^2-h^2)+3(a^2-h^2)^2}{4h(3a^2-h^2)} \\
&= \frac{3[2a^2h^2-h^4+a^4-2a^2h^2+h^4]}{4h(3a^2-h^2)} \\
&= \frac{3(a^4)}{4h(3a^2-h^2)} = \frac{3a^4}{\frac{4.2a^3\rho}{\sigma}} \quad \text{From (2)} \\
\therefore &= \frac{3a\sigma}{8\rho}.
\end{aligned}
$$

Now,

$$GM = OM - OG = \frac{3a\sigma}{8\rho} - \frac{3a}{8} = \frac{3a}{8\rho}(\sigma - \rho).$$

As $\sigma > \rho$, So, M is above G. So the equilibrium is stable.

▶ **Ex. 8.** A paraboloid of revolution floats with its axis vertical and vertex downwards in a liquid, the density of which varies as the depth, the equilibrium will be stable or unstable according as $4c$ is less or greater than $3(m+a)$, where c is the length of the axis, a the length immersed and m the letus rectum of the generating parabola. **[I.A.S. 76, Delhi 61]**

Solution: Since m is the latus rectum of the generating parabola. Hence, equation of the paraboloid is $x^2+y^2=mz$ or $u^2=mz$.

The length $OC = a$ of the vessel is immersed in the liquid.

$\therefore$ density ρ at a height Z above O is given by

$$\rho = \lambda(a-z).$$

Hence, ρ_0 = density at free surface = $\lambda(a-a)=0$.

Hence,

$$A_0K_0^2\rho_0 = 0.$$

Let H, G and M be the centre of buoyancy, centre of gravity and metacentre of the liquid.

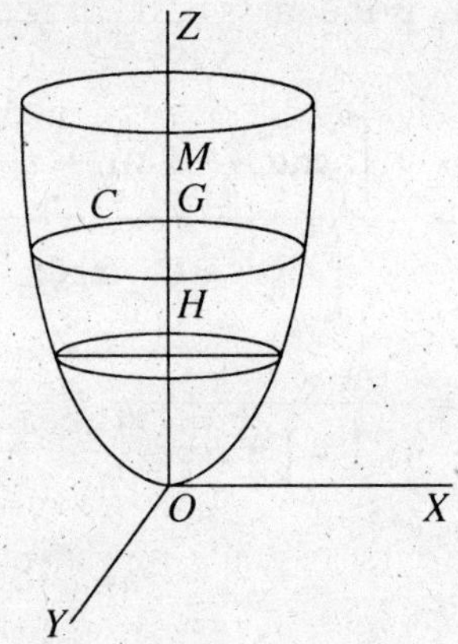

Fig. 9.17

Hence,

$$OG = \text{distance of C.G. of the paraboloid from the vertex}$$

$$= \frac{\int_0^c z\pi u^2 \sigma dz}{\int_0^c \pi u^2 \sigma dz},$$

where σ is the density of the liquid

$$= \frac{\int_0^c zmz\, dz}{\int_0^c mz\, dz} = \frac{\int_0^c z^2 dz}{\int_0^c z\, dz} = \frac{c^3}{3}\frac{2}{c^2} = \frac{2}{3}c$$

and OH = distance of centre of buoyancy of the liquid

$$= \frac{\int_0^a z\pi u^2 \lambda(a-z)\, dz}{\int_0^a \pi u^2 \lambda(a-z)\, dz} = \frac{\int_0^a mz(az-z^2)\, dz}{\int_0^a mz(a-z)\, dz} = \frac{\int_0^a (az^2-z^3)\, dz}{\int_0^a (az-z^2)\, dz}$$

$$= \frac{\left[a\frac{z^3}{3} - \frac{z^4}{4}\right]_0^a}{\left[a\frac{z^2}{2} - \frac{z^3}{3}\right]_0^a} = \frac{\frac{a^4}{3} - \frac{a^4}{4}}{\frac{a^3}{2} - \frac{a^3}{3}} = \frac{a^4}{12} \times \frac{6}{a^3} = \frac{a}{2}.$$

Also

$$HM = \frac{A_0 K_0^2 \rho_0 + \int_{\text{water line}}^{\text{low}} AK^2 d\rho}{M_0},$$

where M_0 is the mass of the liquid displaced

$$= \frac{\int_a^0 \pi u^2 \frac{u^2}{4}\, d[\lambda(a-z)]}{\int_0^a \pi u^2 \lambda(a-z)\, dz} = \frac{1}{4}\frac{\int_0^a m^2 z^2 dz}{\int_0^a mz(a-z)\, dz} = \frac{m}{4}\frac{\int_0^a z^2 dz}{\int_0^a (az-z^2) dz}$$

$$= \frac{m}{4}\frac{\frac{a^3}{3}}{a\left[\frac{a^2}{2} - \frac{a^3}{3}\right]} = \frac{m}{4}\frac{a^3}{3}\frac{6}{a^3} = \frac{m}{2}.$$

Also

$$HG = OG - OH = \frac{2}{3}c - \frac{1}{2}a.$$

Hence, equilibrium will be stable or unstable according as $HM \gtrless HG$

$$\text{or,} \quad \frac{m}{2} \gtrless \frac{2}{3}c - \frac{a}{2},$$
$$\text{or,} \quad \frac{m+a}{2} \gtrless \frac{2}{3}c,$$
$$\text{or,} \quad m + a \gtrless \frac{4}{3}c,$$
$$\text{or,} \quad 3(m+a) \gtrless 4c,$$
$$\text{or,} \quad 4c \gtrless 3(m+a).$$

▶ **Ex. 9.** A solid paraboloid of specific gravity σ floats in a liquid of specific gravity ρ with its axis vertical and vertex downwards. Show that the equilibrium will be stable of $3a > h\left(1 - \sqrt{\frac{\sigma}{\rho}}\right)$, where $4a$ is the latus rectum of the generating parabola and h is the height of the solid. **[C.U. 83H, 85H, 61H]**

Solution: Let the equation of the paraboloid of revolution be

$$x^2 + y^2 = 4az, \text{ i.e., } u^2 = 4az$$

Volume of the paraboloid

$$= \int_0^h \pi u^2 \, dz = \int_0^h \pi 4az \, dz = 2a\pi h^2.$$

If h' be the length of the axis immersed in the liquid, then for equilibrium weight of the paraboloid = weight of the liquid displaced

$$\text{or,} \quad 2\pi a h^2 \sigma = 2\pi a h'^2 \rho,$$
$$\text{or,} \quad \frac{h'^2}{h^2} = \frac{\sigma}{\rho}. \qquad (1)$$

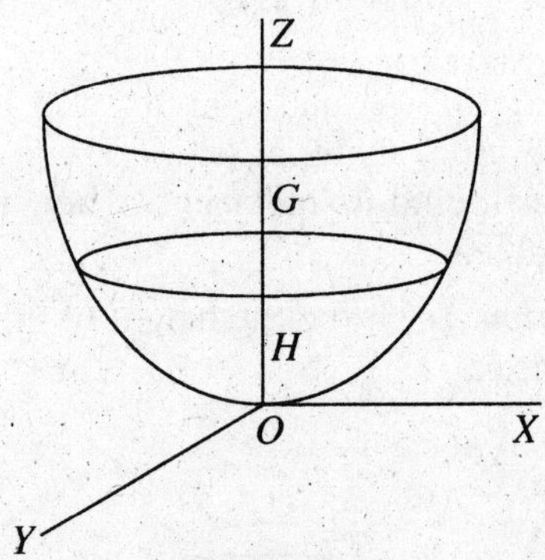

Fig. 9.18

Let H, G and M be the centre of buoyancy, centre of gravity and metacentre of the body.

Now,

$$HM = \frac{AK^2}{V} = \frac{\pi r^2 \frac{1}{4} r^2}{2\pi a h'^2} = \frac{r^4}{8ah'^2},$$

where $r^2 = 4ah'$

$$= \frac{16a^2h'^2}{8ah'^2} = 2a$$

$$\text{OH} = \text{height of the C.G. of the displaced liquid}$$

$$= \frac{\int_0^{h'} z\pi u^2 dz}{\int_0^{h'} \pi u^2 dz} = \frac{\int_0^{h'} z4az\, dz}{\int_0^{h'} 4az\, dz}$$

$$= \frac{h'^3}{3}, \frac{2}{h'^2} = \frac{2}{3}h'$$

$$\text{and} \quad \text{OG} = \text{height of the C.G. of the body} = \frac{2}{3}h.$$

Thus,

$$MG = OG - OH = \frac{2}{3}(h - h').$$

The equilibrium will be stable of HM > HG,

$$\text{or,} \quad 2a > \frac{2}{3}(h - h'),$$

$$\text{or,} \quad 3a > h\left(1 - \frac{h'}{h}\right),$$

$$\text{or,} \quad 3a > h\left(1 - \sqrt{\frac{\sigma}{\rho}}\right) \quad \text{[from (1)]}.$$

▶ **Ex. 10.** A solid cylinder of 3 meters diameter has a height of 3 meters. It is made up of material whose sp. gravity is .8 and is floating in water with its axis vertical. Find its metacentric height and state whether its equilibrium is stable or unstable. [I.A.S. 85]

Solution: Let h′ be the height of the cylinder immersed in the liquid, when the axis is vertical.

$$\therefore \quad h' = .8h = .8 \times 3 = \frac{4}{5} \times 3 = \frac{12}{5}.$$

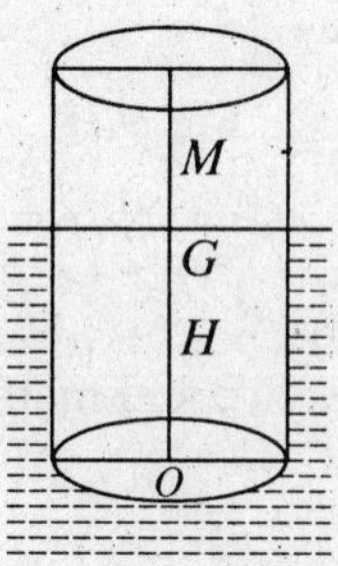

Fig. 9.19

Let in this case, O be the centre of the base, h the C.G. and M the metacentre. Then

$$OG = \frac{h}{2} = \frac{3}{2}$$

$$OH = h' = \frac{12}{5 \times 2} = \frac{6}{5}.$$

Since the section of the cylinder from the surface of the liquid is a circle of radius a, therefore,

$$\text{radius of gyration} = K = \frac{a}{2} = \frac{3}{4}.$$

Since $a = \frac{3}{2}$.

Also $A = \pi a^2 = \frac{9\pi}{4}$.

Also

$$\begin{aligned} V &= \text{the volume of the cylinder immersed in the liquid} \\ &= \pi a^2 h = \pi \times \frac{9}{4} \times \frac{12}{5} = \frac{27\pi}{5}. \end{aligned}$$

Hence,

$$HM = \frac{AK^2}{V} = \frac{9\pi}{4} \times \frac{9}{16} \times \frac{5}{27\pi} = \frac{15}{64}.$$

Now metacentric height

$$\begin{aligned} GM &= OM - OG = OH + HM - OG \\ &= \frac{6}{5} + \frac{15}{64} - \frac{3}{2} = \frac{6}{5} - \frac{81}{64} = \frac{384 - 405}{320} = \frac{-21}{320} = -.066 \text{ m} \\ HG &= OG - OH = \frac{3}{2} - \frac{6}{5} = \frac{3}{10} = 0.3. \end{aligned}$$

Since $HM < HG$. So, equilibrium is unstable.

Exercises

1. Show that a uniform circular cylinder of sp. gr. $\frac{1}{3}$ cannot be in stable equilibrium, when floating upright in water if its length exceeds three quarters of its diameter.

   [Bih. 67H, P.U. 63H, M.U. 65H, C.U. 93H]

2. A rectangular parallelopiped of sp. gravity $\frac{1}{2}$ and edges $2a$, $2b$, $2c$ floats in water. Show that if a^2 and b^2 are both greater than $3c^2$, the position of equilibrium in which the edge are vertical is stable.

3. Find the position of the metacentre in a right cylinder of radius a and length h floating with its axis vertical in a liquid whose density is $\frac{4}{3}$ times that of the cylinder.

   [*Ans.* $\left(\frac{3}{8} + \frac{a^2}{3}\right)$ h above the base]

4. Show that a circular cylinder of radius a and length $\frac{a}{n}$ cannot float up right in stable equilibrium if its sp. gravity lies between

   $$\frac{1}{2}\left[1 - \sqrt{1 - 2n^2}\right]$$
   $$\text{and} \quad \frac{1}{2}\left[1 + \sqrt{1 - 2n^2}\right].$$

   [P.U. 60H, Bih. 62H, L.N.M.U. 96H, C.U. 74]

5. The ratio of the length to the diameter of a uniform solid circular cylinder, specific gravity $\frac{1}{2}$ lies

between $\sqrt{\frac{1}{2}}$ and 1. Prove that the cylinder cannot float in stable equilibrium in water with its axis either horizontal or vertical.

6. Show that in case of a right circular cylinder of radius a and height h floating with its axis vertical in any liquid, the equilibrium will be stable whatever be the specific gravities if $a\sqrt{2} > h$.

[Bih. 73H, P.U. 77H, C.U. 82H]

7. A uniform solid of revolution of sp. gravity S and volume V, floats in water in stable equilibrium with its axis vertical and a volume V' immersed. Prove that a uniform solid of the same shape and of the same volume but of sp. gravity $(1 - S)$ can float in water on stable equilibrium with its axis vertical and a volume $(V - V')$ immersed, and with the same plane of floatation. [P.U. 59H, M.U. 72H]

8. A thin cylindrical stick consists of two parts equal in length. The specific gravity of one part is $\frac{5}{9}$. Show that the least specific gravity of the other part that will permit of the stick floating upright in stable equilibrium in water is $\frac{10}{9}$.

9. A hollow circular cylinder floats in water with its vertical, and three-quarters immersed, prove that for stability the ratio of the radius to the height of the cylinder must be greater than $\sqrt{\frac{3}{8}}$. In vestigate the effect, on the stability, of putting a given quantity of water into the cylinder.

[*Ans.* For stability the depth of water poured in must exceed $\frac{1}{8}$th height of the cylinder]

10. A conical shell, vertex downwards, floats in unstable equilibrium, how much water must be poured in to make the equilibrium stable.

[M.U. 68H]

[*Ans.* If h, α be the height and the semivertical angle of the cone and h' the height of the water inside, then $\left(\frac{z}{h'}\right)^3 > \frac{9h'\sec^2\alpha - 8h}{9z\sec^2\alpha - 8h}$, where z is given by $\frac{z}{8}\rho\pi\tan^2\alpha(z^3 - h'^3) =$ weight of the cone.

11. A uniform solid cylinder of length l and radius a, which can turn round a diameter of its upper end as a fixed axis, is held vertically immersed in water to a depth h, show that this vertical position is not stable unless the ratio of the weight of the cylinder to the water displaced exceeds

$$2 - \frac{h}{l} - \frac{a^3}{2hl}.$$

12. The cross-section of a cylindrical ship is two equal arcs of equal parabolas of latus rectum l which touch at the keel, the common vertex of the two parabolas, so that sides of the ship are concave to the water. The ship is floating up-right with its keel at a depth h. Prove that the height of the metacentre above the keel is

$$h\left(\frac{3}{4} + \frac{h^2}{l^2}\right).$$

[I.A.S. 62, 77]

13. A solid body consists of a right cone joined to a hemisphere on the same base and floats with the spherical portion partly immersed; prove that the greatest height of the cone consistent with stability is $\sqrt{3}$ times the radius of the base.

[P.U. 58H, 62H]

14. Find the form of a uniform solid of revolution such that when floating with its axis vertical the distance between the centre of buoyancy and the metacentre is independent of the length of the axis immersed.

[Bih. 66H, 62H]

[*Ans.* Paraboloid of revolution]

15. A thin cylindrical vessel floats when empty with its axis vertical and its base at a depth h below the surface and the equilibrium is unstable. Show that the vessel can float in stable equilibrium with its axis vertical, if a liquid of volume greater than $\pi a^2 \left(C - \frac{h}{2}\right)$ is poured in, a is the radius of the vessel and C the distance of its centre of gravity from its base.

16. Two equal uniform rods whose density is ρ, are joined together at an angle 2α. If they be immersed with angle downwards in a fluid of density σ, find the poition of equilibrium and prove that surface of buoyancy is a parabola. Also show that the rods cannot float with the line joining their ends inclined to the horizon, unless $\frac{\sigma-\rho}{\sigma+\rho}$ be greater than $\sin^2 \alpha$.

17. A hollow cone whose weight is half that of the water it can contain, floats in stable equilibrium with its axis vertical and vertex downwards. Prove that of α be the semi-vertical angle of cone,

$$\cos^6 \alpha < \frac{729}{1024}.$$

[C.U. 92H, 01H]

18. A hollow cylindrical vessel, of radius r, is supported at the ends of a diameter of the cross-section in which is situated at the centre of gravity of the vessel. The vessel contains water to a depth h. Prove that the vertical position is unstable unless $r < b\sqrt{2}$ and h lies between

$$b \pm \sqrt{b^2 - \frac{r^2}{2}},$$

where b is the distance of the centre of gravity of the vessel from the base. [Cal 91H]

19. Prove that a circular cylinder of radius R and length λR can float upright in stable equilibrium of $\lambda^2 < 2$. [Cal 86H]

20. A homogeneous right circular cone of vertical angle 2α floats with its axis vertical and vertex downwards and is in unstable equilibrium. Prove that the equilibrium becomes stable when a particle of weight w is attach to the vertex if

$$\frac{w}{W} > \left\{ \left(\frac{\rho \cos^6 \alpha}{\sigma} \right)^{1/4} - 1 \right\},$$

where W is the weight of the cone. It is assumed that the base of the cone is not submerged, and that σ is the density of the cone and ρ that of the liquid.

[R.U. 67H, Bhag. 66H, C.U. 94H]

21. If a segment of a sphere of density σ floats in a liquid of density ρ, proe that the position in which the plane face is immersed and horizontal, is a position of stable equilibrium and that the metacentric height is $\frac{\rho-\sigma}{\sigma}$ times the distance of the C.G. of the segment from the centre of the sphere.

22. A uniform elliptic disc of eccentricity e and mass M has a particle of

mass m attached at one focus and focus with its major axis vertical in a liquid of such density that one-fourth of that axis is immersed. Show that the equilibrium will be stable, if

$$\left(\frac{4\pi}{3\sqrt{3}} - 1 - e\right) m > eM.$$

23. A right circular solid cone of semi-vertical angle α floats, wholly immersed with its vertex upwards and axis vertical in a liquid the density of which varies as the depth. If h is the height of the cone, and b the depth of its vertex below the surface, the distance of the meta centre from the vertex is equal to $\frac{3h}{5}\frac{5b+4h-h\tan^2\alpha}{4b+3h}$.

[C.U. 96H, 02H]

24. A solid circular cylinder floating in a liquid with its axis horizontal and lying on the surface, is displaced in the vertical plane through the axis. Prove that the equilibrium is stable if the length of the cylinder is greater than its diameter.

Also discuss this stability of the cylinder of it if floats in the liquid with its axes vertical.

[C.U. 95H, 2000H, 04H]

[*Ans.* $\sqrt{2}\,a > h$, where h is the height of the cylinder and a is the radius of its base]

CHAPTER 10

Gases

10.1 Introduction

Fluids which can be compressed easily and expand indefinitely by increasing the volume are called *Gases.*

Compression is the ratio of the reduction in volume to the original volume, i.e., $-\frac{dv}{V}$.

Air is a gas. It exerts pressure and has weight. Atmosphere is supposed to be spherical shell surrounding the earths surface. Its density gradually increases as one proceeds up from the surface of the earth. If we suppose that the density remains the same through out, i.e., of we suppose that the atmosphere is of homogeneous air and if we take a column of homogeneous air of the same density as at the earths surface and of such a height that the pressure produced by the column at the surface of the earth is the same as the actual pressure of the atmosphere, then this height is called the *height of the homogeneous atmosphere.*

The *barometer* is an instrument which is used for measuring the atmospheric pressure.

There are a *number of corrections* which are applied to an observation with the mercury barometer.

- Correction for capacity.
- Correction for temperature.
- Correction for height, above the sea level.
- Correction for capillarity.
- Correction for vapour pressure.

10.2 Relations between Pressure, Density and Temperature

For a given mass of gas the relation between pressure, density and temperature are expressed by the following two laws which are confirmed by experiment and found to be holding good in a wide range of circumstances.

1. **Boyle's Law**—The law states that if the temperature of a given mass of a gas remains constant, the pressure varies inversely as the volume.

[R.U. 92H, 95H, 97H]

Thus, if V denotes the volume of a given mass of gas and p the pressure, then

$$p \alpha \frac{1}{V}, \quad \text{or,} \quad pV = \text{constant.} \tag{10.1}$$

If m be the given mass of gas and ρ its density, then $m = \rho V$.

Hence, from (10.1)

$$p \cdot \frac{m}{\rho} = \text{constant}, \quad \text{or,} \quad p = k\rho, \quad \text{or,} \quad p \alpha \rho,$$

where k is a constant.

Hence, *temperature remaining the same, the pressure of a given mass of gas is proportional to its density.*

Note 1. *Equation (10.1) is known as the Isothermal Equation of the state.*

Note 2. *This law is also known as Marriotte's Law.*

Note 3. *A gas which obeys Boyle's Law is known as Perfect Gas.*

2. **Charle's Law**—The law states that if the pressure remains constant, the volume increases by a fraction $\frac{1}{273}$ of the volume at $0°C$; for every degree centigrade by which the temperature is raised. [R.U. 95H]

Thus, if V_0 and V be the volumes of the gas at $0°$ and $t°C$, we have

$$\begin{aligned} V &= V_0 + \frac{V_0 t}{273} = V_0\left(1 + \frac{t}{273}\right), \\ \text{or, } V &= V_0(1 + \alpha t), \end{aligned} \tag{10.2}$$

where $\alpha = \frac{1}{273}$.

Again, mass $= \rho V =$ constant,

$$\begin{aligned} \text{or, } \rho V &= \rho_0 V_0, \\ \text{or, } \frac{\rho}{\rho_0} &= \frac{V_0}{V} = \frac{V_0}{V_0(1+\alpha t)} = \frac{1}{1+\alpha t}, \\ \text{or, } \rho_0 &= \rho(1 + \alpha t) \end{aligned} \tag{10.3}$$

which is the *relation between density and temperature.*

If a gas is continually cooled down, we see from eqn. (10.2) if it does not liquify, the volume becomes zero when

$$1 + \alpha t = 0, \quad \text{or,} \quad t = -\frac{1}{\alpha} = -273°\text{C}.$$

This temperature $-273°$C is known as the *Absolute Zero* of the gas thermometer. [R.U. 94H]

The Absolute Temperature T is denoted by

$$t + \frac{1}{\alpha}, \quad \text{or,} \quad 273 + t.$$

[R.U. 93H]

Then from (10.1), we get

$$\begin{aligned} V &= V_0\left(1 + \frac{t}{273}\right) = V_0 \frac{273 + t}{273}, \\ \text{or,}\ V &= \frac{V_0 T}{273}, \quad \text{i.e.,} \quad V \alpha T. \end{aligned}$$

Hence, *volume of a given mass of gas under the same pressure varies directly as its absolute temperature.*

For a given mass V_ρ is constant.

Hence, $\rho \alpha \frac{1}{T}$.

Thus, *if temperature of a given mass of gas, free from any pressure, at a series of points be in A.P., the densities of those points are in H.P.*

When temperature is constant, we have from Boyle's law at 0°C

$$p = K\rho_0 = K\rho(1 + \alpha t) \quad \text{From (10.3)} \tag{10.4}$$

If T be the absolute temperature corresponding to t°C, than

$$T = t + 273 = t + \frac{1}{\alpha}.$$

Hence, (10.4) can be written as

$$\begin{aligned} p &= k\rho\alpha\left(t + \frac{1}{\alpha}\right) = k\rho\alpha T, \\ \text{or,}\ p &= R\rho T, \quad \text{where } R = k\alpha, \end{aligned} \tag{10.5}$$

where T is the absolute temperature and R is a constant depending on the nature of the gas.

Since mass of the gas is constant, $\rho V =$ constant.

From (10.1)

$$\frac{Vp}{RT} = \text{constant}, \quad \text{or}, \quad \frac{pV}{T} = \text{cons. as R is constant.}$$

This is combined Boyle's and Charle's law.

Thus, if a given mass of gas has pressures, volumes and absolute temperatures p, V, T and p', V', T' in two positions, then

$$\frac{pV}{T} = \frac{p'V'}{T'}.$$

10.3 Mixture of Gases

From experimental observations it has been seen that—if two gases are kept in two closed vessels of different sizes at the same pressure and temperature, and if two vessels are connected so that the gas from one may flow into the other and if no chemical action takes place, then the gases form a mixture whose pressure and temperature is the same as of the gases.

This experimental fact enables us to establish the following propositions.

◇ Property 1. *If the pressures of two gases of the same temperature and volume v be p_1 and p_2, the pressure of the two gases, when the combined volume is v, is $p_1 + p_2$, temperature being unaltered.*

Let us suppose that the pressure of the first gas changes from p_1 to p_2 before mixing, then the volume will change. Let it be v_1.

Now, applying Boyle's law, we have

$$p_1 v = p_2 v_1. \tag{10.6}$$

Thus, at the same pressure p_2, we have two gases of combined volume $v + v_1$. Let the pressure of the mixture be p, if the volume changes from $v + v_1$ to v.

Thus, by Boyle's law,

$$\begin{aligned} pv &= p_2(v + v_1) \\ = p_2 v + p_2 v_1 &= p_2 v + p_1 v \quad \text{From (10.6)} \\ &= (p_1 + p_2)v. \end{aligned}$$

Hence, $p = p_1 + p_2$.

Corollary 10.3.1 *If the pressures of gases of the same temperature and volume be $p_1, p_2, \ldots, p_n$ the pressure of the mixture of the gases, when the combined volume is V, is*

$$(p_1 + p_2 + \cdots + p_n).$$

◇ Property 2. *Two volumes v_1 and v_2, of different gases at different pressures p_1 and p_2 are mixed together and put into a vessel of volume V, then the pressure of the mixture is*

$$\frac{p_1 v_1 + p_2 v_2}{V},$$

the temperature being constant.

[M.U. 73, Mith. 81, B.U. 76, P.U. 82, R.U. 79]

Let the pressure of the first gas be changed from p_1 to p before mixing, then the volume will also change. Let it be V'.

Therefore, from Boyle's law

$$p_1 v_1 = p_2 V'. \tag{10.7}$$

Now, there are two gases of volumes V' and v_2 but each at pressure p_2. When these two gases are mixed, the volume of the mixture is $V' + v_2$ at pressure p_2.

If the volume of the mixture be changed to V, then the pressure p_2 will consequently change to P. Therefore, from Boyle's law

$$\begin{aligned} PV &= p_2(V' + v_2) = p_2 V' + p_2 v_2 \\ &= p_1 V_1 + p_2 v_2 \quad \text{From (10.7)} \\ P &= \frac{p_1 V_1 + p_2 v_2}{V}. \end{aligned}$$

Corollary 10.3.2 *If volumes $V_1, V_2, \ldots, V_n$ of different gases at pressures $p_1, p_2, \ldots, p_n$ are mixed together and put into a vessel of volume V, then the pressure of the mixture is*

$$\frac{p_1 V_1 + p_2 V_2 + \cdots + p_n V_n}{V},$$

the temperature remaining constant.

Property 3. *If volumes V_1, V_2 of two gases at pressures p_1 and p_2 and absolute temperatures t_1 and t_2, respectively be mixed, the pressure of the mixture is*

$$\frac{T}{V}\left(\frac{p_1 v_1}{t_1} + \frac{p_1 v_2}{t_2}\right),$$

where V and T are the volume and absolute temperature of the mixture. [M.U. 77, R.U. 75, R.U. 95H]

We suppose that the volumes and absolute temperatures of both the gases be changed to V and T, then their corresponding pressures P_1 and P_2 are given by,

$$\begin{aligned} \frac{P_1 V}{T} &= \frac{p_1 v_1}{t_1} \quad \text{and} \quad \frac{P_2 V}{T} = \frac{p_2 v_2}{t_2}. \\ \therefore \quad P_1 &= \frac{T}{V}\frac{p_1 v_1}{t} \quad \text{and} \quad P_2 = \frac{T}{V}\frac{p_2 v_2}{t_2}. \end{aligned}$$

Now, the temperature remaining constant at T, let the gases be mixed in such a way that the volume of the mixture remains V, then the pressure P of the mixture is given by

$$P = P_1 + P_2 = \frac{T}{V}\left(\frac{p_1 v_1}{t_1} + \frac{p_2 v_2}{t_2}\right).$$

Corollary 10.3.3 *If volumes $V_1, V_2, \ldots, V_n$ of gases at pressures $p_1, p_2, \ldots, p_n$ and absolute temperatures $t_1, t_2, \ldots, t_n$, respectively be mixed, the pressure of the mixture is*

$$\frac{T}{V}\left(\frac{p_1 v_1}{t_1} + \frac{p_2 v_2}{t_2} + \cdots + \frac{p_n v_n}{t_n}\right),$$

where V, T are volume and absolute temperature of the mixture.

10.4 Pressure and Density of the Air at a given Height

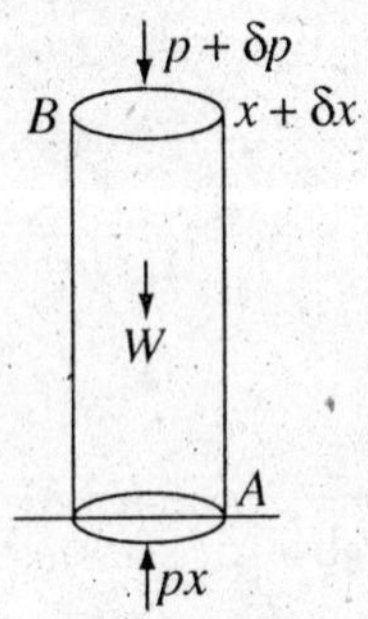

Fig. 10.1

Let p and p+δp be the pressure of the air at heights x and x + δx. We consider a thin vertical cylinder AB through two points, clearly δx is the length of the cylinder. Let α be the cross-sectional area of the cylinder and ρ be the mean density of the air. The cylinder is, in equilibrium under the following forces.

- Weight of the cylinder $W = \alpha\rho g x \delta x$ acting through the C.G. downwards.
- Thrust $(p + \delta p)\alpha$ acting vertically downwards to the upper section.
- Thrust $p\alpha$ acting normally vertically upwards to the lower section.

$$\therefore \quad p\alpha = (p + \delta p)\alpha + \alpha\rho g \delta x$$

$$\text{or,} \quad \delta p + \rho g \delta x = 0, \quad \text{or,} \quad \frac{\delta p}{\delta x} = -\rho g.$$

Taking limit, we get

$$\frac{dp}{dx} = -\rho g. \tag{10.8}$$

But

$$p = k\rho. \tag{10.9}$$

Hence, from (10.8)

$$\frac{dp}{p} = -\frac{g}{k}\,dx$$

$$\int_{p_0}^{p} \frac{dp}{p} = -\frac{g}{k}\int_0^x dx$$

$$\log\frac{p}{p_0} = -\frac{gx}{k}.$$

$$\therefore \quad \frac{p}{p_0} = e^{-\frac{gx}{k}}$$

$$\text{or,}\ p = p_0 e^{-\frac{gx/z}{gx}}.$$

This determines the pressure at a given height. [C.U. 84H]

Also

$$k\rho = k\rho_0 e^{-\frac{gx}{k}} \quad \text{From (10.9)}$$

where ρ_0 is the density of the air at sea-level.

[R.U. 80H, 82H, M.U. 84, Mith. 82, P.U. 82, 83]

This determines the density at a given height.

10.5 If the atmosphere be at rest and if its temperature and force of gravity remains constant, then as altitude increases in A.P., the pressure or density decreases in G.P.

We know that

$$p = p_0 e^{-\frac{gx}{k}} \quad \text{and} \quad \rho = \rho_0 e^{-\frac{gx}{k}}.$$

Let the heights on A.P. above the sea-level be $x, x + a, x + 2a, \ldots$, then the corresponding pressures and densities at these heights are

$$p_0^{-\frac{gx}{k}}, \quad p_0 e^{-\frac{g}{k}(x+a)}, \quad p_0 e^{-\frac{g}{k}(x+2a)}$$

which are clearly in G.P. whose common ratio is $p_0 e^{-\frac{ga}{k}}$ which is less than 1.

Also densities at these heights are

$$\rho_0 e^{-\frac{gx}{k}}, \quad \rho_0 e^{-\frac{g}{k}(x+a)}, \quad \rho_0 e^{-\frac{g}{k}(x+2a)}$$

which are clearly in G.P. whose common ratio is $\rho_0 e^{-\frac{ga}{k}}$ which is less than 1.

10.6 Determination of Heights by Barometers

Let h and h_1 be the heights of barometric observations at the upper and lower stations, ρ and ρ_1 the densities of the air at the two places, p and p_1 be the atmospheric pressures at these two places, respectively. Let x be the difference of the altitudes of the two place.

Then

$$\frac{p_1}{p_2} = \frac{\rho g h_1}{\rho g h_2} = \frac{e^{-\frac{gh_1}{k}}}{e^{-\frac{gh_2}{k}}} = e^{-\frac{g}{k}(h_1 - h_2)} = e^{-\frac{gx}{k}}.$$

$$\log \frac{h_1}{h_2} = -\frac{gx}{k}, \quad \text{or,} \quad \log \frac{h_2}{h_1} = \frac{gx}{k}.$$

$$x = \frac{k}{g} \log \frac{h_2}{h_1}.$$

10.7 The Height of the Homogeneous Atmosphere is the same at all places if the temperature is the same [R.U. 99H]

Let H be the height of the homogeneous atmosphere, p the pressure and ρ the density of the air at the given point. Then

$$p = \rho g H.$$

But by Boyle's law

$$\begin{aligned} p &= k\rho \\ \therefore\ k\rho &= \rho g H \\ H &= \frac{k}{g} = \text{constant; if the temperature is constant.} \end{aligned}$$

10.8 Pressure in an Isothermal Atmosphere

Case I. *When gravity g is constant.*

Let p and ρ denote the pressure and density at a height z above the sea-level, then the pressure equation is

$$dp = -\rho g dz.$$

Also by Boyle's law

$$\begin{aligned} p &= k\rho. \\ \frac{dp}{p} &= -\frac{g}{k}dz. \\ \int_{p_0}^{p}\frac{dp}{p} &= -\frac{g}{k}\int_0^z dz. \\ \log\frac{p}{p_0} &= \frac{-gz}{k}, \\ \text{or,}\ \frac{p}{p_0} &= e^{-\frac{gz}{k}}. \end{aligned}$$

Hence,

$$p = p_0 e^{-\frac{gz}{k}} \tag{10.10}$$

If p_0 be the pressure on the surface of the earth due to a homogeneous atmosphere of density ρ_0 and height H, we have

$$p_0 = \rho_0 g H.$$

Also $p_0 = k\rho_0$

$$\therefore\ k = gH.$$

Hence, from (10.10), we get

$$p = p_0 e^{-\frac{z}{H}}.$$

Case II. *When variation of gravity g is taken into consideration.*

[L.N.M.U. 94H, R.U. 92H]

Let r be the radius of the earth and g the value of gravity at the sea-level. Then at a height z above the sea-level, the value of gravity is

$$\frac{r^2}{(r+z)^2}\cdot g.$$

Therefore, the pressure equation becomes

$$dp = \frac{-r^2}{(r+z)^2}\rho g \cdot dz$$

Also by Boyle's law

$$\begin{aligned} p &= k\rho \\ \frac{dp}{p} &= \frac{-r^2}{(r+z)^2}\frac{g}{k}\,dz. \end{aligned}$$

Integrating, we get

$$\log p = C + \frac{gr^2}{k(r+z)}. \tag{10.11}$$

At the sea-level, $p = p_0$ when $z = 0$.
Hence, from (10.11)

$$\log p_0 = C + \frac{gr}{k}. \tag{10.12}$$

Subtracting eqn. (10.12) from eqn. (10.11) we get

$$\begin{aligned} \log p - \log p_0 &= \frac{gr}{k}\left[\frac{r}{r+z} - 1\right]. \\ \log\frac{p}{p_0} &= -\frac{grz}{k(r+z)}, \\ \text{or, } \frac{p}{p_0} &= e^{-\frac{grz}{k(r+z)}}. \\ p &= p_0 e^{-\frac{grz}{k(r+z)}}. \end{aligned}$$

[C.U. 92H]

Also $p = k\rho$.
Hence,

$$\rho = \rho_0 e^{-\frac{grz}{k(r+z)}},$$

where ρ_0 is the density of the air at sea-level.

10.9 Some Definitions

1. *Thermal capacity:* The thermal capacity of a body is the amount of heat required to raise its temperature by one degree.

2. *Isothermal atmosphere:* Isothermal atmosphere is one in which temperature is supposed to be constant at all heights.

3. *Specific heat:* The specific heat of a gas is the quantity of heat required to raise the temperature of a unit mass of the gas by one degree. If the change takes place at constant volume we have the *specific heat at constant volume*, denoted by C_v and of the change takes place at constant pressure we have the *specific heat at constant pressure* denoted by C_p. It is clear that $C_p > C_v$ for in the first case some heat is used in expanding the gas to keep the pressure constant.

10.10 Prove that $C_p - C_\nu = R$

If a quantity dQ of heat raises the temperature of a unit mass by an amount dt, then

$$dQ = C_\nu dt \tag{10.13}$$

when the volume is constant

$$dQ = C_p dt \tag{10.14}$$

when the pressure is constant.

Further, let the imparting of quantity dQ of heat increase the internal energy of gas by dE and produce an expansion $d\nu$ in its volume, then by the first law of thermodynamics

$$dQ = dE + p d\nu. \tag{10.15}$$

If the volume ν be kept constant, then

$$d\nu = 0 \quad \text{and} \quad dE = dQ = C\nu dt.$$

Hence, from (10.15)

$$dQ = C_\nu dt + p d\nu. \tag{10.16}$$

Also for a perfect gas

$$p\nu = Rt, \text{ or, } p d\nu + \nu dp = R dt$$

Hence, (10.16) becomes

$$dQ = C_\nu dt + R dt - \nu dp. \tag{10.17}$$

Now, let p be kept constant

$$\therefore \quad dp = 0, \quad dQ = C_p dt.$$

Hence, from (10.17)

$$C_p dt = C_\nu dt + R dt, \text{ or, } C_p - C_\nu = R.$$

10.11 Adiabatic Changes

When a change of state takes place in a gas without any gain or less of heat, it is said to be an adiabatic change.

10.12 Relation between Pressure and Volume in an Adiabatic Change

By first law of thermodynamics, we have

$$dQ = dE + p d\nu.$$

For a perfect gas, E is a function of T alone and C_ν is constant, we have by keeping ν constant.

$$dQ = C_\nu dt + p d\nu.$$

In an adiabatic change $dQ = 0$

$$C_\nu dT + pdV = 0. \tag{10.18}$$

Also by Boyle's law and Charle's law $p\nu = RT$,

$$\text{or,} \quad pd\nu + \nu dp = RdT = R\left(-\frac{pd\nu}{c_\nu}\right) \quad \text{From (10.18),}$$

$$\text{or,} \quad pd\nu + \nu dp + \frac{Rpd\nu}{C_\nu} = 0,$$

$$\text{or,} \quad pd\nu\left(1 + \frac{R}{C_\nu}\right) + \nu dp = 0,$$

$$\text{or,} \quad pd\nu(C_\nu + R) + C_\nu \nu dp = 0,$$

$$\text{or,} \quad pd\nu C_p + C_\nu \nu dp = 0,$$

$$\text{or,} \quad pd\nu\frac{C_p}{C_\nu} + \nu dp = 0,$$

$$\text{or,} \quad \frac{dp}{p} + d\nu\frac{\gamma}{\nu} = 0,$$

$$\text{where} \quad \frac{C_p}{C_\nu} = \gamma = \text{constant.}$$

Integrating, we get

$$\log p + \gamma \log \nu = \text{constant,}$$

$$\text{or,} \quad \log p + \log \nu^\gamma = \text{constant,}$$

$$\text{or,} \quad \log(p\nu^\gamma) = \text{constant.}$$

Hence, $p\nu^\gamma = \text{constant}$.

10.13 Work Done in Compressing a Gas

Let the volume of a gas be ν when its pressure is p and let the external atmospheric pressure be π. Then if the volume is decreased by $-d\nu$, the work done to reduce the volume by this amount is $-(p - \pi)\, d\nu$.

Hence, the total work done in reducing the volume from V to V',

$$= -\int_V^{V'} (p - \pi)\, d\nu$$

Case I. *When the change is isothermal.*

In this case, $PV = K$, therefore, the work done

$$= -\int_V^{V'} \left(\frac{K}{V} - \pi\right) dV$$

$$= K \log \frac{V}{V'} - \pi(V - V')$$

$$= PV \log \frac{V}{V'} - \pi(V - V')$$

P being the initial pressure when initial volume is V.

Case II. *When the change is adiabatic.*

In this case, $pV^\gamma = K$, therefore, the work done

$$\begin{aligned}
&= -\int_v^{v'} \left[\frac{K}{V^\gamma} - \pi\right] dV \\
&= \frac{K}{1-\gamma}\left[\frac{1}{V'^{\gamma-1}} - \frac{1}{V^{\gamma-1}}\right] - \pi(V - V') \\
&= \frac{PV^\gamma}{\gamma-1}\left[\frac{1}{V'^{\gamma-1}} - \frac{1}{V^{\gamma-1}}\right] - \pi(V - V').
\end{aligned}$$

P being the initial pressure, when the volume is V.
If P′ be the final pressure, then

$$P'V'^\gamma = PV^\gamma.$$

Hence, the work done

$$\begin{aligned}
&= \frac{1}{\gamma-1}\left[\frac{P'V'^\gamma}{V'^{\gamma-1}} - \frac{PV^\gamma}{V^{\gamma-1}}\right] - \pi(V - V') \\
&= \frac{1}{\gamma-1}[P'V' - PV] - \pi(V - V').
\end{aligned}$$

If T and T′ be the initial and final absolute temperatures, then

$$\frac{PV}{T} = \frac{P'V'}{T'} = R = \frac{P'V' - PV}{T' - T}.$$

$$\therefore \quad \text{The work done} = \frac{R(T' - T)}{\gamma - 1} - \pi(V - V').$$

10.14 Atmosphere in Convective Equilibrium

By convective equilibrium we mean a state in which if equal masses of air at any two stations be interchanged, each would assume the pressure,density and temperature of the other. Thus, the change takes place without loss or gain of heat, i.e., adiabatically.

Case I. *When the gravity is constant.*

Thus, pressure equations

$$dp = -\rho g dz. \tag{10.19}$$

In adiabatic condition

$$\left.\begin{aligned} p &= k\rho^\gamma \\ dp &= k\gamma\rho^{\gamma-1} d\rho \end{aligned}\right\}. \tag{10.20}$$

So, from (10.19)

$$\begin{aligned}
k\gamma\rho^{\gamma-1}\, d\rho &= -\rho g dz \\
\text{or, } k\gamma\rho^{\gamma-2}\, d\rho &= -g dz.
\end{aligned}$$

Integrating, we get

$$\frac{k\gamma\rho^{\gamma-1}}{\gamma-1} = C - gz,$$

$$\text{or,} \quad \frac{k\gamma\rho^{\gamma}}{\rho(\gamma-1)} = C - gz,$$

$$\text{or,} \quad \frac{p}{\rho}\frac{\gamma}{\gamma-1} = C - gz \quad \text{From (10.20).}$$

But

$$p = R\rho T, \quad \text{or,} \quad \frac{p}{\rho} = RT.$$

$$\therefore \quad RT\frac{\gamma}{\gamma-1} = C - gz. \tag{10.21}$$

At sea level, $z = 0$ and $T = T_0$

$$\frac{RT_0\gamma}{\gamma-1} = C.$$

So from (10.21), we get

$$\frac{RT\gamma}{\gamma-1} = \frac{RT_0\gamma}{\gamma-1} - gz,$$

$$\text{or,} \quad \frac{\gamma R}{\gamma-1}(T - T_0) = -gz,$$

$$\text{or,} \quad T - T_0 = -\frac{\gamma-1}{\gamma}\frac{gz}{R},$$

$$\text{or,} \quad \frac{T}{T_0} - 1 = -\frac{\gamma-1}{\gamma}\frac{gz}{RT_0},$$

$$\text{or,} \quad \frac{T}{T_0} = 1 - \frac{\gamma-1}{\gamma}\frac{gz}{RT_0}. \tag{10.22}$$

If the height of the homogeneous atmosphere be H, then

$$p_0 = R\rho_0 T_0 \quad \text{and} \quad p_0 = \rho_0 gH.$$

$$\therefore \quad R\rho_0 T_0 = \rho_0 gH \quad \text{or,} \quad RT_0 = gH.$$

Thus, from (10.22), we have,

$$\frac{T}{T_0} = 1 - \frac{\gamma-1}{\gamma}\frac{Z}{H}$$

[Cal 80H, 94H]

Case II. *Thus, we see that the temperature in an atmosphere in convective equilibrium diminishes as the height increases.*

Case III. *When variation of gravity is taken into account.*

If g be the value of gravity at sea-level and r the earths radius, the value of gravity at a height z is

$$\frac{gr^2}{(r+z)^2}.$$

Then, the pressure equation becomes

$$dp = -\frac{\rho g r^2}{(r+z)^2}\,dz. \tag{10.23}$$

In an adiabatic change

$$p = k\rho^{\gamma} \tag{10.24}$$

$$dp = k\gamma\rho^{\gamma-1}\,d\rho$$

so from (10.23), we have

$$k\gamma\rho\gamma^{-1}\,d\rho = -\frac{\rho g r^2}{(r+z)^2}\,dz,$$

$$\text{or, } k\gamma\rho\gamma^{-2}\,d\rho = -r^2 g\frac{dz}{(r+z)^2}.$$

Integrating, we get

$$\frac{k\gamma\rho^{\gamma-1}}{\gamma-1} = C + \frac{r^2 g}{r+z},$$

$$\text{or, } \frac{k\gamma\rho^{\gamma}}{\rho(\gamma-1)} = C + \frac{r^2 g}{r+z},$$

$$\text{or, } \frac{\gamma}{\gamma-1}\frac{p}{\rho} = C + \frac{r^2 g}{r+z} \quad \text{From (10.24).}$$

Also

$$p = R\rho T, \text{ or, } \frac{p}{\rho} = RT,$$

$$\text{or, } \frac{\gamma}{\gamma-1}RT = C + \frac{r^2 g}{r+z}. \tag{10.25}$$

On earth's surface $z = 0$ and $T = T_0$

$$\frac{\gamma}{\gamma-1}RT_0 = C + rg. \tag{10.26}$$

Subtracting (10.26) from (10.25), we get

$$\frac{\gamma}{\gamma-1}R(T-T_0) = rg\left(\frac{r}{r+z} - 1\right) = -\frac{rgz}{r+z},$$

$$\text{or, } T - T_0 = -\frac{\gamma-1}{\gamma}\frac{rgz}{R(r+z)},$$

$$\text{or, } \frac{T}{T_0} - 1 = -\frac{\gamma-1}{\gamma}\frac{rgz}{(r+z)RT_0},$$

$$\frac{T}{T_0} = 1 - \frac{\gamma-1}{\gamma}\frac{rgz}{(r+z)RT_0}. \tag{10.27}$$

If the height of the homogeneous atmosphere be H, then

$$p_0 = R\rho_0 T_0 \quad \text{and} \quad p_0 = \rho_0 g H.$$

$$\therefore \quad R\rho_0 T_0 = \rho_0 g H, \quad \text{or,} \quad \frac{g}{RT_0} = \frac{1}{H}.$$

Hence, from (10.27), we get

$$\frac{T}{T_0} = 1 - \frac{\gamma - 1}{\gamma}\frac{rz}{(r+z)H}.$$

[C.U. 82H, 86H, 99H, 02H]

Illustrative Examples

▶ **Ex. 1.** A hollow cylinder of height h and open at the top, is inverted and partially immersed so that a length k of it is under water. Prove the air inside it occupies a length x is given by the equation

$$x^2 + x(H + k - h) = Hh,$$

where H is the height of water barometer.

[P.U. 65A, 70A, R.U. 75, M.U. 78H, C.U. 82H, 88H]

Solution: When the cylinder is immersed, with open end downwards the air in side it is at atmospheric pressure. Level of water inside the cylinder will be lower than that of outside.

Let AB, CD be the levels of water outside and inside the cylinder.

Now,

atmospheric pressure $= \pi = \rho g H$ (1)

and pressure at any point on the level

$CD = P = \rho g H + \rho g(k - h + x)$ (2)

x being the length occupied by the air inside the cylinder.

Applying Boyle's law, we get

$$P \cdot Ax = \pi \cdot Ah,$$

where A is the area of cross-section of the cylinder.

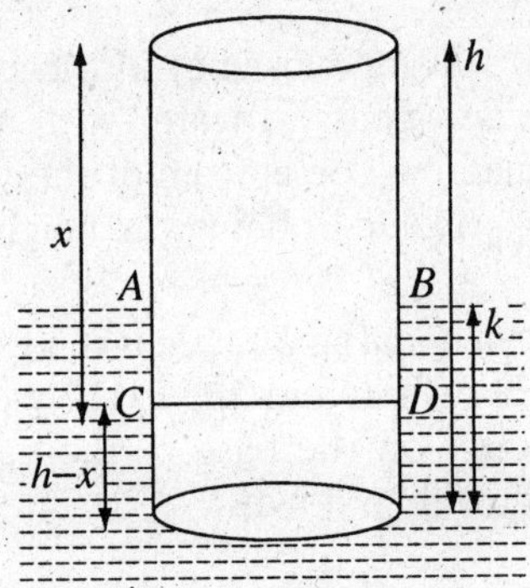

Fig. 10.2

$$P = \frac{\rho g H \cdot h}{x} \quad \text{From (1)}$$

$$\frac{\rho g H \cdot h}{x} = \rho g H + \rho g(k - h + x), \text{ From (2)}$$

or, $$Hh = xH + x(k - h + x),$$

or, $$x^2 + x(H + k - h) - Hh = 0$$

which gives true value of x as the positive square root.

► **Ex. 2.** Assuming the height of water barometer to be h, find to what depth a small inverted, conical glass must be lowered so that the water may rise half up.

[M.U. 61A, Mith. 78A, 66A, 71A, 75A, R.U. 79, 64A, Mith. 78A, L.N.M.U. 94H]

Solution: Let r be the radius of the base of the cone and y its height.

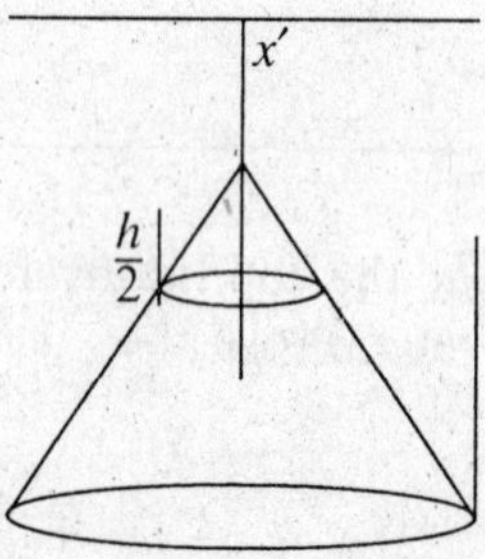

Fig. 10.3

When the rim is just on the surface of water, then the volume of the air within the cone is $\frac{1}{3}\pi r^2 y$, and it is at a pressure of water head h ft. high.

When the water rises half-way up within the cone, let the middle point of the axis of the cone sink to adepth x ft. and the radius of the surface of the water inside the cone is $\frac{r}{2}$, hence, volume of air inside the cone $\frac{1}{3}\pi\left(\frac{r}{2}\right)^2 \frac{y}{2}$.

This is at a pressure of water head $(h + x)$ ft.

Hencc, by Boyle's law

$$\frac{1}{3}\pi r^2 y h = \frac{1}{3}\pi\left(\frac{r}{2}\right)^2 \frac{y}{2}(h + x),$$

or, $$8h = h + x, \quad 7h = x.$$

► **Ex. 3.** A thin closed cylindrical vessel, of height a, contains air at atmospheric pressure and floats in water with axis vertical and length b immersed. If a small hole is made in the bottom of the vessel, show that water leaks in until there is a depth $\frac{ab}{h+b}$ inside, h being the height of the water barometer. **[I.A.S. 86]**

Solution: When a small hole is made, the water will rush in and due to the weight of the water on the base of the vessel, the vessel will go down.

Let y be the depth of the base from the free surface in this position and let x be the height of water in the vessel.

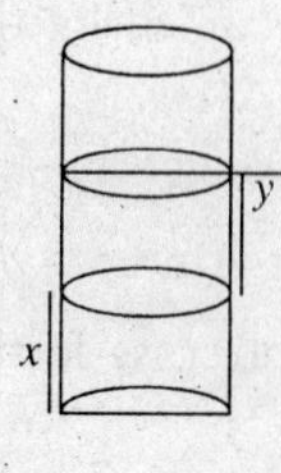

Fig. 10.4

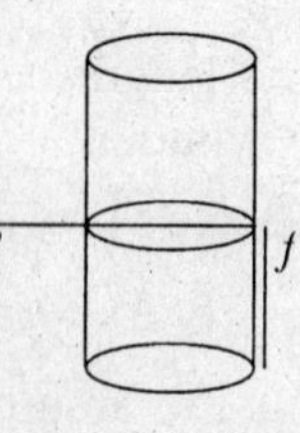

Fig. 10.5

The pressure at any point on the surface of water in the cylinder

$$= \rho g h + (y - x)\rho g = (h + y - x)\rho g.$$

Let A be the area of cross-section of the cylinder. So the volume of the air in the second position will be $(a - x)A$.
Then from Boyle's law, we have

$$A a \rho g h = A(a - x)(h + y - x)\rho g,$$

$$\text{or,} \quad (h + y - x)(a - x) = ah. \tag{1}$$

Also we know that weight of the body = weight of the fluid displaced.
The weight of the gas in the vessel will remain the same in both the cases

$$\therefore \quad b\rho g = (y - x)\rho g \quad \text{or, } y - x = b.$$

From (1), we get

$$(h + b)(a - x) = ah,$$

$$\text{or,} \quad a(h + b) - x(h + b) - ah,$$

$$\text{or,} \quad ab = x(h + b). \quad \therefore \quad x = \frac{ab}{h + b}.$$

▶ **Ex. 4.** Show that if $P_1, \rho_1, t_1; P_2, \rho_2, t_2; P_3, p_2, t_3$ be the three corresponding pressures, densities and absolute temperature of a perfect gas

$$t_1\left(\frac{P_2}{\rho_2} - \frac{P_3}{\rho_3}\right) + t_2\left(\frac{P_3}{\rho_3} - \frac{P_1}{\rho_1}\right) + t_3\left(\frac{P_1}{\rho_1} - \frac{P_2}{\rho_2}\right) = 0.$$

[Bag. U. 65A, 69A; M.U. 76A, Bih. 78A, P.U. 80A, R.U. 79H, 97H]

Solution: From Charle's law

$$p = k\rho(1 + \alpha t)$$

we have

$$P_1 = k\rho_1(1 + \alpha t_1),$$
$$P_2 = k\rho_2(1 + \alpha t_2),$$
$$P_3 = k\rho_3 = k\rho_3(1 + \rho t_3).$$

$$\therefore \quad \frac{P_1}{\rho_1} = k(1 + \alpha t_1), \frac{P_2}{\rho_2} = k(1 + \alpha t_2), \frac{P_3}{\rho_3} = k(1 + \alpha t_3).$$

$$\begin{aligned}
\text{LHS} &= t_1\left(\frac{P_2}{\rho_2} - \frac{P_3}{\rho_3}\right) + t_2\left(\frac{P_3}{\rho_3} - \frac{P_1}{\rho_1}\right) + t_3\left(\frac{P_1}{\rho_1} - \frac{P_2}{\rho_2}\right) \\
&= t_1[k(1 + \alpha t_2) - k(1 + \alpha t_3)] + t_2[k(1 + \alpha t_3) - k(1 + \alpha t_1)] \\
&\quad + t_3[k(1 + \alpha t_1) - k(1 + \alpha t_2)] \\
&= k\alpha[t_1(t_2 - t_3) + t_2(t_3 - t_1) + t_3(t_1 - t_2)] \\
&= k\alpha[t_1 t_2 - t_1 t_3 + t_2 t_3 - t_1 t_2 + t_3 t_1 - t_2 t_3] \\
&= 0 = \text{R.H.S.}
\end{aligned}$$

▶ **Ex. 5.** The readings of a faulty barometer in which there is some air are a and b, when the true readings are α and β, find the true reading when the faulty barometer reads c. [P.U. 67A, Ranj. 80H, 81, M.U. 68A, Bih. 68A, Bhag. 69H]

Solution: Let x be the true reading of the barometer, when c is the reading of the faulty barometer and l the length of the barometric tube.

In case of faulty barometer:

- The volume of the enclosed air is proportional to the length of the barometer tube—faulty reading.
- The pressure of the enclosed air is proportional to the true reading—faulty reading.

In the first case, the volume of the enclosed air

$$= (l-a)k \text{ and its pressure } = (\alpha - a)k',$$

where k and k' are constants.

In the second case, the volume of the enclosed air

$$= (l-b)k \text{ and its pressure } = (\beta - b)k'.$$

In the third case, the volume of the enclosed air

$$= (l-c)k \text{ and its pressure } = (x - c)k'.$$

Therefore, by Boyle's law

$$(l-a)(\alpha-a) = (l-b)(\beta-b) = (l-c)(x-c)$$

since

$$(l-a)(\alpha-a) = (l-b)(\beta-b).$$

Hence, we have

$$l[(\alpha-a)-(\beta-b)] = a(\alpha-a) - b(\beta-b)$$

$$\Rightarrow \quad l = \frac{a(\alpha-a)-b(\beta-b)}{(\alpha-a)-(\beta-b)}.$$

$$\therefore \quad l-a = \frac{a(\alpha-a)-b(\beta-b)}{(\alpha-a)-(\beta-b)} - a$$

$$= \frac{a(\alpha-a)-b(\beta-b)-a(\alpha-a)+a(\beta-b)}{(\alpha-a)-(\beta-b)}$$

$$= \frac{(\beta-b)(a-b)}{(\alpha-a)-(\beta-b)} \tag{1}$$

$$\text{and} \quad l-c = \frac{a(\alpha-a)-b(\beta-b)}{(\alpha-a)-(\beta-b)} - c$$

$$= \frac{a(\alpha-a)-b(\beta-b)-c(\alpha-a)+c(\beta-b)}{(\alpha-a)-(\beta-b)}$$

$$= \frac{(\alpha-a)(a-c)-(\beta-b)(b-c)}{(\alpha-a)-(\beta-b)}. \tag{2}$$

Again,

$$(l-a)(\alpha-a)=(l-c)(x-c).$$

$$\therefore \quad x-c=\frac{(l-a)(\alpha-a)}{l-c},$$

$$\text{i.e.,} \quad x=c+\frac{(l-a)(\alpha-a)}{l-c}$$

$$=c+\frac{(\beta-b)(a-b)(\alpha-a)}{(\alpha-a)(a-c)-(\beta-b)(b-c)}.$$

▶ **Ex. 6.** Assuming that a change from 30″ to 27″ in the height of a barometer corresponds to a rise in the attitude of 2700 ft, find the altitude corresponding to the height 21.87″ of the barometer.

[P.U. 72A, R.U. 66A, 74A, M.U. 70A, 74A, Bhag. 72H, 75A, Bih. 74A, 76A]

Solution: Let h be the height of the first station and p_1 the pressure at the point from which the height is measured.

Therefore, from the formula

$$p=p_1e^{-\frac{gh}{k}}$$

we have

$$30 = p_1e^{-\frac{gh}{k}} \tag{1}$$

$$\text{and} \quad 27 = p_1e^{-\frac{g}{h}(h+2700)}. \tag{2}$$

Let h be the attitude corresponding to the height 21.87" of the barometer.

Therefore

$$21.87=p_1e^{-\frac{g}{k}(h+h')}. \tag{3}$$

Dividing (1) by (2), we have

$$\frac{30}{27}=e^{\frac{2700g}{k}}$$

$$\log\frac{30}{27}=\frac{2700g}{k}$$

$$\log 30-\log 27=\frac{2700g}{k},$$

$$\text{or,} \quad \log 10+\log 3-3\log 3=\frac{2700g}{k},$$

$$\text{or,} \quad \log 10-2\log 3=\frac{2700g}{k},$$

$$\text{or,} \quad 1-2\log 3=\frac{2700g}{k}. \tag{4}$$

Again, dividing (1) by (3), we get

$$\frac{30}{21.87}=e^{\frac{gh'}{k}} \Rightarrow \log\frac{30}{21.87}=\frac{gh'}{k}$$

$$\begin{aligned}\frac{gh'}{k} &= \log 10 + \log 3 - (\log 2187 - \log 10^2)\\ &= 1 + \log 3 - (\log 3^7 - 2)\\ &= 1 + \log 3 - 7\log 3 + 2\\ &= 3 - 6\log 3. \qquad (5)\end{aligned}$$

Dividing (5) by (4), we get

$$\begin{aligned}\frac{h'}{2700} &= \frac{3 - 6\log 3}{1 - 2\log 3} = 3\\ h' &= 8100 \text{ ft.}\end{aligned}$$

▶ **Ex. 7.** A cylindrical vessel of depth h and section α is maintained at constant temperature. If ρ_0 and ρ_1 are the densities of the air at the top and bottom, show that the total amount of air contained is

$$\frac{h\alpha(\rho_1 - \rho_0)}{\log \rho_1 - \log \rho_0}.$$

[I.A.S. 73A, R.U. 73H, 83H, M.U. 74H, Bhag. 63H, 68H]

Solution: Let the origin be on the upper surface of the cylindrical well, so that when $z = 0, \rho = \rho_0$ when, $z = h, \rho = \rho_1$.

At a depth z, the pressure equation is

$$dp = \rho g dz.$$

Also $p = k\rho$

$$\begin{aligned}\therefore \quad dp &= k d\rho\\ \rho g dz &= k d\rho, \text{ or, } \frac{d\rho}{\rho} = \frac{g}{k} dz.\end{aligned}$$

Integrating, we get

$$\begin{aligned}\log \rho &= \log c + \frac{g}{k} z,\\ \text{or, } \log \rho &= \log\left(c e^{\frac{gz}{k}}\right)\\ \rho &= c e^{\frac{gz}{k}}\end{aligned}$$

when $z = 0, \rho = \rho_0$

$$\therefore \quad c = \rho_0, \ \rho = \rho_0 e^{\frac{gz}{k}}$$

when $z = h, \rho = \rho_1$ at the top.

$$\begin{aligned}\rho_1 &= \rho_0 e^{\frac{gh}{k}}\\ \frac{\rho_1}{\rho_0} &= e^{\frac{gh}{k}}\\ \text{or, } \log \frac{\rho_1}{\rho_0} &= \frac{gh}{k} \qquad (1)\\ k &= \frac{gh}{\log \rho_1 - \log \rho_0}. \qquad (2)\end{aligned}$$

$\therefore$ the mass of the liquid contained in the well, area of whose base is α

$$\begin{aligned}
&= \int_0^h \alpha\rho dz = \alpha \int_0^h \rho_0 e^{\frac{gz}{k}} dz \\
&= \alpha \frac{\rho_0 k}{g} \cdot \left[e^{\frac{gz}{k}}\right]_0^h = \frac{\alpha\rho_0 k}{g}\left[e^{\frac{gh}{k}} - 1\right] \\
&= \frac{\alpha\rho_0 h}{\log\rho_1 - \log\rho_0}\left(\frac{\rho_1}{\rho_0} - 1\right) \quad \text{From (1) and (2)} \\
&= \frac{\alpha h(\rho_1 - \rho_0)}{\log\rho_1 - \log\rho_0}.
\end{aligned}$$

▶ **Ex. 8.** A gas satisfying Boyle's law $p = k\rho$ is acted on by forces

$$X = \frac{-y}{x^2+y^2}, \qquad Y = \frac{x}{x^2+y^2}$$

show that the density varies as $e^{\frac{\theta}{k}}$, where $\tan\theta = \frac{y}{x}$.

[I.A.S. 76, M.U. 63H, Bih. 66H, Bhag. 67H, C.U. 04H]

Solution: The pressure equation is given by

$$\begin{aligned}
dp &= \rho(Xdx + Ydy), \\
\text{or, } dp &= \rho\left(\frac{-y\,dx + x\,dy}{x^2+y^2}\right).
\end{aligned}$$

Also $p = k\rho$

$$\frac{dp}{p} = \frac{(x\,dy - y\,dx)}{k(x^2+y^2)} = \frac{1}{k}\frac{x\,dy - y\,dx}{x^2\left(1+\frac{y^2}{x^2}\right)} = \frac{1}{k}\frac{d\left(\frac{y}{x}\right)}{1+\frac{y^2}{x^2}}.$$

Integrating, we get

$$\begin{aligned}
\log p &= \frac{1}{k}\tan^{-1}\frac{y}{x} + \log A = \frac{\theta}{k} + \log A \left[\tan\theta = \frac{y}{x}\right] \\
\log p &= \log A e^{\frac{\theta}{k}}, \quad \text{or,} \quad p = Ae^{\frac{\theta}{k}}, \quad \text{or,} \quad p \propto e^{\frac{\theta}{k}}.
\end{aligned}$$

▶ **Ex. 9.** If the law connecting the pressure and density of the air were $p = k\rho^n$, prove that neglecting variations of gravity and temperature, the height of the atmosphere would be $\frac{n}{n-1}$ times the homogeneous atmosphere.

[M.U. 66H, 79H, R.U. 71H]

Solution: The pressure equation is

$$dp = -\rho g dz. \tag{1}$$

Given $p = k\rho^n$

$$dp = kn\rho^{n-1}\,d\rho. \tag{2}$$

From (1) and (2)

$$\begin{aligned} -\rho g dz &= kn\rho^{n-1}\, d\rho, \\ \text{or,}\quad -g dz &= kn\rho^{n-2}\, d\rho. \end{aligned}$$

Integrating, we get

$$C - gz = \frac{kn\rho^{n-1}}{n-1}. \tag{3}$$

At the sea level $z = 0, \rho = \rho_0$.
So from (2)

$$C = \frac{kn\rho_0^{n-1}}{n-1}.$$

Hence, from (3), we get

$$\begin{aligned} \frac{kn\rho_0^{n-1}}{n-1} - gz &= \frac{kn\rho^{n-1}}{n-1} \\ gz &= \frac{kn}{n-1}(\rho_0^{n-1} - \rho^{n-1}). \end{aligned} \tag{4}$$

If h be the height of the atmospheric pressure, then $\rho = 0$ at $z = h$.
Thus, from (4), we have

$$gh = \frac{kn}{n-1}\rho_0^{n-1}. \tag{5}$$

If H be the height of the homogeneous atmosphere, then

$$\begin{aligned} & p_0 = \rho_0 g H \text{ and } p_0 = k\rho_0^n. \\ \therefore\quad & k\rho_0^n = \rho_0 g H \\ \therefore\quad & k\rho_0^{n-1} = gH. \end{aligned}$$

Hence, from (5) we get

$$\begin{aligned} gh &= \frac{ngH}{n-1}. \\ h(n-1) &= nH. \\ \therefore\quad h &= \frac{n}{n-1}H. \end{aligned}$$

► **Ex. 10.** In an atmosphere at rest under gravity the height H of the equivalent homogeneous atmosphere above any point at height h measured in terms of air of density at height h, is given as a function of h. Show that the pressure p at a height h is given by

$$\frac{p}{p_0} = e^{-\int_0^h \frac{dh}{H}},$$

where p_0 is the pressure at $h = 0$. [C.U. 96H]

Solution: Let at height h, the density of air be ρ.

Then $p = k\rho_0$.

The pressure equation is

$$dp = -\rho g dh.$$

$$\therefore \quad \frac{dp}{p} = -\frac{g}{k} dh. \tag{1}$$

Also

$$\rho_0 = H\rho_0 g \quad \text{and} \quad p_0 = k\rho_0$$

$$k\rho_0 = H\rho_0 g, \; k = gH.$$

From (1), we get

$$\frac{dp}{p} = -\frac{dh}{H}.$$

$$\int_{\rho_0}^{p} \frac{dp}{p} = -\int_0^h \frac{dh}{H}.$$

$$\therefore \quad \log\frac{p}{\rho_0} = -\int_0^h \frac{dh}{H}$$

$$\frac{p}{\rho_0} = e^{-\int_0^h \frac{dh}{H}}.$$

► **Ex. 11.** A hollow gas-tight sphere containing hydrogen requires a force mg to prevent it from rising when the lowest point touches the ground, the total mass of the sphere and hydrogen is M. Show that the sphere can float in equilibrium with its lowest point at a height h above the ground, where

$$h = \frac{k}{g} \log \frac{M + m}{M}$$

k being the ratio of the pressure of the atmosphere to its density.

[P.U. 79H, 82H, Bhag. 65H, R.U. 69H, M.U. 77H, 81H]

Solution: When the sphere touches the ground, then the two forces namely the weight Mg of the sphere and the hydrogen and the force mg act on the sphere. These two forces prevent the sphere from rising. Let V be the volume of the sphere and ρ_0 be the density at the ground.

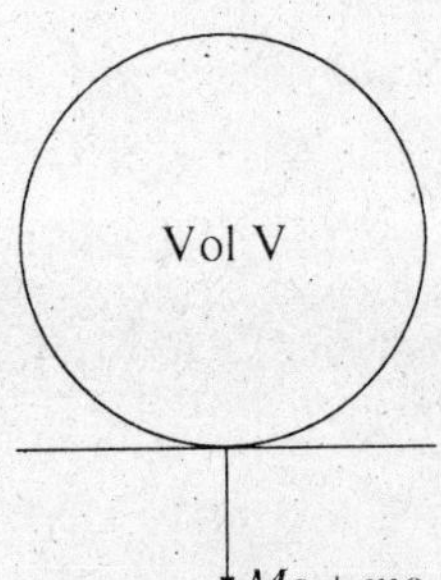

Fig. 10.6

Then, for equilibrium

$$V\rho_0 g = Mg + mg. \tag{1}$$

Since the sphere floats in equilibrium at a height h, then

$$V\rho_1 g = Mg. \tag{2}$$

Dividing (1) by (2), we get

$$\frac{\rho_0}{\rho_1} = \frac{M+m}{M}. \tag{3}$$

Also $p = k\rho$.

The pressure equation is

$$\begin{aligned} dp &= -g\rho dz \\ \frac{dp}{p} &= -\frac{g}{k}\, dz \\ \int_{\rho_0}^{p} \frac{dp}{p} &= -\frac{g}{k}\int_0^z dz \\ \log\frac{p}{\rho_0} &= -\frac{gz}{k} \\ \frac{p}{\rho_0} &= e^{-\frac{gz}{k}}. \end{aligned}$$

$$\begin{aligned} \therefore\quad p &= \rho_0 e^{-\frac{gz}{k}} \\ k\rho &= k\rho_0 e^{-\frac{gz}{k}} \quad [\because\ \rho_0 = k\rho_0,\ p = k\rho] \\ \frac{\rho}{\rho_0} &= e^{-\frac{gz}{k}} \end{aligned}$$

when $z = h$, $\rho = \rho_1$

$$\begin{aligned} \frac{\rho_1}{\rho_0} &= e^{-\frac{gh}{k}}. \\ \therefore\quad \frac{\rho_0}{\rho_1} &= e^{\frac{gh}{k}}. \end{aligned}$$

Therefore from (3), we get

$$\begin{aligned} e^{\frac{gh}{k}} &= \frac{M+m}{M} \\ \frac{gh}{k} &= \log\frac{M+m}{M}. \end{aligned}$$

Hence,

$$h = \frac{k}{g}\log\frac{M+m}{M}.$$

▶ **Ex. 12.** If near the earth's surface gravity be assumed to be constant, and the temperature in the atmosphere to be given by

$$t = t_0 \left(1 - \frac{z}{nH}\right),$$

where H is the height of the homogeneous atmosphere, show that the pressure in the atmosphere will be given by

$$p = \rho_0 \left(1 - \frac{z}{nH}\right)^n.$$

[Bih. 69H, 65H, Bhag. 66H, 72H, R.U. 67H, P.U. 79H, C.U. 88H, 01H]

Solution: We have $\rho_0 = \rho_0 gH,\ p = R\rho t$.

Therefore, pressure equation is $dp = -\rho g dz$

$$\therefore \quad \frac{dp}{p} = -\frac{gdz}{Rt} = -\frac{gdz}{Rt_0\left(1 - \frac{z}{nH}\right)}$$

$$\int_{p_0}^{p} \frac{dp}{p} = -\frac{g}{Rt_0}\int_0^z \frac{dz}{1 - \frac{z}{nH}}$$

$$\log \frac{p}{p_0} = \frac{gnH}{Rt_0} \log\left(1 - \frac{z}{nH}\right). \qquad (1)$$

Now, $p_0 = R\rho_0 t_0$ and $p_0 = \rho_0 gH$, $Rt_0 = gH$.

Therefore, from (1), we get

$$\log \frac{p}{p_0} = n \log\left(1 - \frac{z}{nH}\right)$$

$$\frac{p}{p_0} = \left(1 - \frac{z}{nH}\right)^n, \quad \text{or,} \quad p = p_0\left(1 - \frac{z}{nH}\right)^n.$$

▶ **Ex. 13.** If the absolute temperature T, diminishes upwards in the atmosphere according to the law $T = \frac{T_0}{1+\beta z}$, where β is a constant, show that the pressure at a height z is given by

$$p = p_0 e^{-\left(\frac{z}{H} + \frac{\beta z^2}{2H}\right)}$$

H being the height of the homogeneous atmosphere. **[I.A.S. 74]**

Solution: The pressure equation is

$$dp = -g\rho dz.$$

Also $p = R\rho T$

$$\frac{dp}{p} = -\frac{g}{RT}dz = -\frac{g}{RT_0}(1 + \beta z)\, dz.$$

Integrating, we get

$$\log p = -\frac{g}{RT_0}\left(z + \frac{\beta}{2}z^2\right) + C. \qquad (1)$$

When $z = 0,\ p = p_0$

So from (1), $\log p_0 = C$

So from (1), we get

$$\log p = -\frac{g}{RT_0}\left(z + \frac{\beta}{2}z^2\right) + \log p_0,$$

$$\text{or,} \quad \log p - \log p_0 = -\frac{g}{RT_0}\left(z + \frac{\beta}{2}z^2\right).$$

$$\therefore \quad \log \frac{p}{p_0} = -\frac{g}{RT_0}\left(z + \frac{\beta}{2}z^2\right). \qquad (2)$$

Also $p = \rho g H$, $p = R\rho T$, $p_0 = \rho_0 g H$, $p_0 = R\rho_0 T_0$, $RT_0 = gH$.

Hence, from (2), we get

$$\log \frac{p}{p_0} = -\frac{1}{H}\left(z + \frac{\beta}{2}z^2\right)$$

$$\frac{p}{p_0} = e^{-\left(\frac{z}{H} + \frac{\beta z^2}{2H}\right)}.$$

$$\therefore \quad p = p_0 e^{-\left(\frac{z}{H} + \frac{\beta z^2}{2H}\right)} \quad \text{Proved.}$$

▶ **Ex. 14.** A gaseous atmosphere in equilibrium is such that $p = k\rho^{y} = R\rho T$. Where p, ρ, T are pressure, density and temperature and K, γ, R are constants. Prove that the temperature decreases upwards at a constant rate α, so that

$$\frac{dT}{dz} = -\alpha = -\frac{g}{R}\frac{\gamma - 1}{\gamma}$$

[L.N.M.U. 86H, P.U. 65H, R.U. 68H, C.U. 98H]

In a certain atmosphere of uniform composition $T = T_0 - \beta z$, where T_0, β are constants and $\beta < \alpha$. Find the pressure and density and show that they both vanish at height $\frac{T_0}{\beta}$. [P.U. 61H, Big. 62H, I.A.S. 75]

Solution: Given

$$\left.\begin{aligned} p &= k\rho^{y} \\ dp &= k\gamma\rho^{y-1}\,d\rho \end{aligned}\right\}. \qquad (1)$$

Also the pressure equation is

$$dp = -g\rho dz.$$

$$\therefore \quad k\gamma\rho^{y-1}\,d\rho = -g\rho dz$$

$$k\gamma\rho^{y-2}\,d\rho = -gdz.$$

Integrating we get

$$\frac{k\gamma\rho^{y-1}}{\gamma - 1} = c - gz, \quad \text{or,} \quad \frac{k\gamma\rho^{y}}{\rho(\gamma - 1)} = c - gz.$$

From (1), we get

$$\frac{p\gamma}{\rho(\gamma - 1)} = c - gz.$$

Also $p = R\rho T$

$$\frac{RT\gamma}{\gamma - 1} = c - gz.$$
$$T = \frac{\gamma - 1}{\gamma R}(c - gz).$$
$$\therefore \quad \frac{dT}{dz} = -\frac{\gamma - 1}{\gamma} \cdot \frac{g}{R} = -\alpha.$$

For the second case $p = R\rho T$ and the prėssure equation is $dp = -g\rho dz$.

$$\therefore \quad \frac{dp}{p} = -\frac{g}{RT}\, dz$$
$$\frac{dp}{p} = -\frac{g}{R(T_0 - \beta z)}\, dz \quad [\because T = T_0 - \beta z].$$

Integrating, we get

$$\log p = C + \frac{g}{R\beta} \log(T_0 - \beta z). \tag{2}$$

At $z = 0$, $p = p_0$

$$\log p_0 = c + \frac{g}{R\beta} \log T_0. \tag{3}$$

Subtracting (3) from (2), we get

$$\log p - \log p_0 = \frac{g}{R\beta} [\log(T_0 - \beta z) - \log T_0]$$
$$\log \frac{p}{p_0} = \frac{g}{R\beta} \log \frac{T_0 - \beta z}{T_0}$$
$$= \frac{g}{R\beta} \log\left(1 - \frac{\beta z}{T_0}\right) = \log\left(1 - \frac{\beta z}{T_0}\right)^{\frac{g}{R\beta}}$$
$$\frac{p}{p_0} = \left(1 - \frac{\beta z}{T_0}\right)^{\frac{g}{R\beta}}$$
$$p = p_0\left(1 - \frac{\beta z}{T_0}\right)^{\frac{g}{R\beta}}.$$

At height $\frac{T_0}{\beta}$, $p = 0$.

Also $p = k\rho^{\gamma}$

$$\therefore \quad k\rho^{\gamma} = 0. \text{ Hence, } \rho = 0.$$

▶ **Ex. 15.** In a vertical column of perfect gas, the pressure and absolute temperature at any height z are p and T, prove that

$$z = \frac{p_0}{\rho_0 g T_0} \int_p^{p_0} \frac{T}{p}\, dp,$$

where p_0, ρ_0, T_0 are pressure, density and absolute temperature at the bottom and g has its usual meaning. [R.U. 66H, 70H, 98H, M.U. 76H]

Solution: The pressure equation is

$$dp = -\rho g dz$$

Also

$$\begin{aligned} p &= R\rho T \\ dz &= -\frac{dp}{\rho g} = -\frac{dp}{\frac{g \cdot p}{Rt}} \\ dz &= -dp\frac{RT}{pg} \\ \int_0^z dz &= -\frac{R}{g}\int_{p_0}^{p} \frac{Tdp}{p} \\ z &= \frac{R}{g}\int_{p}^{p_0} \frac{Tdp}{p}. \end{aligned} \tag{1}$$

Now $p = R\rho T$, $p_0 = R\rho_0 T_0$.

Therefore, from (1) we get

$$z = \frac{p_0}{\rho_0 T_0 g}\int_{p}^{p_0} \frac{Tdp}{p}.$$

Exercises

1. A hollow cylinder of certain height closed at the bottom and open at the top is inverted and pushed down, vertically in water until water rises 2 ft. in it. If water barometer be 34 ft. height, find the height of the cylinder. [R.U. 67A, Bih. 75A]

   [$\left(2 + \frac{68}{d}\right)$ **ft, d being the depth of water surface within the cylinder**]

2. A cylinder tube 9 feet long with one end closed is lowered into water with open end downwards until its top is 10 ft. below the surface, find how high the water rises in the interior. [P.U. 71A, R.U. 68A, M.U. 74S]

   [*Ans.* 2.9 ft nearly]

3. A hollow cone of height h and mouth downwards is immersed vertically in water. Show that the water within the cone is at a depth $7H - \frac{h}{2}$ below the surface of water, where H is the height of the barometer. [R.U. 74H, 83H]

4. A closed air-tight cylinder of length $2a$ is half full of water and

half full of air at atmospheric pressure, which is equal to that of a column of height h of water is introduced without letting the air escape so as to fill an additional height K of the cylinder, and the pressure of the base is doubled. Prove that

$$K = a + h - \sqrt{ah + h^2}$$

[M.U. 69, 68H, Bih. 69, Bhag. 63A]

5. A cylinder contains two gases which are separated from each other by a movable piston, the gases are both at 0°C and the volume of one gas is double that of the other. If the temperature of the first be raised t°C, prove that the piston will move through a space $\frac{2l\alpha t}{9 + 6\alpha t}$, where l is the length of the cylinder, and α is the coefficient of expansion per 1°C. [P.U. 62A]

6. A hollow cone of height 4 ft is immersed mouth downwards in water. The surface of water is exposed to atmospheric pressure. Show how far it should be depressed so that water within the cone may rise half way up if the height of the water barometer be equal to 34 ft. [*Ans.* 236 ft]

7. A conical wine-glass 4 inches in height is lowered, mouth downwards, into water till the level of the water inside is 34 ft. below the surface, the height of the water barometer being 34 ft, what is the height of the part of the cone now occupied by air. [*Ans.* $2\sqrt[3]{4}$ inches]

8. A hollow closed conical vessel of height h floats partially immersed in water with vertex downwards and axis vertical. A hole is then made very near the vertex and water allowed to come into the vessel so that no air escapes from within. If the vertex was originally at a depth b and H is the height of the water barometer, prove that the new depth c of the vertex is given by

$$c^3 - b^3 = \left[c - \frac{H(c^3 - b^3)}{h^3 - (c^3 - b^3)}\right]^3.$$

9. A thin conical surface of weight w just sinks to the surface of water. When immersed with its vertex downwards a weight mw must be placed inside it to make it sink to the same depth as before, if h be the height of a water barometer, prove that the height of the cone is $hm(1+m)^{\frac{1}{3}}$. [Luck. 76, 79, 80]

10. Prove that if volumes V_1 and V_2 of atmospheric air are forced into vessels of volume v_1 and v_2 and a communication is opened between them, a mass of air of volume $\frac{V_1 v_2 - V_2 v_1}{V_1 + V_2}$ at atmospheric pressure will pass from one vessel to the other.

11. If a body floats in a liquid with volumes v_1, v_2 and v_3 above the surface when the barometric heights are h_1, h_2 and h_3, prove that

$$\begin{aligned} &h_1 v_1 (v_2 - v_3) \\ &+ h_2 v_2 (v_3 - v_1) \\ &+ h_3 v_3 (v_1 - v_2) = 0. \end{aligned}$$

[Bih. 68A, M.U. 6xA, R.U. 69A, Mith. 74]

12. A barometer stands at 30 inches and the space occupied by the Torricellian vacuum is then 2 inches. If now a bubble of air which would at atmospheric pressure occupy half an inch of the tube to be introduced into it, show that the surface of mercury will be lowered by 3 inches.

13. The height of the Torricellian vacuum in a barometer is a inches and the instrument indicates a pressure of b inches of mercury when the true reading is c inches. If the faulty readings are due to an imperfect vacuum, prove that the true reading corresponding to an apparent reading of d inches is

$$d + \frac{a(c-b)}{a+b-d}.$$

[R.U. 62A, Bih. 67A, 72A, 77A, Bhag. 71A, P.U. 79A]

14. When the reading of a true barometer is 30″, the reading of a faulty one is 28″, its tube containing a quantity of air of length $3\frac{1}{8}''$. What will be the reading of a correct barometer when the faulty of indicates $27\frac{1}{8}''$? [*Ans.* 29″ nearly]

15. A barometer stands at 30 inches, and the space occupied by the Torricellian vacuum is then 2 inches, if now a bubble of air which would at atmospheric pressure occupy half an inch of the tube be introduced into it, show also that the height of a correct barometer, when the in correct one stands at α inches is

$$\alpha + \frac{15}{32-\alpha} \quad \text{inches.}$$

16. The readings of a faulty barometer in which there is some air are 29.8″ and 29.4″, the true readings are 30.4 and 29.8 inches, respectively, find the true reading when the barometer reads 29 inches.

[P.U. 60S, 61A, M.U. 69A]

[*Ans.* 29.3″]

17. A box is filled with a heavy gas at uniform temperature. Prove that if a is the altitude of the highest point the lowest, and p, p' be the pressures at these points, the relation of the pressure to the density at any point is equal to

$$\frac{ag}{\log \frac{p'}{p}}.$$

[Bih. 80H, 67H, Bhag. 62H, 66A, M.U. 74, R.U. 78, 81, P.U. 80H, 81H, Luck. 76, 80]

18. A heavy gas at constant temperature is confined in a vertical cylinder of height h. If ρ be the density at the base, prove that the mean density is

$$\frac{k\rho}{gh}\left(1 - e^{-ghlk}\right).$$

[Agra 81, Luck. 74, 77, 82, Bih. 79H, R.U. 82H, M.U. 87H, 90H, B.P.S.C. 82]

19. A closed straight tube of fine uniform bore contains mercury in the middle and air at each end. When the tube is vertical, the portions occupied by air are of lengths a and b, respectively which be come a', b' when the tube is inverted. Prove that the lengths, when the tube is horizontal are

$$\frac{aa'(b+b')}{ab+a'b'}, \frac{bb'(a+a')}{ab+a'b'}.$$

20. A hollow gas-tight balloon containing helium weights w lbs. When its lowest point touches the ground, it requires a force of W lbs. to prevent it from rising. Show that it can float in equilibrium at a height $H \log\left(1 + \frac{w}{W}\right)$, where H is the height of the homogeneous atmosphere.

[P.U. 62H, 79H, M.U. 73H, R.U. 74H, C.U. 85H, I.A.S. 72]

21. If the law connecting the pressure and density of the air were $p = k\rho^2$, prove that neglecting changes

of temperature and gravity, the height of the atmosphere will be twice the height of the homogeneous atmosphere.

[P.U. 67H, 79H, L.N.M.U. 96H]

22. If the pressure of the air varies as $\left(1+\frac{1}{m}\right)$ the power of the density, show that neglecting variation of temperature and gravity, the height of the atmosphere should be equal to $(m+1)$ times the height of the homogeneous atmosphere.

23. A balloon carrying a self-registering barometer records pressures equivalent to h and h_1 inches of mercury, when it ascends to heights equal to fractions α and α_1 of the earth's radius, respectively. Prove that (taking into account the variation of gravity with height and assuming the temperature constant)

$$\frac{ag_0}{k}\frac{\alpha_1-\alpha}{(1+\alpha)(1+\alpha_1)} = \log\frac{h}{h_1}+2\log\frac{1+\alpha_1}{1+\alpha},$$

where a is the earth's radius.

[P.U. 64H, 78H, Bih. 63H, R.U. 62H, M.U. 64H, C.U. 92H]

24. Prove that, if the temperature in the atmosphere fall uniformly with the height ascended, the height of a station above sea-level is given by

$$z=a\left\{1-\left(\frac{h}{h_0}\right)^m\right\},$$

where h, h_0 are the readings of the barometer at the station and the sea-level, and a, m, are constants.

[P.U. 60H, 62H, 66H, 77H, R.U. 64H, 66H, 70H, Bih. 76H, Mith. 87H]

25. A thin uniform circular tube of radius a contains air and rotates with uniform angular velocity w about an axis in its plane, distance c from the centre, find the pressure at any point neglecting the weight of the air. If $c < a$ and if p and p' the greatest and least pressures, prove that

$$\log\frac{p}{p'}=\frac{W^2}{2k}(a+c)^2.$$

[Bih. 66H, 68H, M.U. 64H]

26. Assuming the temperature of the air to diminish uniformly with the height, prove that the difference of level between two stations is

$$H\frac{T_0-T_1}{273}\frac{\log\frac{h_0}{h_1}}{\log\frac{T_0}{T_1}},$$

where H is the height of homogeneous atmosphere at $0°C$. T_0 and T_1 are the absolute temperatures at the stations and h_0, h_1 the barometric heights reduced to $0°C$ (consider gravity as constant). Prove also that, if as an approximation the temperature were taken constant and equal to $\frac{1}{2}(T_0+T_1)$ the calculated height would be too great by a fraction of its true value equal roughly to

$$\frac{1}{3}\left(\frac{T_0-T_1}{T_0+T_1}\right)^2.$$

[Bih. 61H, 68H, R.U. 63H, Bhag. 70H]

27. Masses m and m' of two gases in which the ratio of the pressure to the density are respectively k and k', are mixed at the same temperature. Prove that the ratio of the pressure to density in the compound is $\frac{mk+m'k'}{m+m'}$.

[M.U. 66A, 68S, Bhag. 64A, Bih. 81, R.U. 76, 82,I.A.S. 99, R.U. 93H, 96H]

28. As an aeroplane ascends the temperature and the pressure are simultaneously recorded and a curve is drawn plotting the absolute temperature against the logarithm of the pressure. Prove that the height ascended between true reading is

$$-\frac{k}{g}\int_{x_1}^{x_2} T dx, \quad \text{where } x = \log p.$$

[P.U. 63H]

29. The height of a balloon is calculated from the barometer pressure reading (p) on the assumption that the pressure of the air varies as the density. Show that of the pressure actually varies as the nth power of the density, there will be an error.

$$h_0\left[\frac{n}{n-1}\left\{1-\left(\frac{p}{p_0}\right)^{\frac{n-1}{n}}\right\}-\log\frac{p_0}{p}\right]$$

is the calculated height, where h_0 is the height, of the homogeneous atmosphere and p_0 is the pressure at the surface of the earth.

[R.U. 65H, M.U. 67H, I.A.S. 94]

30. If the absolute temperature T at a height z is a given function f(z) of the height, show that the ratio of pressure at two heights z_1 and z_2 is given by

$$\log\frac{p_2}{p_1} = -\frac{g}{R}\int_{z_2}^{z_1}\frac{dz}{f(z)}$$

R being the constant in the equation $p = R\rho T$. [C.U. 96H]

31. Find the pressure of the atmosphere at a distance r from the earth's centre, having given that the temperature of the atmosphere varies inversely as the nth power of the distance from the centre of the earth, its radius being a.

[Bih. 71H]

[*Ans.* $\log\frac{p}{p_0} = \frac{g}{Rk}\frac{r^{n+1}-a^{n+1}}{n+1}$]

32. A bent tube of uniform bore, the arms of which are at right angles, revolves with constant angular velocity ω about the axis of one of its arms, which is vertical and has its extremity immersed in water. Prove that the height to which water h as risen in the vertical arm is

$$\frac{\pi\left(1-e^{-\frac{\omega^2 a}{2k}}\right)}{\rho g}$$

a being the length of the horizontal arm π, the atmospheric pressure, and ρ the density of the water, and k the ratio of the pressure of the atmosphere to its density.

[I.A.S. 78, Delhi 66, Agra 63, Vik ,3]

33. One half of a circular tube, radius a, of fine uniform bore, is occupied by mercury, and the other half by air at atmosphere pressure. The tube is made to rotate about a vertical diameter with uniform angular velocity w. Show that if the air in the tube remain homogeneous, and $a^2w^2 > 2g(a+h)$ h being the height of the mercury barometer, there will be a vacuum of length $2a\theta$ at the bottom of the tube, such that

$$a^2w^2\cos 2\theta = \frac{2gh\pi}{\pi - 2\theta} + 2ag(\sin\theta + \cos\theta)$$

[C.U. 94H]